A Second Opinion....

"Jay Benford designs fascinating anachronisms for people who spurn the world of plast cs and mass production."

Pacific Yachting

"Jay Benford is one of those freethinking designers who is just as happy working on a 14ft (4.27m) worldgirdler as a 131ft (40m) luxury cruiser. His magic is such that he can make both boats, and those in between, individual, attractive and practical."

Yachting World

"Jay Benford is a widely known yacht designer with a far-ranging imagination and great versatility, unbound by prejudices on materials or methods of construction."

WoodenBoat

"The designs of Jay R. Benford are usually easy to recognize but difficult to describe precisely. They definitely have a traditional look about them yet don't resemble any particular type of traditional craft. Often they look like caricatures of boats — the kind of boats you see in children's books.

No insult is intended. On the contrary, Benford's boats all seem to have some sort of magical quality, a unique character all their own that defies definition."

Danny Greene, **Cruising Sailboat Kinetics**

"...let me say that I have spent **hours** perusing **Small Ships**...to my utter delight. Here is Benford at his best with an infinite variety of arrangements and layouts on several series of working craft hulls including his Florida Bay Coaster 40, 50, and 65 footers built by Reuben Trane....

"...The varieties are endless, and if one of these doesn't start you dreaming of retiring to a life of floating luxury with your beloved, you're reading the wrong magazine....

"...original, fun, and thought provoking which is what makes this book worth owning, not just reading. It's a book which will be referred to again and again."

Ted Jones, **Coastal Cruising**

It was while working on the first **Mariner's Catalog**—a period of nautical despair for those who liked traditional boats made of traditional materials—that we first encountered the design works of J. Benford Associates. We could not believe it, a firm devoted almost exclusively to boats so salty that they made our eyes rust. So, like the girl who dreamed away childhood over knights on white chargers and screamed rape when one finally showed up, we didn't list them. Peculiar the human mind....

Volume 6, **The Mariner's Catalog**

The Benford 30 shows what a clever and uninhibited designer can do in a relatively small boat (indeed, the smallest one in this book).

The Proper Yacht by Arthur Beiser

Puffin's *launching. Designer is holding bow line, and builder Howard Cain is holding stern line.*

Puffin *has a boom crutch that can be used in 3 positions, on centerline and off center port or starboard. It's shown in the port position here.*

Puffin *winter sailing off Seattle.*

BY THE SAME AUTHOR:

CRUISING BOATS, SAIL & POWER, 4 editions in 1968, 1969, 1970 & 1971. Design catalog and article reprints. OP*

PRACTICAL FERRO-CEMENT BOATBUILDING, with Herman Husen, 3 editions in 1970, 1971 & 1972. Best-selling construction handbook, a how-to on ferro-cement. OP*

DESIGNS & SERVICES, 7 editions, 1971, 1972, 1987, 1988, 1990, 1993 & 1996. Catalog of plans & services of our firm. This material is now all online at www.benford.us.

BOATBUILDING & DESIGN FORUM, 1973. A monthly newsletter with more information on ferro-cement boatbuilding & other boatbuilding information. OP*

THE BENFORD 30, 3 editions in 1975, 1976 & 1977. An exposition on the virtues of this design and general philosophy on choosing a cruising boat. OP*

CRUISING DESIGNS, 1975, 1976, 1993, 1996 & 2003. A catalog of plans and services and information about boats and equipment.

DESIGN DEVELOPMENT OF A 40 METER SAILING YACHT, 1981. A technical paper presented to the Society of Naval Architects & Marine Engineers, at the fifth Chesapeake Sailing Yacht Symposium, and in the bound transactions of that meeting.

CRUISING YACHTS, 1983. A hard cover book with a selection of Benford designs covered in detail, including several complete sets of plans, a lot of information about the boats and how they came to be. Eight pages of color photos. OP*

THE FLORIDA BAY COASTERS, A FAMILY OF SMALL SHIPS, 1988. A book of study plans of these Benford designed freighter yachts. OP*

SMALL CRAFT PLANS, 1990, 1991, 1997 & 2005. A book with 15 sets of full plans for 7'-3" to 18'-0" dinghies and tenders.

SMALL SHIPS, 5 editions, 1990, 1995, 1995, 1997 & 2002. A book of study plans for Benford designs for tugs, freighters (like the Florida Bay Coasters), ferries, excursion boats, trawler yachts, houseboats & fishing vessels. Eight pages of color photos.

POCKET CRUISERS & TABLOID YACHTS, Volume 1, 1992 & 1996. A book with 6 complete sets of plans (11 boats including the different versions) for boats from 14' to 25', including 14' & 20' Tug Yachts, 17' & 25' Fantail Steam Launches, a 14' Sloop, a 20' Catboat, and 20' Supply Boat & Cruiser.

* OP = Out of print

CATBOATS

FROM THE BOARDS OF THE

BENFORD DESIGN GROUP

www.benford.us

ISBN 10 – 1-888671-25-4
ISBN 13 – 978-1-888671-25-4

Graphic design and production by:
Scribe, Inc., 842 S. 2nd St., Philadelphia, PA 19147

Printed in the USA by:
Victon Graphics, 1211 Bernard Drive, Baltimore, MD 21223

Questions regarding the content of this book should be addressed to:

Tiller Publishing
29663 Tallulah Lane
Easton, MD 21601

410-745-3750
www.tillerbooks.com

Dedication:

To the builders and owners of our catboat designs, who shared their photos with us, to Roy Montgomery, who did a masterful job of photographing ***Puffin***, and to the crew who skilfully put ***Puffin*** together.

And, of course, Dona....

Table of Contents

Note: Some of the drawings are cut-and-paste assemblies to fit the page sizes of the book. There are some details that are not included here, but are part of the actual building plan sets. We have striven to reproduce all these plans accurately so that they can be scaled with an architect's scale. Most will be found to be at 1/4" = 1'-0" except the 22' Cruising Cat drawings, which will be 3/16" = 1'-0". Some of the details were that done at larger scales originally will be here at larger scales than 1/4" = 1'-0".

The drawings in this book were scanned from originals that were done mostly in pencil, with a few done in ink. Thus their line quality varies somewhat, depending on the darkness of the originals and their age....

Introduction — What Is a Catboat?

The classical definition of a catboat is a sailing boat rigged with only a mainsail – no headsails as working sails – with the mast set nearly in the bow. Using the traditional plumb stem gives deeper socketing of the mast – usually unstayed – and lets the mast get right into the eyes of the boat. It is usually relatively beamy, often a centerboarder, and thus has both good initial stability and the ability to sail in shallow waters. Sometimes a cat will have a smaller headsail that is set flying and/or on a bowsprit.

A catboat should not be confused with a catamaran – which has two hulls – and is often also called a cat. A catamaran could also be a catboat, though, since the catboat designation relates specifically to the rig.

Since the region where the catboat primarily was developed is an area of shoal water, most of the original working catboats were centerboarders. Many of the yacht versions have emulated this arrangement. However, in a small cruising boat, the trunk makes quite an intrusion into the living space in the cabin. This has led us to develop shallow keel versions of our cat designs which still have modest draft and yet do not have the centerboard trunk subdividing the cabin.

In our rendition of these types of designs, we have taken advantage of the roominess of this sort of beamy boat to make good cruising accommodations so that the crews will be kept comfortable.

There are several types of construction represented by these designs, from very simple sheet plywood to the complexity of ferrocement and one-off fiberglass. There are additional hours required to get a one-off 'glass hull really well faired, which is a disincentive to build that way. It's the classic trade-off of time versus expense....

There are also some first cousins to catboats. They are known as cat ketches, cat schooners and cat yawls. They share many of the virtues of ease of handling of the smaller cats. Check out some of our other design books – **Cruising Designs** and **Small Craft Plans** – for some designs that also have cat-rigged variations.

What material would I recommend? My own prejudices lean towards wooden construction, either strip/cold-molded or traditional carvel. A well built and cared for wooden boat will live for generations. And, were I doing one of the Gunkholers, I'd have no problem with the plywood and epoxy structure they use.

Additional Reading:

For an authoritative read on how to get the best performance out of your Catboat, get a copy of Bill Welch's **The Competitive Cat: Racing Small Gaff-Rigged Catboats**.

*19' **Catspaw** and **Sylvester** bult by Paul Miller.*

Chapter 1

Puffin & *Liberty*

17' Catboat *Puffin* & 17' Cruising Cat *Liberty*
Design number 67 & 67R
1970 & 1972

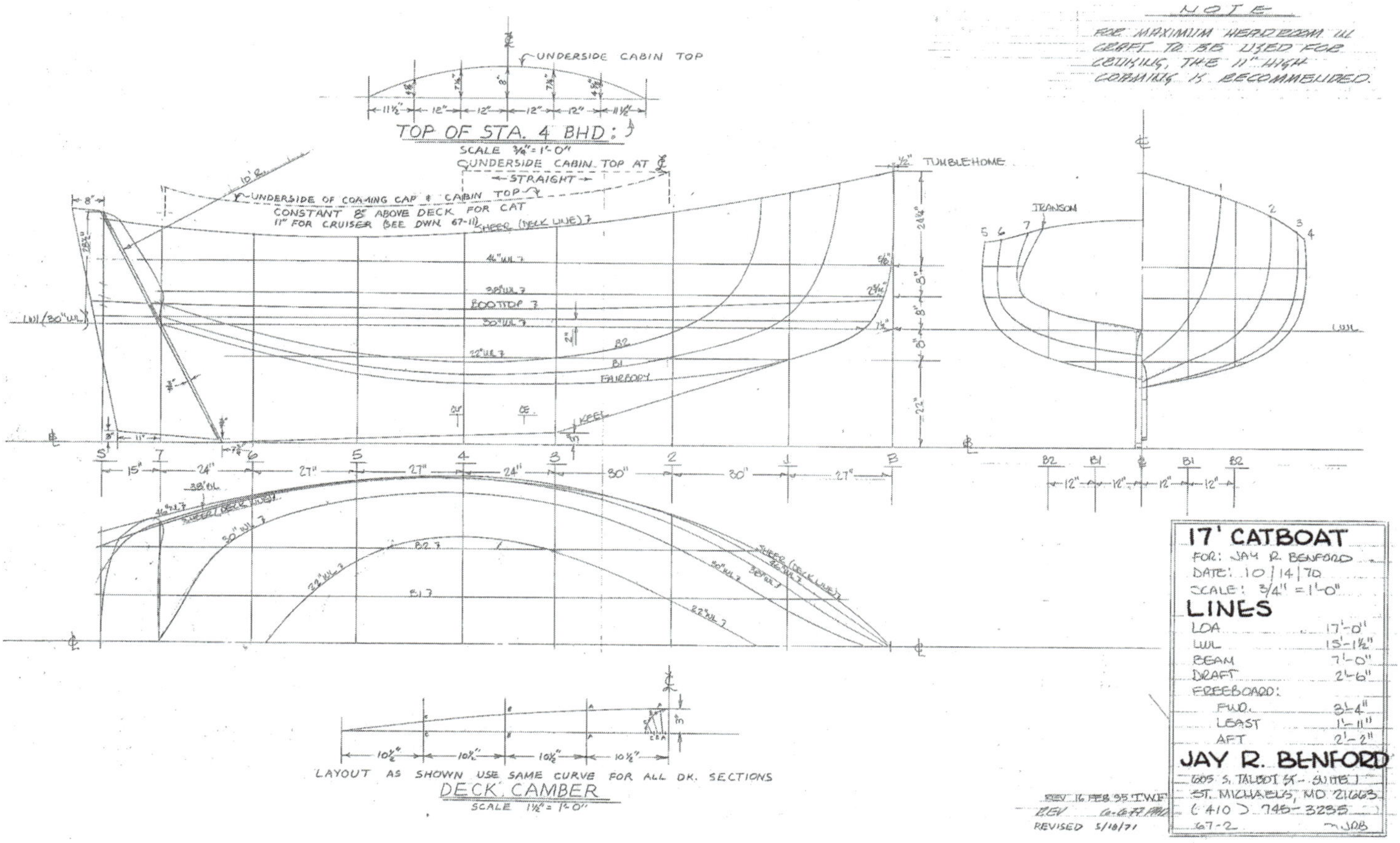

When this design was being done, I was living aboard and cruising the 50' **Kiyi**, *an elegant fantail yacht built in 1926 by Schertzer Brothers in Seattle. This lines plan was done on the drawing board in her pilothouse and additional information was added several times over the years. Note the slight hollow in her bow waterlines and easy run. Her firm bilges help give her good form stability.*

This handsome small cruiser accommodates two, either in a V-berth/double or singles. She has a head, stove, and plenty of storage space. The sink, counter, lounging seat and hanging locker for oilskins are additions in Cruising Cat version, which has a shorter cockpit and an optional deeper keel with room for an inboard engine. Inboard or outboard power of 5 to 10 horsepower is all that is needed. She has a deep, self-draining cockpit, she performs well in light airs and is able to carry full sail in 20 knots of wind. A safe, sane investment—an ideal first boat for the amateur builder. The alternate rigs, taken from the 18' Canoe Yawl and 20' Gaff Sloop, can be used on the basic 17' Catboat hull. Optional interior details can be expanded by way of joinery details and sections on a custom basis to order. An alternative Marconi sloop rig is also available, as are gaff cutter and topsail sloop rigs.

The 17' Cat design was originally done as a conceptual illustration when we were working on writing the first draft of our **Practical Ferro-Cement Boatbuilding** book. This was in the fall of 1970, during the peak of the interest in building ferro-cement boats. The hype surrounding the applications of ferro at the time said that it wasn't practical to use the building method to build something smaller than a 30-footer, which I thought was singularly uninformed. Of course, this "conventional wisdom" ignored the fact that some of the original work done in developing the ferro-cement method—more than a century before—had been to build a small rowing boat. . . .

This "why not?" thinking got further along when I started looking at the preliminary that became this 17' cat. A smaller boat ought to have thinner planking and we put together some test panels to prove out this theory. The lab results indicated that the theory would be borne out in practice.

Knowing this impractical under 30' theory to be a fallacy, we fleshed out this 17' design and eventually saw hundreds of plans sold for her. (In fact, we later designed and built a ferro-cement 12-footer, whose plans are shown in our **Small Craft Plans** book.) Over the following decades, we've also created wood and 'glass versions of this design. The wood ones seem to be the ones most often built.

The myth that ferro-cement was not practical for boats under 30 feet masqueraded as fact too long, and is at least part of the reason many people over-extend themselves. If a person were bitten by the ferro-cement bug, he was offered little choice but to start building a larger boat—often one larger than he needed—there were simply no designs available for small ferro-cement boats. However, there is really no good reason for this to be true. With careful design and attention to detail on the part of the designer AND the builder, the weight can be kept down to reasonable amounts, and the small boat becomes practical.

Living and cruising aboard the 50' fantail motoryacht ***Kiyi*** at the time, I'd had taken out the seating in the small pilothouse—she had only 10' beam—and had a drawing board in its place with a couple file cabinets, drawing files and a bookcase under it. This is where ***Puffin*** started. Having always admired Charlie Witholz's catboat designs—and been given a partial set of prints of one of them—I pulled them out to look at and started on a smaller one. To make her a practical cruising boat, I made her a bit deeper, so she had more freeboard—and gained good sitting headroom inside. Living on Puget Sound, shoal draft wasn't as big an issue, so she gained a long, shallow keel and lost the centerboard trunk that all too often intrudes in the middle of the cabin.

To go along with the publication of our book, we hired some friends to work on building one as

Kiyi *underway on Puget Sound*

a demo to show at the Seattle and Vancouver Boat Shows. We took it to the Seattle show with the square welded mesh not fully applied, and worked on adding more mesh on her during the show. At Vancouver, we showed her with the mesh completed, but not yet plastered. It made quite a contrast to the chicken wire boats that were so common there. Thinner shell, but stronger—and lighter!

What we started as a little cruising boat for us to sail and enjoy was sold well before we finished it. However, I did get to cruise her a couple times, bringing her down from Sidney, BC, to Seattle for the next year's show and returning her back there—in January! The trip down to Seattle included some strong winds that eventually led us to reefing down, after watching the unstayed wooden mast gracefully bending as the stronger gusts hit us. And since this was mid-winter, we were very grateful for the wood burning stove that provided much needed warmth and hot meals.

The folks who bought her wanted to have an inboard diesel, so we added an angled shaft and off centerline prop, since she'd been built without a stern tube or aperture for a prop. Seeing this as a desirable addition, we created the ***Liberty*** version, which has a deeper keel and wineglass sections. These changes gave more room for the inboard engine, provided some additional displacement to carry the added weight and more cruising accommodations. Thus, the ***Liberty*** version came to be, with her wineglass section, deeper keel hull which gave good room for the engine. She has the cabin lengthened and the cockpit shortened to give more interior room.

As we went about creating the design, we gave some thought to what we might name the boat. When I'd probably had one Pepsi® too many, I suggested it should be called the Sphinx—a stone cat. . . .

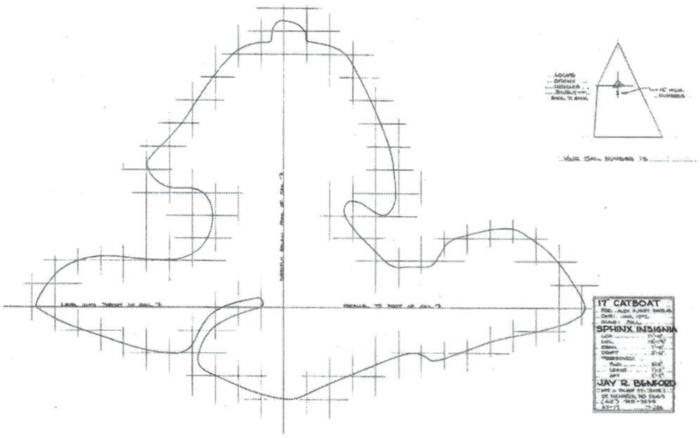

The buyers ultimately decided to name her ***Puffin*** and we've continued using ***Puffin*** as her design name, but the Sphinx logo (see above illustration) is visible on her sail and a template for it is part of her plans. . . .

Too many people who got enthused about ferro-cement bit off more than they could chew, both in the way of man-hours and dollars to see the vessel through to completion. Many of their abandoned projects can still be seen around any coastal city. This small cruising cat is a much better investment of time and money, and will get you cruising much sooner—whatever your material choice is for building her.

Since the original cat was designed, we've done a gaff sloop version, a yawl rigged version and gaff and Marconi cutter versions. And a powerboat version, using the same hull with a shallower keel. The construction plans have been modified for construction in wood, either carvel planked or strip planked with cold-molded veneers over, and for building in foam cored fiberglass. For quickest construction, wood is the way to go. And, it has the most aesthetically pleasing results . . .

Particulars:	***17' Puffin***	***17' Puffin***	***17' Liberty***	***17' Liberty***
Units	Imperial	Metric	Imperial	Metric
Length overall	17'-0"	5.18 m	17'-0"	5.18 m
Length designed waterline	15'-1½"	4.61 m	15'-1½"	4.61 m
Beam	7'-0"	2.13 m	7'-0"	2.13 m
Draft	2'-6"	0.76 m	3'-6"	1.07 m
Freeboard: Forward	3'-4"	1.02 m	3'-4"	1.02 m
Least	1'-11"	0.59 m	1'-11"	0.59 m
Aft	2'-2"	0.66 m	2'-2"	0.66 m
Displacement*, cruising trim.	2,950 Lbs.	1,338 kg.	3,600 Lbs.	1,633 kg.
Displacement-length ratio	381		464	
Ballast, pounds	850	386 kg.	1,200	544 kg.
Ballast ratio	29%		41%	
Sail area, Catboat	200 sq. ft.	18.58 m^2	200 sq. ft.	18.58 m^2
Sail area-displacement ratio	15.56		13.62	
Sail area, Cat rig with jib	260 sq. ft.	24.15 m^2	260 sq. ft.	24.15 m^2
Sail area-displacement ratio	20.22		17.71	
Sail area, Cutter	220 sq. ft.	20.44 m^2	220 sq. ft.	20.44 m^2
Sail area-displacement ratio	17.11		14.99	
Sail area, Cruising Gaff Sloop	210 sq. ft.	19.51 m^2	210 sq. ft.	19.51 m^2
Sail area-displacement ratio	16.34		14.30	
Sail area, Gaff sloop	260 sq. ft.	24.15 m^2	260 sq. ft.	24.15 m^2
Sail area-displacement ratio	20.22		17.71	
Sail area, Marconi sloop	214 sq. ft.	19.88 m^2	214 sq. ft.	19.88 m^2
Sail area-displacement ratio	16.65		14.58	
Sail area, Cruising Yawl rig	233 sq. ft.	21.65 m^2	233 sq. ft.	21.65 m^2
Sail area-displacement ratio	18.12		15.87	
Prismatic coefficient	.53		.53	
Pounds per inch immersion	356		356	
Entrance half-angle	26°		26°	
Water tankage, gallons	20	76 liters	20	76 liters
Fuel tankage, gallons	10	38 liters	10	38 liters
Headroom	4'-4"	1.32 m	4'-10"	1.47 m

***CAUTION:** The figure for displacement quoted here is for the boat in cruising trim. That is, with the fuel and water tanks filled, the crew on board, as well as the crews' gear and stores in the lockers. This should not be confused with the "shipping weight" often quoted as "displacement" by some manufacturers. This should be taken into account when comparing figure and ratios between this and other designs.

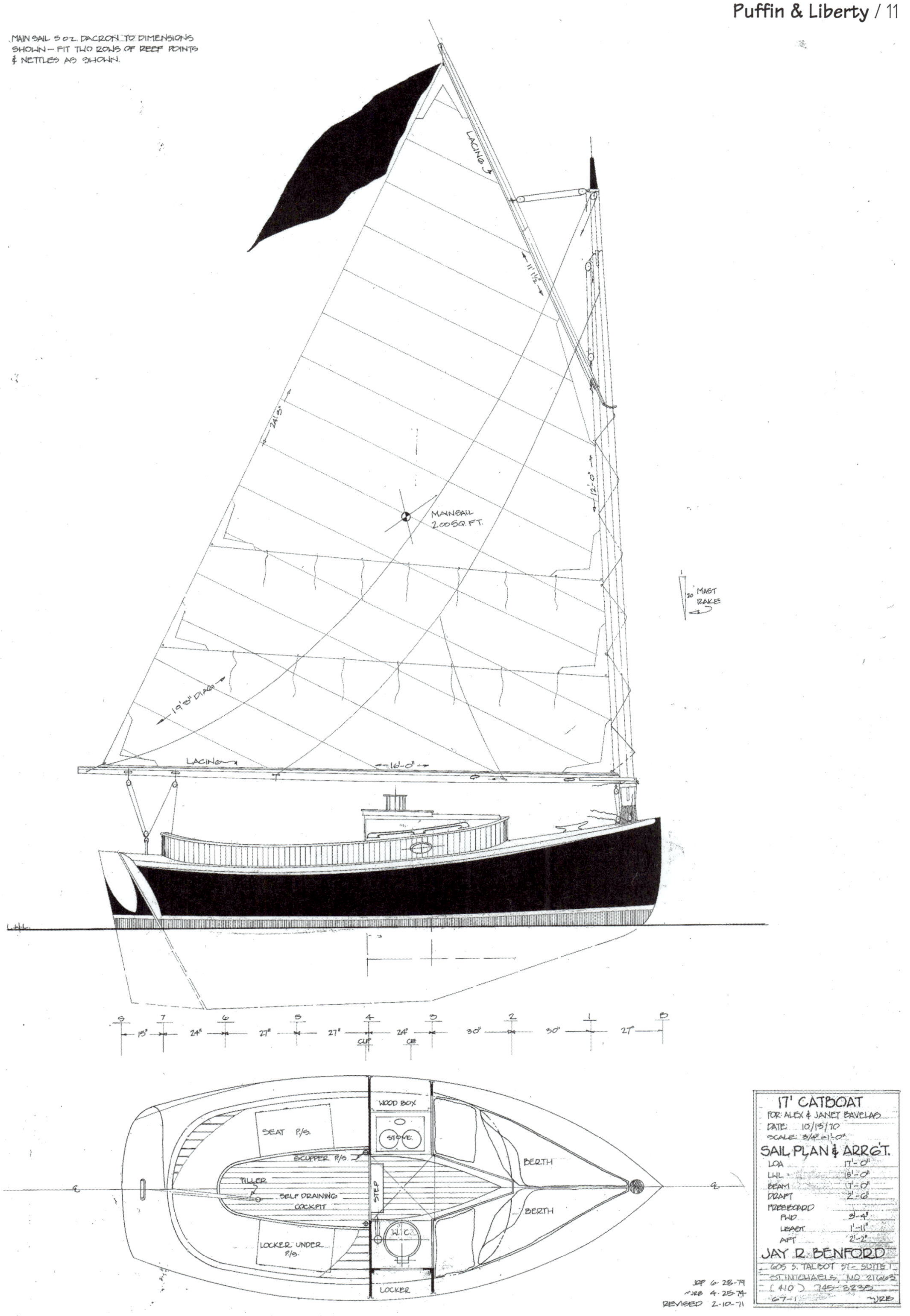
MAIN SAIL 5 OZ. DACRON TO DIMENSIONS SHOWN — FIT TWO ROWS OF REEF POINTS & NETTLES AS SHOWN.
LACING
11'1½"
24'3"
12'-0"
MAINSAIL 200 SQ. FT.
20° MAST RAKE
19'8" DIAG.
LACING
16'-0"
L.W.L.
WOOD BOX
SEAT P/S.
STOVE
SCUPPER P/S.
BERTH
TILLER
SELF DRAINING COCKPIT
STEP
BERTH
W.C.
LOCKER UNDER P/S.
LOCKER
17' CATBOAT
FOR: ALEX & JANET BAVELAS
DATE: 10/13/70
SCALE: 3/4" = 1'-0"
SAIL PLAN & ARRG'T.
LOA 17'-0"
LWL 15'-0"
BEAM 7'-0"
DRAFT 2'-6"
FREEBOARD
FWD 3'-4"
LEAST 1'-11"
AFT 2'-2"
JAY R. BENFORD
605 S. TALBOT ST – SUITE 1
ST. MICHAELS, MD 21663
(410) 745-3235
67-1
JRB
REVISED 2-10-71

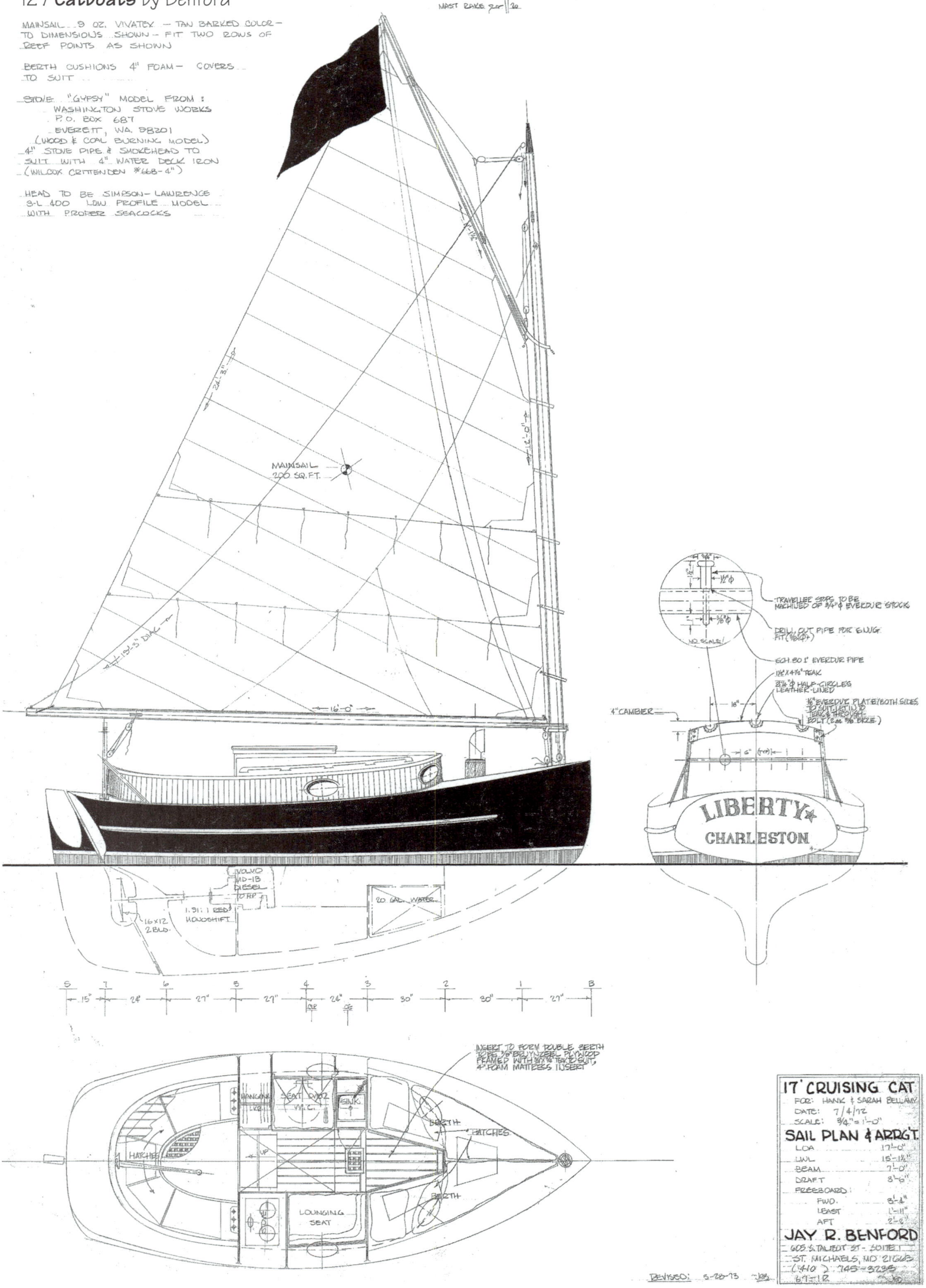
MAINSAIL 9 OZ. VIVATEX — TAN BARKED COLOR — TO DIMENSIONS SHOWN — FIT TWO ROWS OF REEF POINTS AS SHOWN
BERTH CUSHIONS 4" FOAM — COVERS TO SUIT
STOVE "GYPSY" MODEL FROM: WASHINGTON STOVE WORKS P.O. BOX 687 EVERETT, WA. 98201 (WOOD & COAL BURNING MODEL)
4" STOVE PIPE & SMOKEHEAD TO SUIT WITH 4" WATER DECK IRON (WILCOX CRITTENDEN #668-4")
HEAD TO BE SIMPSON-LAWRENCE S-L 400 LOW PROFILE MODEL WITH PROPER SEACOCKS
MAST RAKE
MAINSAIL 200 SQ.FT.
24'-3"
12'-0"
16'-0"
4" CAMBER
TRAVELLER STOPS TO BE MACHINED OF 3/4" EVERDUR STOCK
DRILL OUT PIPE FOR SNUG FIT
NO SCALE
SCH.80 1" EVERDUR PIPE
1½" X 4½" TEAK
LEATHER-LINED
LIBERTY
CHARLESTON
VOLVO MD-1B DIESEL 10 HP
1.91:1 RED. HOWOSHIFT
16X12 2BLD.
20 GAL. WATER
15" 24" 27" 27" 24" 30" 30" 27"
HANGING LKR.
SEAT OVER W.C.
SINK
BERTH
HATCHES
UP
LOUNGING SEAT
17' CRUISING CAT
FOR: HANK & SARAH BELLAMY
DATE: 7/4/72
SCALE: 3/4" = 1'-0"
SAIL PLAN & ARRG'T.
LOA 17'-0"
LWL 15'-1½"
BEAM 7'-0"
DRAFT 3'-6"
FREEBOARD:
FWD. 3'-4"
LEAST 1'-11"
AFT 2'-8"
JAY R. BENFORD
605 S. TALBOT ST. - SUITE 1
ST. MICHAELS, MD 21663
(410) 745-3235
REVISED: 5-28-73

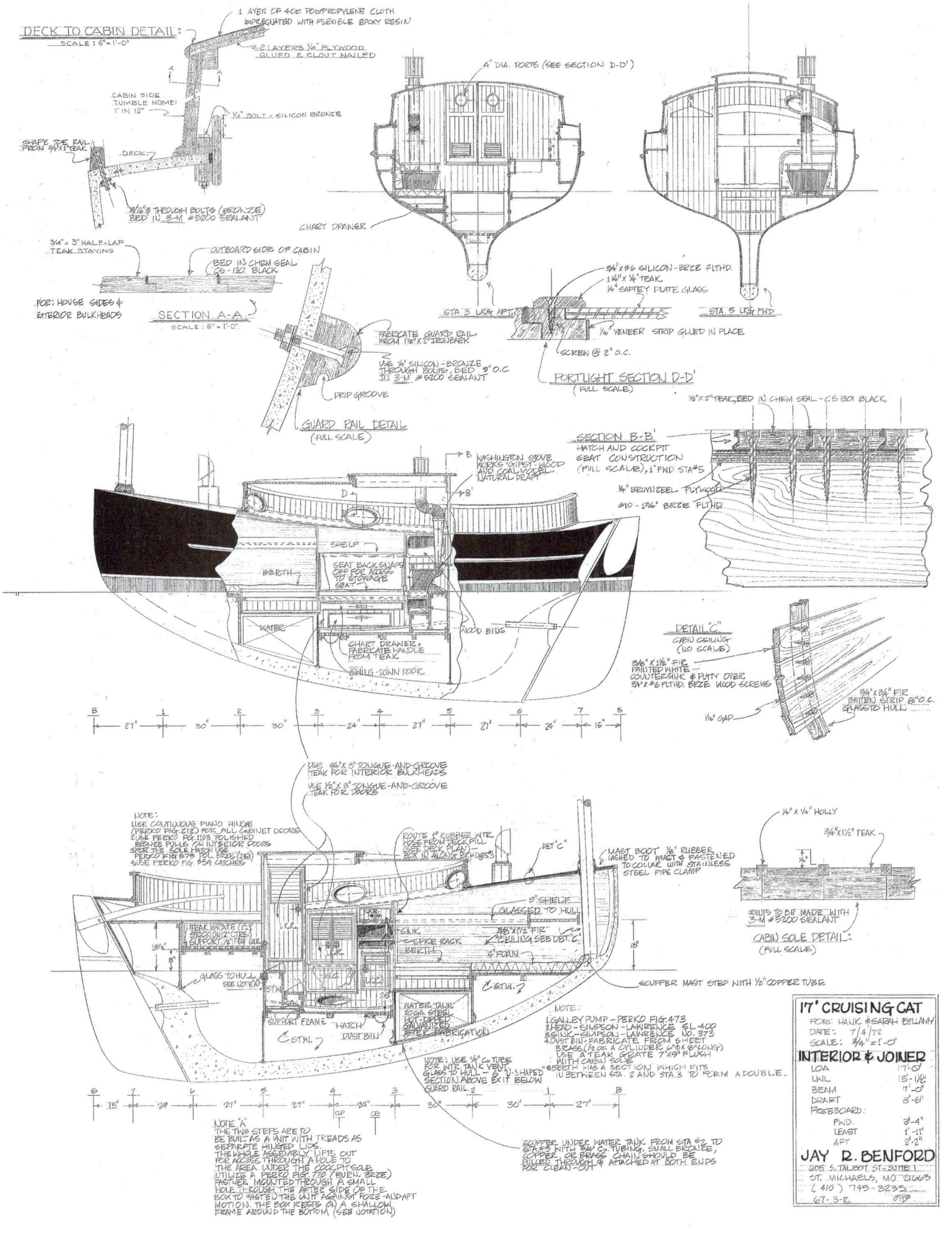
DECK TO CABIN DETAIL:
SCALE: 6" = 1'-0"
1 AYER OF 40oz POLYPROPYLENE CLOTH IMPREGNATED WITH FLEXIBLE EPOXY RESIN
2 LAYERS 1/4" PLYWOOD GLUED & CLOUT NAILED
CABIN SIDE TUMBLE HOME: 1" IN 12"
1/4" BOLT - SILICON BRONZE
SHAPE TOE RAIL FROM 3/4"x1" TEAK
DECK
3/16" ϕ THROUGH BOLTS (BRONZE) BED IN 3-M #5200 SEALANT
3/4" x 3" HALF-LAP TEAK STAVING
OUTBOARD SIDE OF CABIN
BED IN CHEM SEAL CS-130 BLACK
FOR: HOUSE SIDES & EXTERIOR BULKHEADS
SECTION A-A
SCALE: 6" = 1'-0"
FABRICATE GUARD RAIL FROM 1 1/2" x 2" IRONBARK
USE 1/4" SILICON-BRONZE THROUGH BOLTS, BED 9" O.C. IN 3-M #5200 SEALANT
DRIP GROOVE
GUARD RAIL DETAIL
(FULL SCALE)
4" DIA. PORTS (SEE SECTION D-D')
CHART DRAWER
STA. 3 LKG. AFT.
STA. 5 LKG. FWD.
3/4" x #6 SILICON-BRZE FLTHD.
1 1/4" x 1/4" TEAK
1/4" SAFTEY PLATE GLASS
1/16" VENEER STRIP GLUED IN PLACE
SCREW @ 2" O.C.
PORTLIGHT SECTION D-D'
(FULL SCALE)
1/2" x 2" TEAK, BED IN CHEM SEAL - CS 1301 BLACK
SECTION B-B'
HATCH AND COCKPIT SEAT CONSTRUCTION
(FULL SCALE), 1' FWD STA #5
1/4" BRUYNZEEL PLYWOOD
#10 - 1 3/4" BRZE FLTHD.
WASHINGTON STOVE WORKS "GIPSY" WOOD AND COAL MODEL NATURAL DRAFT
SHELF
SEAT BACK SNAPS OFF FOR ACCESS TO STOWAGE SEAT
BERTH
WATER
WOOD BINS
CHART DRAWER, FABRICATE HANDLE FROM TEAK
SWING-DOWN DOOR
DETAIL "C"
CABIN CEILING
(NO SCALE)
3/8" x 1 1/2" FIR PAINTED WHITE COUNTERSINK & PUTTY OVER 3/4" x #6 FLTHD. BRZE WOOD SCREWS
3/4" x 3/4" FIR BATTEN STRIP 8" O.C. GLASS TO HULL
1/16" GAP
B 1 2 3 4 5 6 7 S
27" 30" 30" 24" 27" 27" 24" 16"
USE 3/4" x 3" TONGUE-AND-GROOVE TEAK FOR INTERIOR BULKHEADS
USE 1/2" x 3" TONGUE-AND-GROOVE TEAK FOR DOORS
NOTE:
1. USE CONTINUOUS PIANO HINGE (PERKO FIG. 212) FOR ALL CABINET DOORS
2. USE PERKO FIG. 1103 POLISHED BRONZE PULLS ON INTERIOR DOORS
3. FOR THE SOLE HATCH USE PERKO FIG 575 POL. BRZE (2EA)
4. USE PERKO FIG. 554 CATCHES
ROUTE 1" RUBBER WTR. HOSE FROM DECK FILL (SEE DECK PLAN) BOX IN ALONG BLKHDS.3
DET "C"
MAST BOOT 1/8" RUBBER LASHED TO MAST & FASTENED TO COLLAR WITH STAINLESS STEEL PIPE CLAMP
9" SHELF GLASSED TO HULL
SINK
SPICE RACK
BERTH
3/8" x 1 1/2" FIR CEILING SEE DET. C
4" FOAM
TEAK GRATE (1"x1" STOCK ON 2" CTRS.) SUPPORT 1/2" FOR GLE
GLASS TO HULL SEE NOTE "A"
SCUPPER MAST STEP WITH 1/2" COPPER TUBE
WATER TANK 20 GA. STEEL HOT-DIPPED GALVANIZED AFTER FABRICATION
SUPPORT FRAME
HATCH
DUST BIN
NOTE:
1. GALLEY PUMP - PERKO FIG. 473
2. HEAD - SIMPSON - LAWRENCE SL-400
3. SINK - SIMPSON - LAWRENCE NO. 373
4. DUST BIN - FABRICATE FROM SHEET BRASS (1/2 OF A CYLINDER 6"ϕ x 8" LONG) USE A TEAK GRATE 7"x9" FLUSH WITH CABIN SOLE
5. BERTH HAS A SECTION WHICH FITS IN BETWEEN STA. 2 AND STA. 3 TO FORM A DOUBLE.
NOTE: USE 1/4" Cu. TUBE FOR WTR. TANK VENT, GLASS TO HULL - & U-SHAPED SECTION ABOVE EXIT BELOW GUARD RAIL.
S 7 6 5 4 3 2 1 B
15" 24" 27" 27" 24" 30" 30" 27"
NOTE "A"
THE TWO STEPS ARE TO BE BUILT AS A UNIT WITH TREADS AS SEPARATE HINGED LIDS. THE WHOLE ASSEMBLY LIFTS OUT FOR ACCESS THROUGH A HOLE TO THE AREA UNDER THE COCKPIT SOLE. UTILIZE A PERKO FIG. 770 (BURN. BRZE) FASTNER MOUNTED THROUGH A SMALL HOLE THROUGH THE AFTER SIDE OF THE BOX TO FASTEN THE UNIT AGAINST FORE-AND-AFT MOTION. THE BOX RESTS ON A SHALLOW FRAME AROUND THE BOTTOM. (SEE NOTATION)
SCUPPER UNDER WATER TANK FROM STA #2 TO STA #3 WITH 3/4" Cu. TUBING. SMALL BRONZE, COPPER, OR BRASS CHAIN SHOULD BE PULLED THROUGH & ATTACHED AT BOTH ENDS FOR CLEAN-OUT.
1/4" x 1/4" HOLLY
3/4" x 1 1/2" TEAK
JOINTS TO BE MADE WITH 3-M #5200 SEALANT
CABIN SOLE DETAIL:
(FULL SCALE)
17' CRUISING CAT
FOR: HANK & SARAH BELLAMY
DATE: 7/4/78
SCALE: 3/4" = 1'-0"
INTERIOR & JOINER
LOA 17'-0"
LWL 15'-1 1/2"
BEAM 7'-0"
DRAFT 3'-6"
FREEBOARD:
FWD. 3'-4"
LEAST 1'-11"
AFT 2'-2"
JAY R. BENFORD
605 S. TALBOT ST - SUITE 1
ST. MICHAELS, MD 21663
(410) 745-3235
67-3-8

STATION	S	7	6	5	4	3	2	1
	HEIGHTS							
℄ TO KEEL		(0'-3 1/4")	0'-0"	0'-1 1/4"	0'-4"	0'-8 1/4"	1'-4 3/4"	2'-6"
" " B1	(4'-1 1/2")	3'-8 1/2"	3'-1 1/16"	2'-7"	2'-4"	2'-4 1/4"	2'-9 1/4"	3'-10"
" " B2	(4'-5 7/8")	4'-0 1/8"	3'-4 1/4"	2'-10 5/8"	2'-8 3/4"	2'-10 5/16"	3'-4 13/16"	—
LWL TO BOOTTOP	0'-5 5/8"	0'-5 1/8"	0'-4 5/8"	0'-4 3/8"	0'-4 1/4"	0'-4 1/2"	0'-5 1/4"	0'-6 1/4"
" " SHEER	2'-2 5/8"	2'-0 3/4"	1'-10 15/16"	1'-10 1/2"	1'-11 5/8"	2'-1 5/8"	2'-5 3/8"	2'-10 5/8"
	HALF-BREADTHS							
℄ TO KEEL HALF-SIDING	—	—	0'-3/4"	0'-3 1/4"	0'-4"	0'-3 3/4"	0'-2 1/4"	0'-1/2"
" " 34" WL	—	—	0'-3 3/4"	1'-10"	2'-2 3/4"	1'-11 5/16"	1'-1 3/16"	0'-2 1/4"
" " 42" WL	—	0	2'-4 9/16"	3'-1 5/16"	3'-2 1/8"	2'-11 3/8"	2'-1"	0'-8 11/16"
" " 50" WL	2'-0 3/4"	2'-5 15/16"	3'-0 5/8"	3'-4 7/8"	3'-5 7/8"	3'-3 1/4"	2'-6 3/8"	1'-2 3/8"
" " 58" WL	2'-3 3/4"	2'-7 7/8"	3'-1 1/4"	3'-5"	3'-6 3/8"	3'-4"	2'-8 3/4"	1'-5 1/16"
" " SHEER	2'-0 1/4"	2'-5 5/8"	3'-0 1/4"	3'-4 7/16"	3'-5 3/4"	3'-3 13/16"	2'-8 1/4"	1'-8 1/2"
	DIAGONALS							
DIAG. A	—	—	3 1/2"	9"	12 5/8"	12 7/16"	8 1/8"	2 7/16"
DIAG. B	—	—	1 3/8"	5 5/16"	7 1/8"	6 3/4"	5 5/8"	—

NOTES:

LINES MUST BE LOFTED FULL SIZE- DO NOT SCALE PRINTS - LINES & OFFSETS TO THE OUTSIDE OF THE SHELL. DEDUCT APPROPRIATE SHELL THICKNESS FOR MOLDS, WEB FRAMES BHDS. ETC.

SHELL THICKNESS 3/8"

THESE PLANS ARE THE PROPERTY OF THE DESIGNER & MAY BE USED ONLY AS AUTHORIZED. NO MORE THAN ONE BOAT WILL BE BUILT FROM THESE PLANS WITHOUT PERMISSION OF THE DESIGNERS.

ANY ALTERATION FROM THESE PLANS RELIEVES THE DESIGNER OF FURTHER RESPONSIBILITY.

DECK CAMBER IS 3" IN 7'-0"

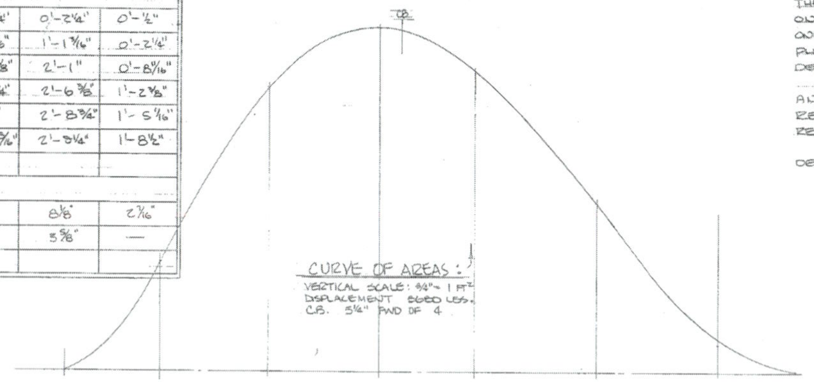

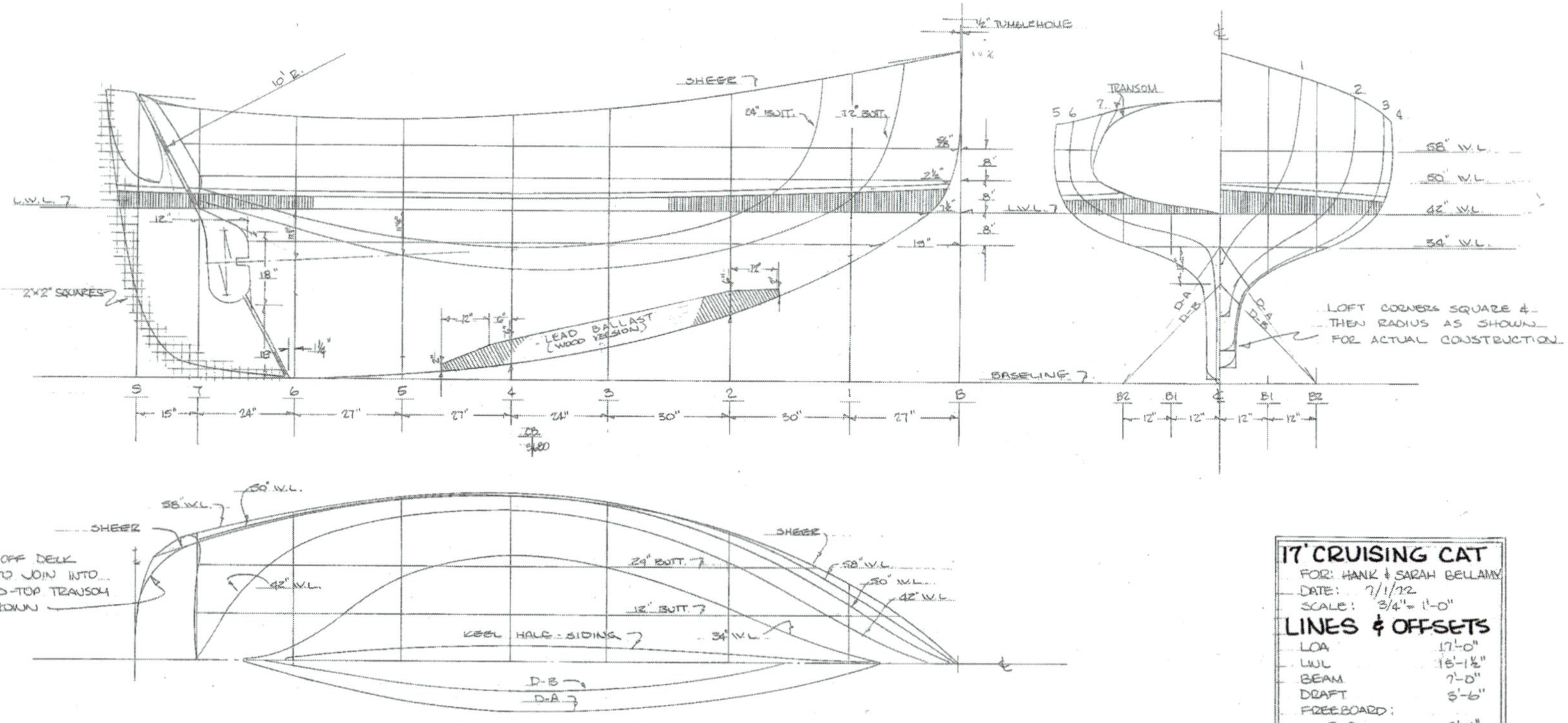

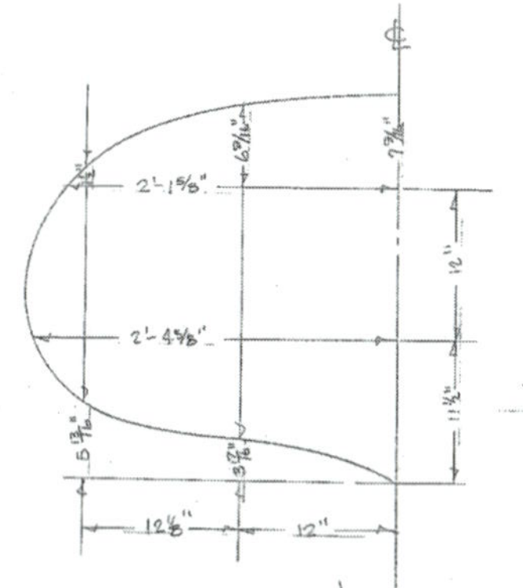

EXPANDED TRANSOM TO INSIDE OF SHELL (OUTSIDE OF FRAME)

17' CRUISING CAT
FOR: HANK & SARAH BELLAMY
DATE: 7/1/72
SCALE: 3/4" = 1'-0"

LINES & OFFSETS

LOA	17'-0"
LWL	15'-1½"
BEAM	7'-0"
DRAFT	3'-6"
FREEBOARD:	
FWD	3'-4"
LEAST	1'-11"
AFT	2'-2"

JAY R. BENFORD
605 S. TALBOT ST - SUITE 1
ST. MICHAELS, MD 21663
(410) 745-3235
67-2R ~JRB

REVISED:
16 FEB 95 T.W.F.
3-3-75 ~JRB
2-2-74 ~JRB
2-13-73 RMP
11-27-72 RMP
8-18-72 ~JRB
8-16-72 ~JRB

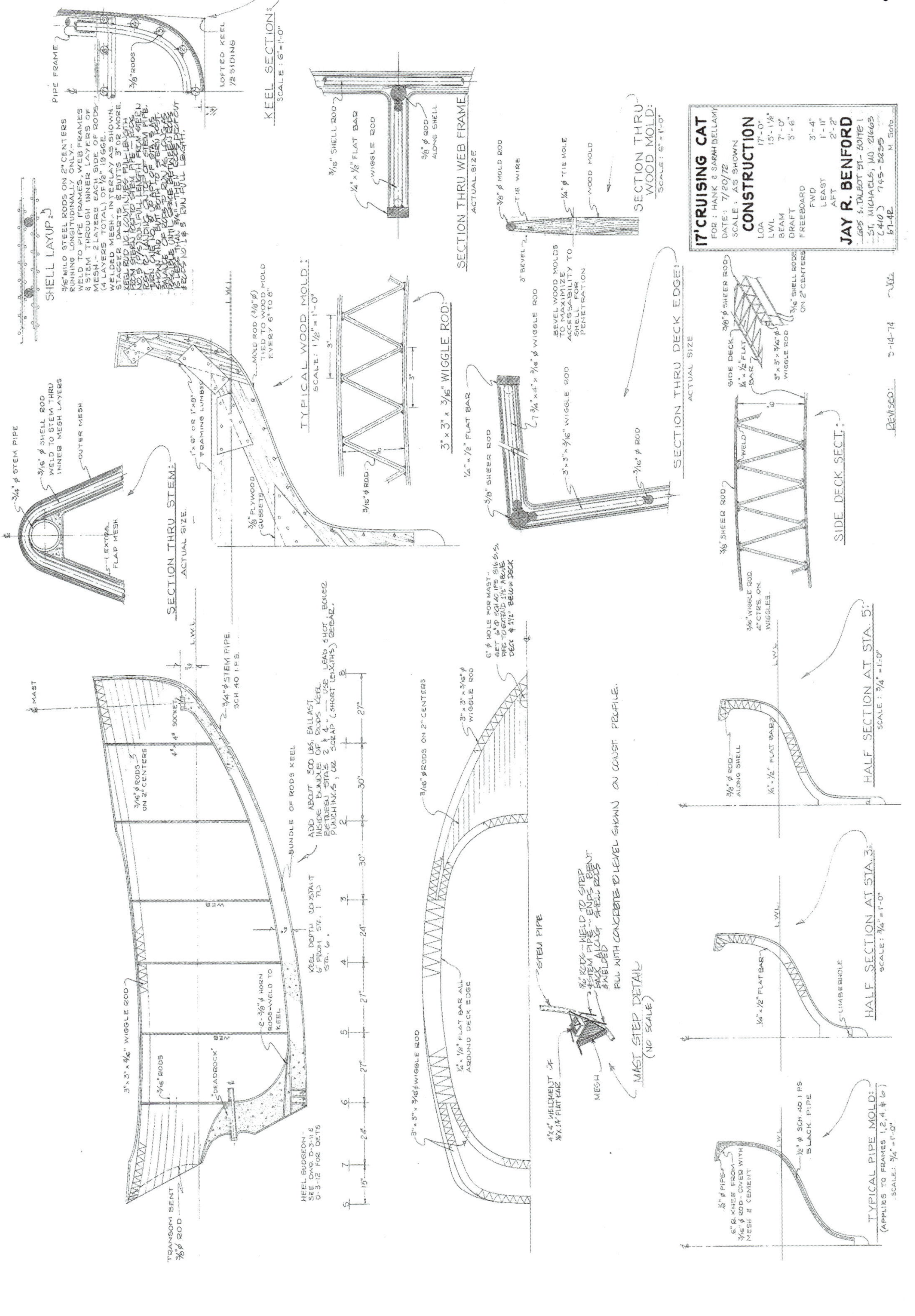
17' CRUISING CAT
FOR: HANK & SARAH BELLAMY
DATE: 7/20/72
SCALE: AS SHOWN
CONSTRUCTION
LOA 17'-0"
LWL 15'-1 1/2"
BEAM 7'-0"
DRAFT 3'-6"
FREEBOARD FWD 3'-4"
LEAST 1'-11"
AFT 2'-2"
JAY R. BENFORD
REVISED: 3-14-74
KEEL SECTION: SCALE: 6" = 1'-0"
SHELL LAYUP:
SECTION THRU WEB FRAME ACTUAL SIZE
SECTION THRU WOOD MOLD: SCALE: 6" = 1'-0"
TYPICAL WOOD MOLD: SCALE: 1 1/2" = 1'-0"
3" x 3" x 3/16" WIGGLE ROD:
SECTION THRU STEM: ACTUAL SIZE
SECTION THRU DECK EDGE: ACTUAL SIZE
SIDE DECK SECT.:
HALF SECTION AT STA. 5: SCALE: 3/4" = 1'-0"
HALF SECTION AT STA. 3: SCALE: 3/4" = 1'-0"
TYPICAL PIPE MOLD: (APPLIES TO FRAMES 1, 2, 4, & 6) SCALE: 3/4" = 1'-0"
MAST STEP DETAIL (NO SCALE)
FILL WITH CONCRETE TO LEVEL SHOWN ON CONST. PROFILE.
BUNDLE OF RODS KEEL
KEEL DEPTH CONSTANT 6" FROM STA. 1 TO STA. 6
HEEL GUDGEON - SEE DWG. D-3-11 & D-3-12 FOR DETS
TRANSOM BENT 3/8" ROD
3/4" ø STEM PIPE SCH 40 I.P.S.
1/2" ø SCH. 40 I.P.S. BLACK PIPE
3" x 3" x 3/16" ø WIGGLE ROD
1/4" x 1/2" FLAT BAR ALL AROUND DECK EDGE
3/16" ø RODS ON 2" CENTERS
LWL

The interior for ***Puffin*** *going together. The thru-hull openings for the seacocks for the head can be seen here. The lift trap in the berth flat provides access to the stowage under, right.*

SCANTLINGS:

(1) LEAD BALLAST — 1200 LBS — TAKE SHAPE FROM LINES & LOFTING — 7 - ½"ø BRONZE KEEL BOLTS ON 12" CENTERS LOCATED AS SHOWN.

(2) STEM, KNEES, KEEL, & STERNPOST FROM "NOMINAL" 4" STOCK SIZES AS NOTED — CAUTION — A "4×4" IS ACTUALLY ABOUT 3½" × 3½" & SO ON — ACTUAL SIZED PIECES ARE TO BE USED & TRIMMED TO FIT AS SHOWN — ALL TO BE CLEAR FIR.

(3) PLANKING TO BE RED CEDAR (OR PHIL. MAHOG. OR WHITE CEDAR) 5/8" THICK STRIPS 3/4" WIDE WITH TOPS & BOTTOMS MILLED WITH HOLLOW & CROWN AS SHOWN — EDGE NAIL (ANCHORFAST 1½"-13GGE) ON STAGGERED PATTERN AS ILLUSTRATED.

(4) TOP OF LEAD KEEL TO BE RABBET LINE — RABBET FWD. & AFT OF KEEL AS SHOWN FAIRS INTO STEM & STERNPOST.

(5) 1⅛" × 1½" ALASKA CEDAR DECK BEAMS SAWN TO 4" IN 7'-0" CAMBER — NOTCH FOR HARPIN & CARLIN & BOLT TOGETHER — AT STA'S 1, 2, & 7

(6) KNEES SHAPED AS SHOWN SAWN FROM 2 PCS. OF 1⅛" × 3½" ALASKA CEDAR AT ALL STAS. — AT STA'S. 4 & 6 FIT DIRECTLY TO HARPIN & CARLIN (NO DECK BEAM AT 4 & 6)

(7) 3/4" × 3½" ALASKA CEDAR HARPIN SAWN TO SHAPE AS SHOWN — SCARF & BOLT AS SHOWN

(8) 3/4" × 2½" ALASKA CEDAR CARLIN SAWN TO SHAPE AS SHOWN — SCARF & BOLT AS SHOWN

(9) TOERAIL FROM 3/4" × 1⅛" TEAK — TAPER TOP TO 5/8" — THRU-BOLT TO HARPIN (3/16"ø BOLTS ON 9 TO 12" CENTERS)

(10) SHEER GUARD FROM 3/4" × 1½" IRONBARK — CUT DRIP GROOVE AS SHOWN — SCREW THRU INTO HARPIN.

(11) NOTE SQUARE CORNER ON TRANSOM-SHEER INTERSECTION — USE EXPANDED TRANSOM SHAPE ON SHEET 67-3 FOR TRANSOM FRAME — NOTE THIS SHAPE IS FOR AFT FACE OF FRAME & FRAME BEVELS OUT FOR GREATER WIDTH AS IT GOES FORWARD — TAKE BEVELS FROM LOFT FLOOR.

(12) 3/4" IRONBARK STEM CAP & WORM GUARD ALONG BOTTOM EDGE OF KEEL — ROUND OFF AS SHOWN & TO FAIR INTO KEEL SECTIONS

(13) NOTE: ALL PLYWOOD BULKHEADS TO STOP 1" CLEAR OF KEEL TO KEEP BILGE WATER OUT OF END GRAIN & TO PROVIDE LIMBERS.

(14) 16 GGE. STEEL FUEL TANK TO FIT SPACE SHOWN — LEAVE STEEL BARE INSIDE & EPOXY PAINT OUTSIDE — APPX. 16 TO 20 GALS. — PROVIDE FILL, VENT, & SUPPLY LINE FITTINGS TO SUIT ENGINE.

(15) BULKHEADS ½" MARINE PLYWOOD (FIR) OR 3/8" BRUYNZEEL PLY

(16) BULKHEAD FRAME TO BE 1⅛" × 1½" YELLOW (ALASKA) CEDAR SAWN TO SHAPE — ON FWD. SIDE OF BHD'S. FWD. OF STA. 4 & ON AFT SIDE OF BHD'S. AFT OF STA. 4

(17) 1½" ALASKA CEDAR FLOORS AT ENGINE BEDS — AFT FLOOR CENTERED AT AFT ENG. MOUNT & FWD. FLOOR CENTERED 1¾" FWD. OF FWD. ENG. MOUNT — CARRY FLOORS UP TO TOP OF BEDS & NOTCH DOWN TO PROVIDE ½" CLEARANCE TO ENGINE

(18) ENGINE BEDS 2½" × 3½" W. OAK (OR FIR) FITTED TO HULL AS SHOWN — NOTCH FWD. FLOOR FOR BED & LAG. FASTEN — BEDS TO BE 4½" OFF ℄ TO INSIDE FACE & RUN 2½" FWD. OF FWD. ENGINE MOUNT — GET NARROW MOUNTS ON VOLVO MD-1B (1.91:1 MONOSHIFT) DIESEL.

(19) CABIN & COAMING 11" HIGH OFF DECK ALL AROUND — HEIGHT CONSTANT & 1" IN 12" TUMBLEHOME ALL AROUND — 3/4" × 3½" TEAK STAVES RUN VERTICALLY — HALF-LAP SEAMS & FILL WITH KNIFING GRADE OF THIOKOL PER DETAILED SECTION — NOTE OUTBOARD SEAM TO BE WIDER WITH MORE THIOKOL & INBOARD SEAM SET TIGHT.

(20) CABIN FRAME ON DECK FROM 1¼" × 1¼" ALASKA CEDAR SAWN TO SHAPE AT ENDS & BEVELED OUTBOARD FOR TUMBLEHOME OF HOUSE & COAMING — THRU-BOLT (3/16"ø GALV.) TO CARLIN ON 8" CTRS. & BED IN THIOKOL.

(21) FACIA OF ½" × 3" TEAK — TO PROJECT ABOVE CABIN FRAME AS SHOWN TO FORM HAND-GRAB.

(22) HOUSETOP FRAME FROM 1⅛" × 1½" ALASKA CEDAR — BEVEL FOR HOUSETOP

(23) ½" × 1¼" TEAK MOLDINGS — CUT BEVELS AS SHOWN & BED IN THIOKOL.

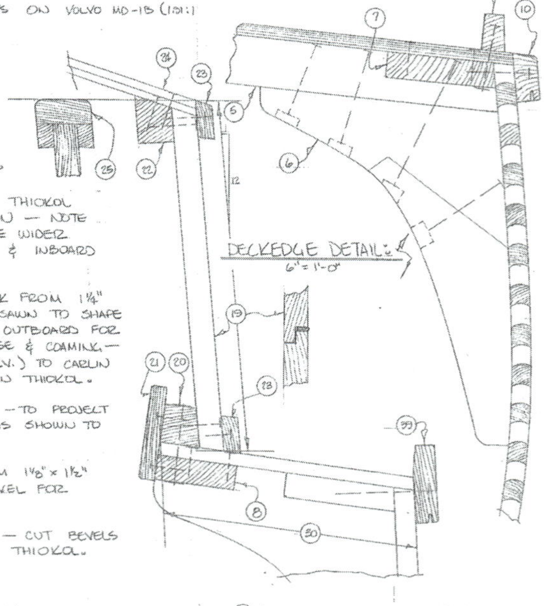

(24) HOUSETOP 2 LAYERS ¼" BRUYNZEEL PLYWOOD CLOUT NAILED & GLUED TOGETHER — STAGGER BUTTS FROM ONE LAYER TO NEXT — COVER WITH 2 LAYERS OF 3/4 OZ. FIBERGLASS MAT & RESIN — RUN MAT OVER EDGES & END UNDER MOLDING.

(25) COCKPIT COAMING TO FORM CONTINUATION OF LINE OF HOUSETOP AS SHOWN — CAP TO BE 3/4" × 1¾" & CHEEKS ½" × 3/4" EACH SIDE — ALL TEAK

(26) DECK 3/8" BRUYNZEEL OR ½" FIR PLYWOOD — ANCHORFAST NAIL TO DECK FRAMING ON 6 TO 8" CENTERS.

(27) BOTTOM EDGES OF DECK BEAMS MAY BE CUT AS SHOWN IF DESIRED — KNEES ALSO

(28) 1½" × 11"+ BREASTHOOK & MAST BLOCK — RABBET OUT FOR HARPIN & SAW CAMBER IN TOP — THRU-BOLT TO HARPIN — ALASKA CEDAR

(29) 3/4" ALASKA CEDAR KNEES — THRU-BOLT TO HARPIN

(30) INCREASE TO 8" FROM STA. 3 TO STA 6 & FAIR INTO CURVES AT ENDS

(31) 1⅛" × 2½" ALASKA CEDAR TRANSOM FRAME SAWN TO SHAPE & CAMBER — SEE NOTE (11)

(32) 3/4" × 3½" TEAK TRANSOM PLANKING — HALF-LAP ETC. PER NOTE (19)

(33) 1⅛" × 1½" ALASKA CEDAR FRAMING FOR COCKPIT SEAT

(34) COCKPIT SEAT FLATS 3/8" BRUYNZEEL OR ½" FIR PLYWOOD — HATCHES BY OWNER TO SUIT

(35) FIR BLOCK FOR MAST STEP BUILT UP FROM 4" STOCK TO 8" ABOVE LWL AS SHOWN 3" × 3" × 2" DEEP SOCKET FOR MAST HEEL (NOTE — REVISE SIZE OF MAST HEEL TO FIT)

(36) 15 GAL. COLLAPSIBLE PLASTIC WATER TANK FROM WEST PRODUCTS OR EQUAL — CONNECT TO SINK PUMP & FILL THRU DECK OR IN CABIN TO SUIT

(37) CUT SLOT IN TRANSOM FOR TILLER PER SIZING ON SHEET 67-6R

(38) REVISE COCKPIT SHAPE AS SHOWN

(39) FOR MORE SIDEDECK WIDTH, TOERAIL MAY BE OMITTED & SHEER GUARD EXTENDED 1" ABOVE DECK — TAPER OFF INBOARD SIDE TO MAKE TOP 5/8" WIDE — TOTAL SIZE OF SHEER GUARD BECOMES 3/4" × 2½"

(40) CABIN SOLE BEAMS 1⅛" × 1½" ALASKA CEDAR — SOLE ½" PLYWOOD — PROVIDE LIFT OUT SECTION FOR BILGE ACCESS TO SUIT

(41) BUILD COMPANIONWAY STEPS TO LIFT OFF FOR ACCESS TO ENGINE

(42) PLANKING TO BE ½" THICK IF POLYPROPELENE CLOTH & FLEXIBLE RESIN SHEATHING OF HULL IS TO BE DONE — SHEATHING TO BE 2 LAYERS INSIDE & OUT — DOUBLED OVER OUTSIDE OF KEEL.

(43) SEE DETAIL DRAWING D-55 FOR ALTERNATIVE COLD-MOLDED VEENER APPLICATION OVER A THINNER STRIP-PLANKED HULL.

The wooden construction plan is shown on these two pages. The 17' **Liberty** *sistership shown left, was built in Sweden and the* **Puffin** *sistership* **Melinda** **Lee** *was built in Port Moody, British Columbia, by Melvin Bottaro. They are typical of the results of hundreds of the plans sold.*

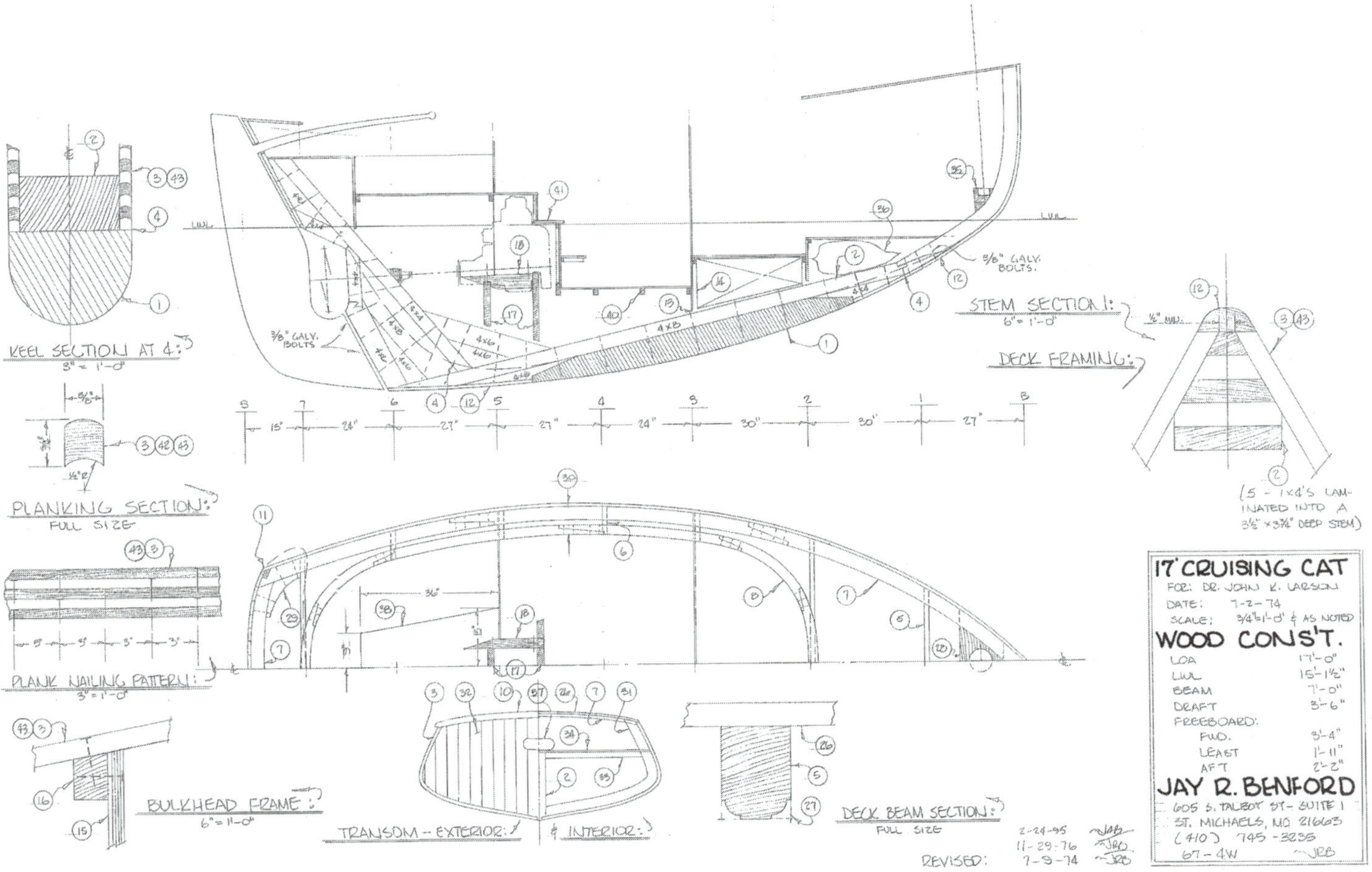

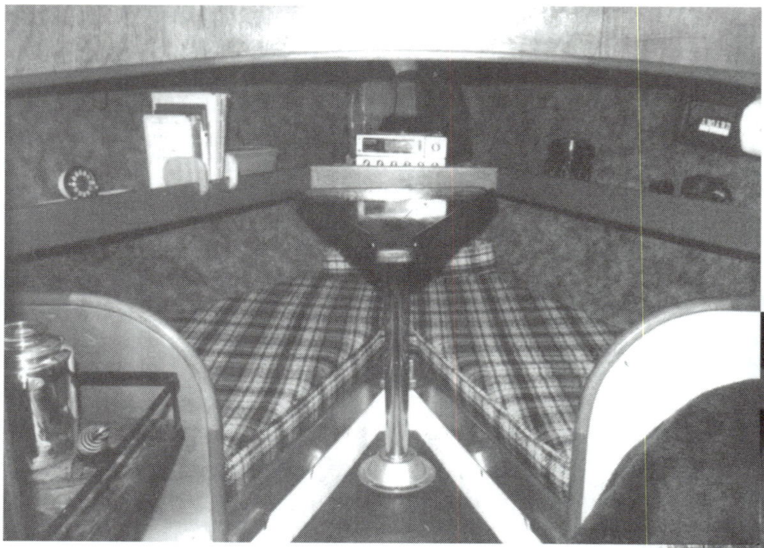

Upper left, ***Melinda Lee****'s v-berth insert is here transformed into a table. Lower right, note the ventilators on the companionway doors and the shelves behind the stove and outboard of the berths.*

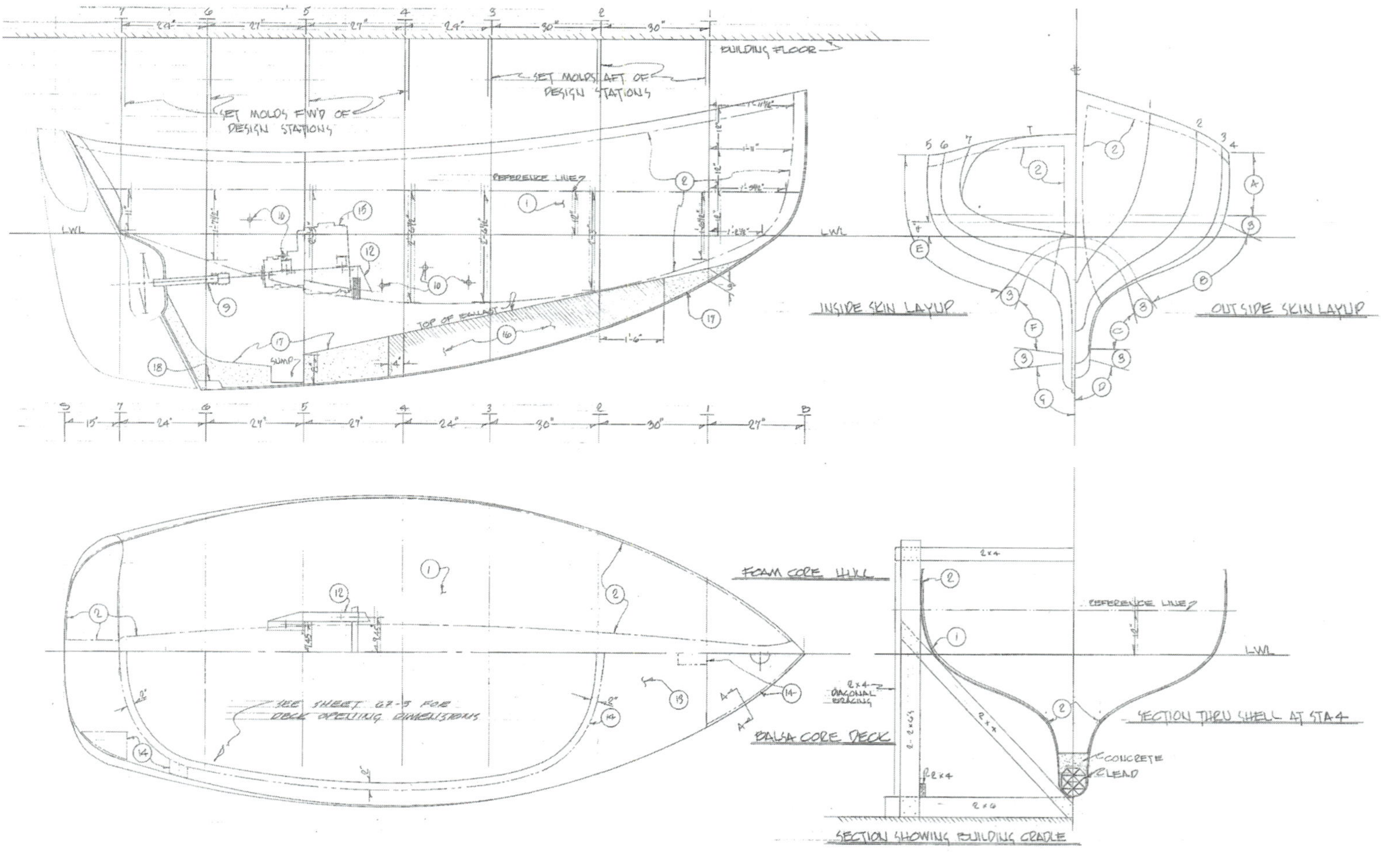

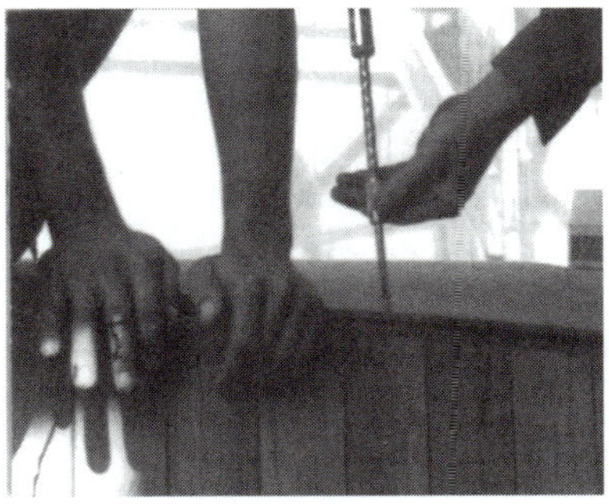

Building **Puffin***'s cabin with vertical teak staves and laminated plywood top.*

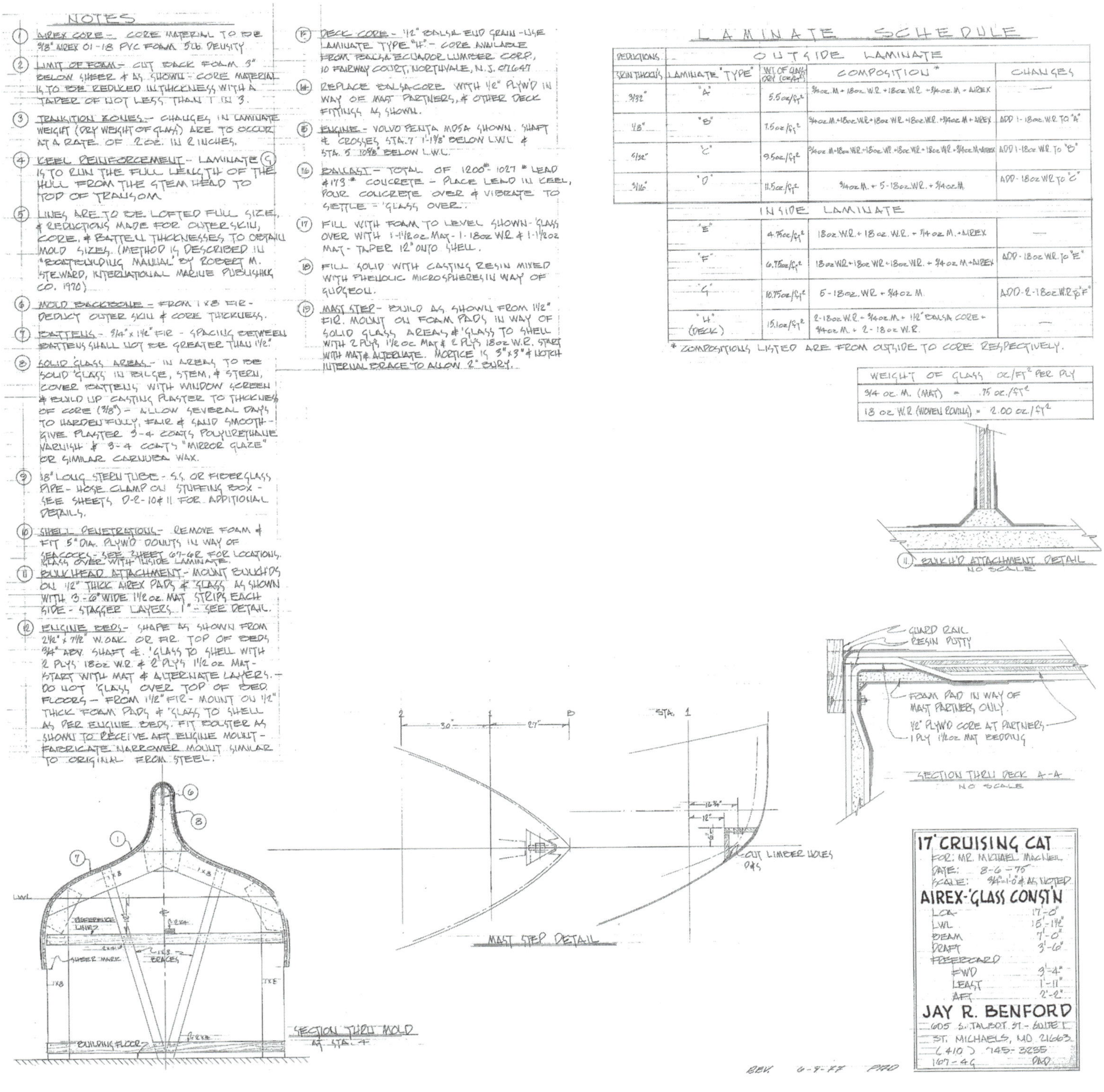

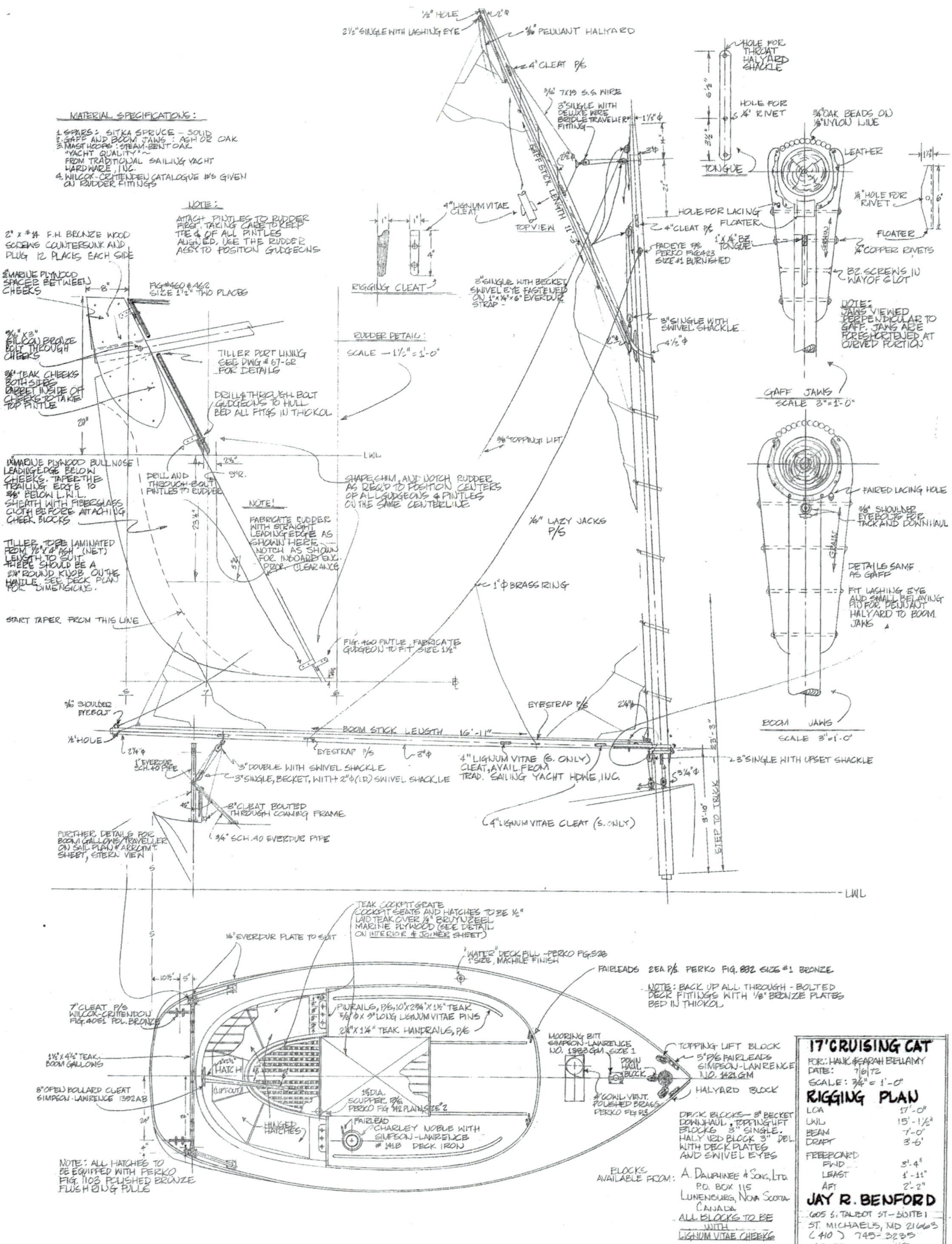

MATERIAL SPECIFICATIONS:
1. SPARS: SITKA SPRUCE – SOLID
2. GAFF AND BOOM JAWS: ASH OR OAK
3. MAST HOOPS: STEAM-BENT OAK "YACHT QUALITY" ~ FROM TRADITIONAL SAILING YACHT HARDWARE, INC.
4. WILCOX-CRITTENDEN CATALOGUE #'S GIVEN ON RUDDER FITTINGS
NOTE: ATTACH PINTLES TO RUDDER FIRST, TAKING CARE TO KEEP THE ℄ OF ALL PINTLES ALIGNED. USE THE RUDDER ASS'Y TO POSITION GUDGEONS
2" x #14 F.H. BRONZE WOOD SCREWS COUNTERSUNK AND PLUG 12 PLACES EACH SIDE
MARINE PLYWOOD SPACER BETWEEN CHEEKS
FIG #460 & 462 SIZE 1½" TWO PLACES
BRONZE BOLT THROUGH CHEEKS
¾" TEAK CHEEKS BOTH SIDES. RABBET INSIDE OF CHEEKS TO TAKE TOP PINTLE
TILLER PORT LINING SEE DWG # 67-6R FOR DETAILS
DRILL & THROUGH-BOLT GUDGEONS TO HULL. BED ALL FITGS IN THIOKOL
MARINE PLYWOOD BULLNOSE LEADING EDGE BELOW CHEEKS. TAPER THE TRAILING EDGE TO ⅜" BELOW L.W.L. SHEATH WITH FIBERGLASS CLOTH BEFORE ATTACHING CHEEK BLOCKS
DRILL AND THROUGH-BOLT PINTLES TO RUDDER
SHAPE, SHIM, AND NOTCH RUDDER AS REQ'D TO POSITION CENTERS OF ALL GUDGEONS & PINTLES ON THE SAME CENTERLINE
NOTE! FABRICATE RUDDER WITH STRAIGHT LEADING EDGE AS SHOWN HERE. NOTCH AS SHOWN FOR INBOARD EXT. PROP. CLEARANCE
TILLER, TO BE LAMINATED FROM ½" X 4" ASH (NET) LENGTH TO SUIT. THERE SHOULD BE A 2¼" ROUND KNOB ON THE HANDLE. SEE DECK PLAN FOR DIMENSIONS.
START TAPER FROM THIS LINE
FIG. 460 PINTLE, FABRICATE GUDGEON TO FIT. SIZE 1½"
RUDDER DETAIL: SCALE – 1½" = 1'-0"
RIGGING CLEAT
4" LIGNUM VITAE CLEAT
TOPVIEW
LWL
½" HOLE
2½" SINGLE WITH LASHING EYE
⁵⁄₁₆" PENDANT HALYARD
4" CLEAT P/S
3/16" 7X19 S.S. WIRE
3" SINGLE WITH DELUXE WIRE BRIDLE TRAVELER FITTING
GAFF STICK LENGTH
HOLE FOR LACING
4" CLEAT P/S
PADEYE P/S PERKO FIG 423 SIZE #1 BURNISHED
3" SINGLE WITH BECKET SWIVEL EYE FASTENED ON 1" x ¼" x 6" EVERDUR STRAP
3" SINGLE WITH SWIVEL SHACKLE
⅜" TOPPING LIFT
¼" LAZY JACKS P/S
1" ⌀ BRASS RING
EYESTRAP P/S
BOOM STICK LENGTH 16'-11"
⅝" SHOULDER EYEBOLT
½" HOLE
EYESTRAP P/S
3" DOUBLE WITH SWIVEL SHACKLE
3" SINGLE, BECKET, WITH 2" ⌀ (I.D.) SWIVEL SHACKLE
1" EVERDUR SCH. 40 PIPE
8" CLEAT BOLTED THROUGH COAMING FRAME
¾" SCH. 40 EVERDUR PIPE
FURTHER DETAILS FOR BOOM GALLOWS/TRAVELLER ON SAIL PLAN & ARR'G'M'T SHEET, STERN VIEW
4" LIGNUM VITAE (S. ONLY) CLEAT, AVAIL. FROM TRAD. SAILING YACHT HDWE, INC.
4" LIGNUM VITAE CLEAT (S. ONLY)
3" SINGLE WITH UPSET SHACKLE
23'-3"
STEP TO TRUCK
LWL
HOLE FOR THROAT HALYARD SHACKLE
HOLE FOR ¼" RIVET
TONGUE
¾" OAK BEADS ON ⅛" NYLON LINE
LEATHER
¼" HOLE FOR RIVET
FLOATER
1" X ¼" BZ. TONGUE
¼" COPPER RIVETS
BZ. SCREWS IN WAY OF SLOT
NOTE: JAWS VIEWED PERPENDICULAR TO GAFF. JAWS ARE FORESHORTENED AT CURVED PORTION
GAFF JAWS SCALE 3" = 1'-0"
FAIRED LACING HOLE
⅜" SHOULDER EYEBOLTS FOR TACK AND DOWNHAUL
DETAILS SAME AS GAFF
FIT LASHING EYE AND SMALL BELAYING PIN FOR PENDANT HALYARD TO BOOM JAWS
BOOM JAWS SCALE 3" = 1'-0"
TEAK COCKPIT GRATE, COCKPIT SEATS AND HATCHES TO BE ½" LAID TEAK OVER ¼" BRUYNZEEL MARINE PLYWOOD (SEE DETAIL ON INTERIOR & JOINER SHEET)
⅛" EVERDUR PLATE TO SUIT
"WATER" DECK FILL – PERKO FIG. 528 1" SIZE, MACHINE FINISH
FAIRLEADS 2EA P/S. PERKO FIG. 882 SIZE #1 BRONZE
NOTE: BACK UP ALL THROUGH-BOLTED DECK FITTINGS WITH ⅛" BRONZE PLATES BED IN THIOKOL
7" CLEAT P/S WILCOX-CRITTENDEN FIG. 4051 POL. BRONZE
PINRAILS, P/S, 10" X 2¾" X 1½" TEAK ⅝" ⌀ X 9" LONG LIGNUM VITAE PINS
2¼" X 1¼" TEAK HANDRAILS, P/S
1½" X 4½" TEAK BOOM GALLOWS
8" OPEN BOLLARD CLEAT SIMPSON-LAWRENCE 1392AB
HATCH
CUTOUT
1½" DIA. SCUPPER P/S PERKO FIG 442 PLAIN SIZE 2
FAIRLEAD
CHARLEY NOBLE WITH SIMPSON-LAWRENCE # 141B DECK IRON
HINGED HATCHES
MOORING BITT SIMPSON-LAWRENCE NO. 1383 GM SIZE 1
DOWN HAUL BLOCK
4" COWL VENT. POLISHED BRASS PERKO FIG 513
TOPPING LIFT BLOCK
5" P/S FAIRLEADS SIMPSON-LAWRENCE NO. 1421 GM
HALYARD BLOCK
DECK BLOCKS – 3" BECKET DOWNHAUL, TOPPING LIFT BLOCKS 3" SINGLE. HALYARD BLOCK 3" DBL. WITH DECK PLATES AND SWIVEL EYES
NOTE: ALL HATCHES TO BE EQUIPPED WITH PERKO FIG. 1103 POLISHED BRONZE FLUSH RING PULLS
BLOCKS AVAILABLE FROM: A. DAUPHINEE & SONS, LTD. P.O. BOX 115 LUNENBURG, NOVA SCOTIA CANADA
ALL BLOCKS TO BE WITH LIGNUM VITAE CHEEKS
17' CRUISING CAT
FOR: HANK & SARAH BELLAMY
DATE: 7/6/72
SCALE: ¾" = 1'-0"
RIGGING PLAN
LOA 17'-0"
LWL 15'-1½"
BEAM 7'-0"
DRAFT 3'-6"
FREEBOARD FWD 3'-4"
LEAST 1'-11"
AFT 2'-2"
JAY R. BENFORD
605 S. TALBOT ST - SUITE 1
ST. MICHAELS, MD 21663
(410) 745-3235
67-5R

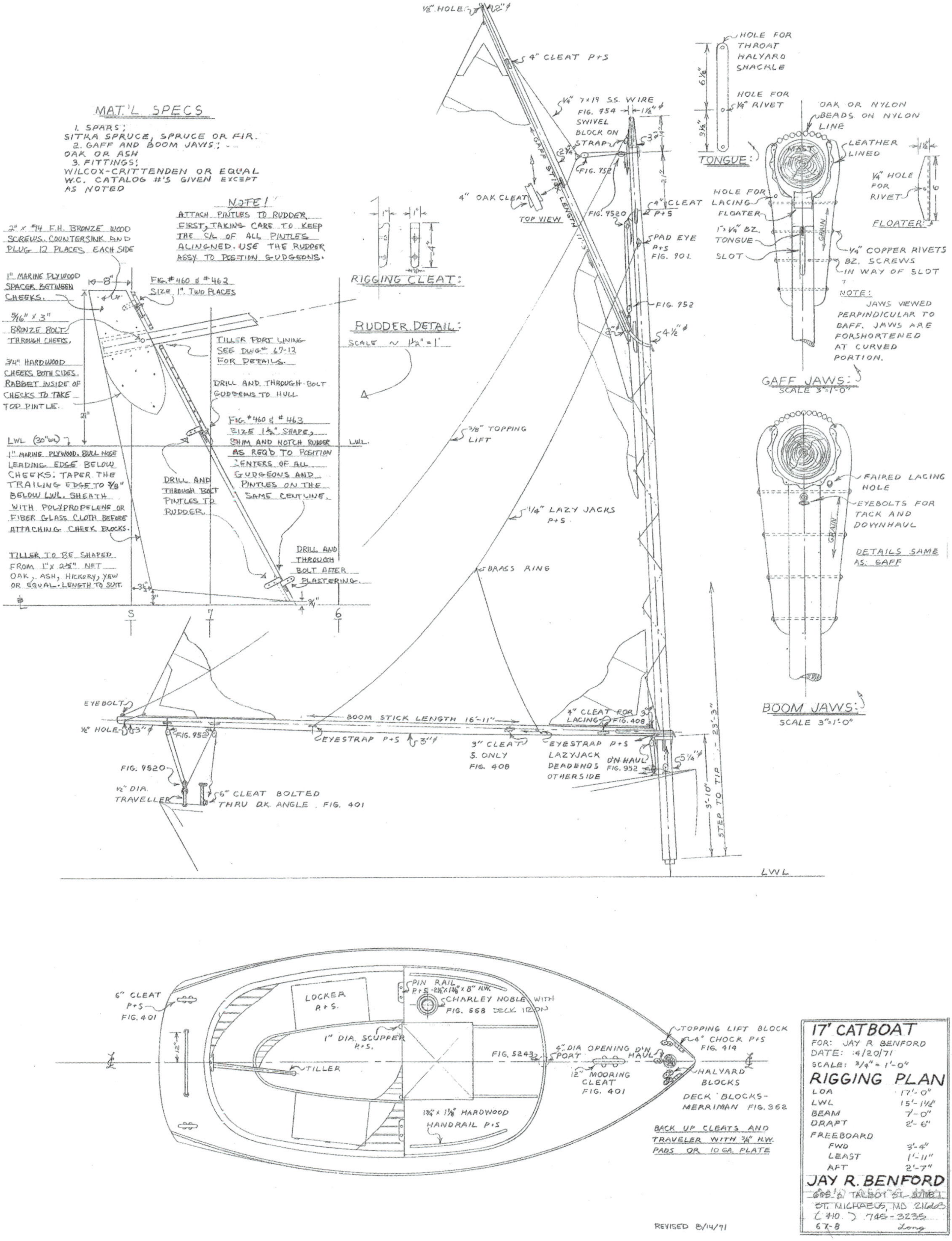
MAT'L SPECS
1. SPARS:
SITKA SPRUCE, SPRUCE OR FIR.
2. GAFF AND BOOM JAWS:
OAK OR ASH
3. FITTINGS:
WILCOX-CRITTENDEN OR EQUAL
W.C. CATALOG #'S GIVEN EXCEPT AS NOTED
NOTE!
ATTACH PINTLES TO RUDDER FIRST, TAKING CARE TO KEEP THE C/L OF ALL PINTLES ALINGNED. USE THE RUDDER ASSY TO POSITION GUDGEONS.
2" x #14 F.H. BRONZE WOOD SCREWS, COUNTERSINK AND PLUG 12 PLACES EACH SIDE
1" MARINE PLYWOOD SPACER BETWEEN CHEEKS.
5/16" x 3" BRONZE BOLT THROUGH CHEEKS.
3/4" HARDWOOD CHEEKS BOTH SIDES. RABBET INSIDE OF CHEEKS TO TAKE TOP PINTLE.
FIG. #460 & #463 SIZE 1", TWO PLACES
TILLER PORT LINING SEE DWG # 67-12 FOR DETAILS.
DRILL AND THROUGH-BOLT GUDGEONS TO HULL
FIG. #460 & #463 SIZE 1½", SHAPE, SHIM AND NOTCH RUDDER AS REQ'D TO POSITION CENTERS OF ALL GUDGEONS AND PINTLES ON THE SAME CENTERLINE.
LWL (30" WL)
1" MARINE PLYWOOD, BULL NOSE LEADING EDGE BELOW CHEEKS: TAPER THE TRAILING EDGE TO 3/8" BELOW LWL. SHEATH WITH POLYPROPELENE OR FIBER GLASS CLOTH BEFORE ATTACHING CHEEK BLOCKS.
DRILL AND THROUGH BOLT PINTLES TO RUDDER.
DRILL AND THROUGH BOLT AFTER PLASTERING.
TILLER TO BE SHAPED FROM 1" x 2½" NET OAK, ASH, HICKORY, YEW OR EQUAL. LENGTH TO SUIT.
RIGGING CLEAT:
RUDDER DETAIL:
SCALE ~ 1½" = 1'
TOP VIEW
1/2" HOLE
4" CLEAT P+S
1/4" 7x19 SS. WIRE
FIG. 954
SWIVEL BLOCK ON STRAP
FIG. 952
4" OAK CLEAT
GAFF STICK LENGTH
FIG. 9520
4" CLEAT P+S
PAD EYE P+S FIG. 901.
3/8" TOPPING LIFT
1/4" LAZY JACKS P+S
BRASS RING
EYEBOLT
1/2" HOLE
BOOM STICK LENGTH 16'-11"
EYESTRAP P+S
3" CLEAT S. ONLY FIG. 408
4" CLEAT FOR LACING FIG. 408
EYESTRAP P+S LAZYJACK DEADENDS OTHERSIDE
D'N HAUL FIG. 952
1/2" DIA. TRAVELLER
6" CLEAT BOLTED THRU O.K. ANGLE, FIG. 401
STEP TO TIP 23'-3"
LWL
HOLE FOR THROAT HALYARD SHACKLE
HOLE FOR 1/4" RIVET
TONGUE:
OAK OR NYLON BEADS ON NYLON LINE
LEATHER LINED
MAST
HOLE FOR LACING
FLOATER
1" x 1/4" BZ. TONGUE
SLOT
GRAIN
1/4" COPPER RIVETS BZ. SCREWS IN WAY OF SLOT
1/4" HOLE FOR RIVET
FLOATER:
NOTE: JAWS VIEWED PERPENDICULAR TO GAFF. JAWS ARE FORSHORTENED AT CURVED PORTION.
GAFF JAWS:
SCALE 3" = 1'-0"
FAIRED LACING HOLE
EYEBOLTS FOR TACK AND DOWNHAUL
DETAILS SAME AS GAFF
BOOM JAWS:
SCALE 3" = 1'-0"
6" CLEAT P+S FIG. 401
LOCKER P+S.
PIN RAIL P+S
CHARLEY NOBLE WITH FIG. 668 DECK IRON
1" DIA. SCUPPER P+S.
TILLER
FIG. 5243
4" DIA OPENING PORT
12" MOORING CLEAT FIG. 401
D'N HAUL
TOPPING LIFT BLOCK
4" CHOCK P+S FIG. 414
HALYARD BLOCKS
DECK BLOCKS - MERRIMAN FIG. 362
1¾" x 1⅛" HARDWOOD HANDRAIL P+S
BACK UP CLEATS AND TRAVELER WITH 3/4" H.W. PADS OR 10 GA. PLATE
17' CATBOAT
FOR: JAY R BENFORD
DATE: 4/20/71
SCALE: 3/4" = 1'-0"
RIGGING PLAN
LOA 17'-0"
LWL 15'-1½"
BEAM 7'-0"
DRAFT 2'-6"
FREEBOARD
FWD 3'-4"
LEAST 1'-11"
AFT 2'-7"
JAY R. BENFORD
ST. MICHAELS, MD 21663
(410) 745-3235
67-8
REVISED 8/14/71

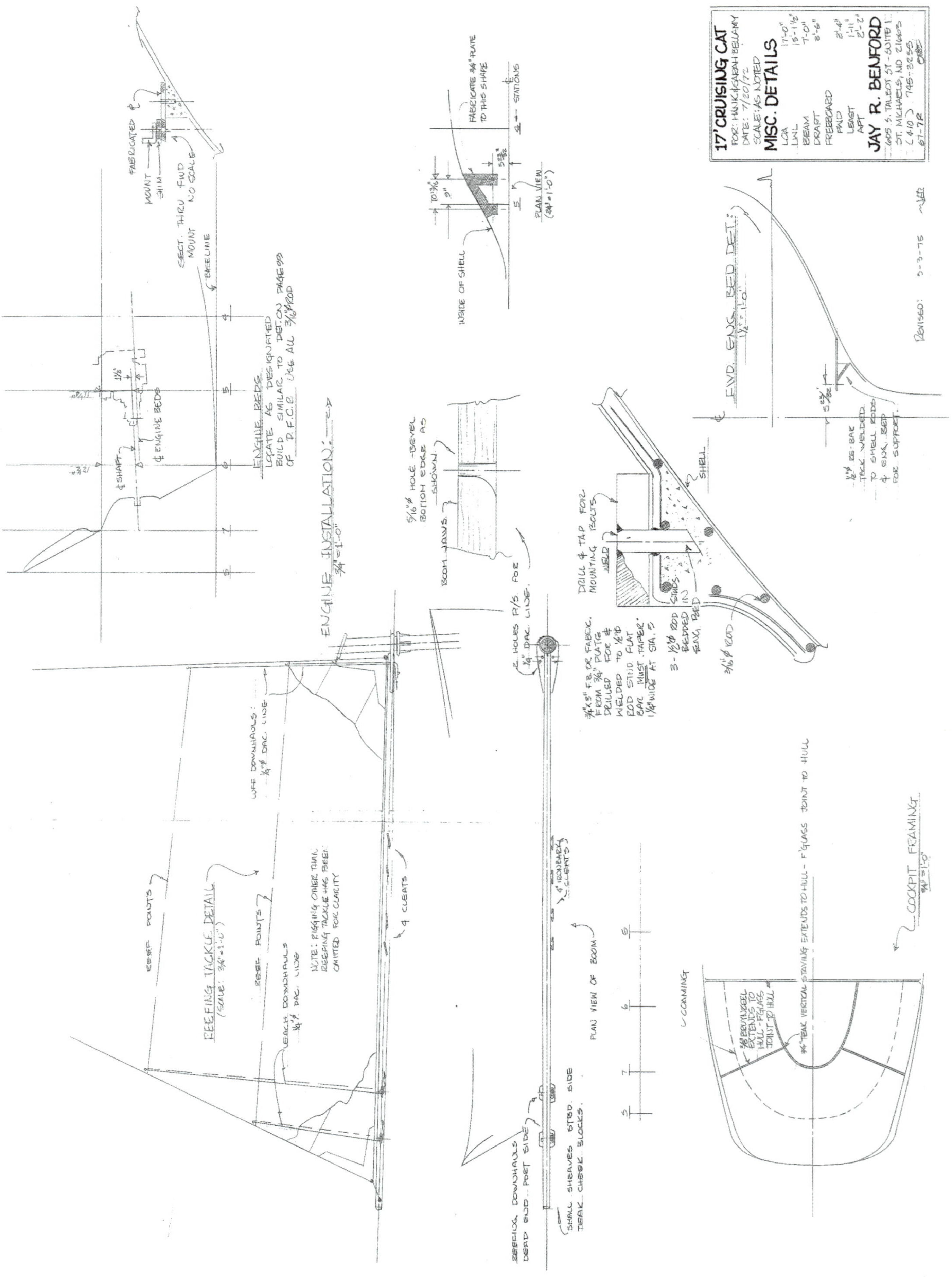
17' CRUISING CAT
FOR: HANK & SARAH BELLAMY
DATE: 7/20/72
SCALE: AS NOTED
MISC. DETAILS
LOA 17'-0"
LWL 15'-11 1/2"
BEAM 7'-0"
DRAFT 3'-6"
JAY R. BENFORD
ENGINE INSTALLATION
3/4"=1'-0"
ENGINE BEDS
FWD. ENG. BED DET.
1 1/2"=1'-0"
PLAN VIEW
3/4"=1'-0"
INSIDE OF SHELL
SHELL
FABRICATED ℄
MOUNT
SHIM
SECT. THRU FWD MOUNT
NO SCALE
BASELINE
DRILL & TAP FOR MOUNTING BOLTS
BOOM JAWS
REEFING TACKLE DETAIL
(SCALE: 3/4"=1'-0")
REEF POINTS
LUFF DOWNHAULS
LEACH DOWNHAULS
4 CLEATS
PLAN VIEW OF BOOM
SMALL SHEAVES STBD. SIDE
TEAK CHEEK BLOCKS
COCKPIT FRAMING
3/4"=1'-0"
COAMING
REVISED: 5-3-75

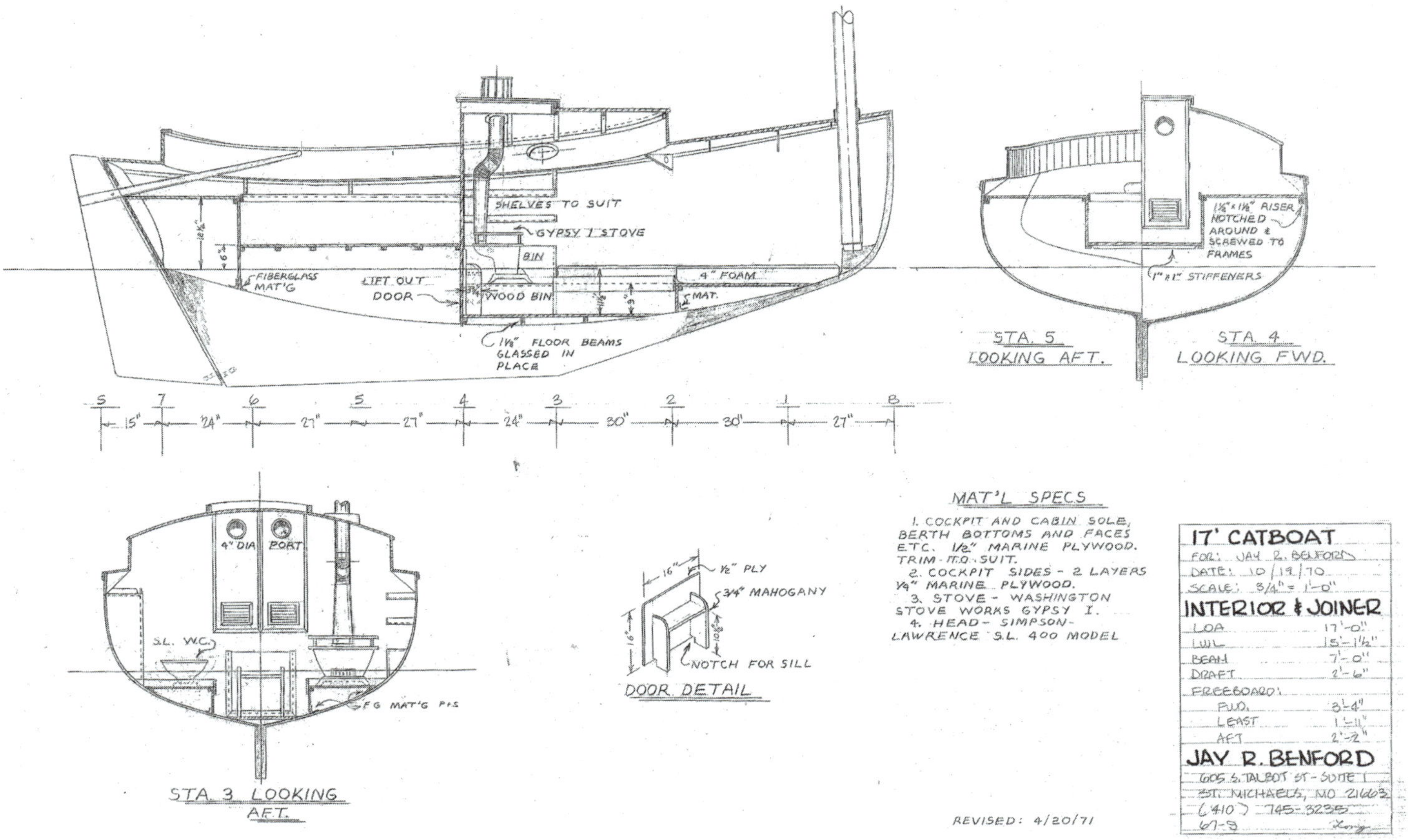
SHELVES TO SUIT
GYPSY I STOVE
BIN
FIBERGLASS MAT'G
LIFT OUT DOOR
WOOD BIN
4" FOAM
MAT.
1½" FLOOR BEAMS GLASSED IN PLACE
1½" x 1½" RISER NOTCHED AROUND & SCREWED TO FRAMES
1" x 1" STIFFENERS
STA. 5 LOOKING AFT.
STA. 4 LOOKING FWD.
S 7 6 5 4 3 2 1 B
15" 24" 27" 27" 24" 30" 30" 27"
4" DIA. PORT
S.L. W.C.
F.G. MAT'G PTS
STA. 3 LOOKING AFT.
16" ½" PLY
3/4" MAHOGANY
NOTCH FOR SILL
DOOR DETAIL
MAT'L SPECS
1. COCKPIT AND CABIN SOLE, BERTH BOTTOMS AND FACES ETC. ½" MARINE PLYWOOD. TRIM - T.O. SUIT.
2. COCKPIT SIDES - 2 LAYERS ¼" MARINE PLYWOOD.
3. STOVE - WASHINGTON STOVE WORKS GYPSY I.
4. HEAD - SIMPSON-LAWRENCE S.L. 400 MODEL
REVISED: 4/20/71
17' CATBOAT
FOR: JAY R. BENFORD
DATE: 10/19/70
SCALE: 3/4" = 1'-0"
INTERIOR & JOINER
LOA 17'-0"
LWL 15'-1½"
BEAM 7'-0"
DRAFT 2'-6"
FREEBOARD:
FWD. 3'-4"
LEAST 1'-11"
AFT 2'-2"
JAY R. BENFORD
605 S. TALBOT ST - SUITE 1
ST. MICHAELS, MD 21663
(410) 745-3235
67-9

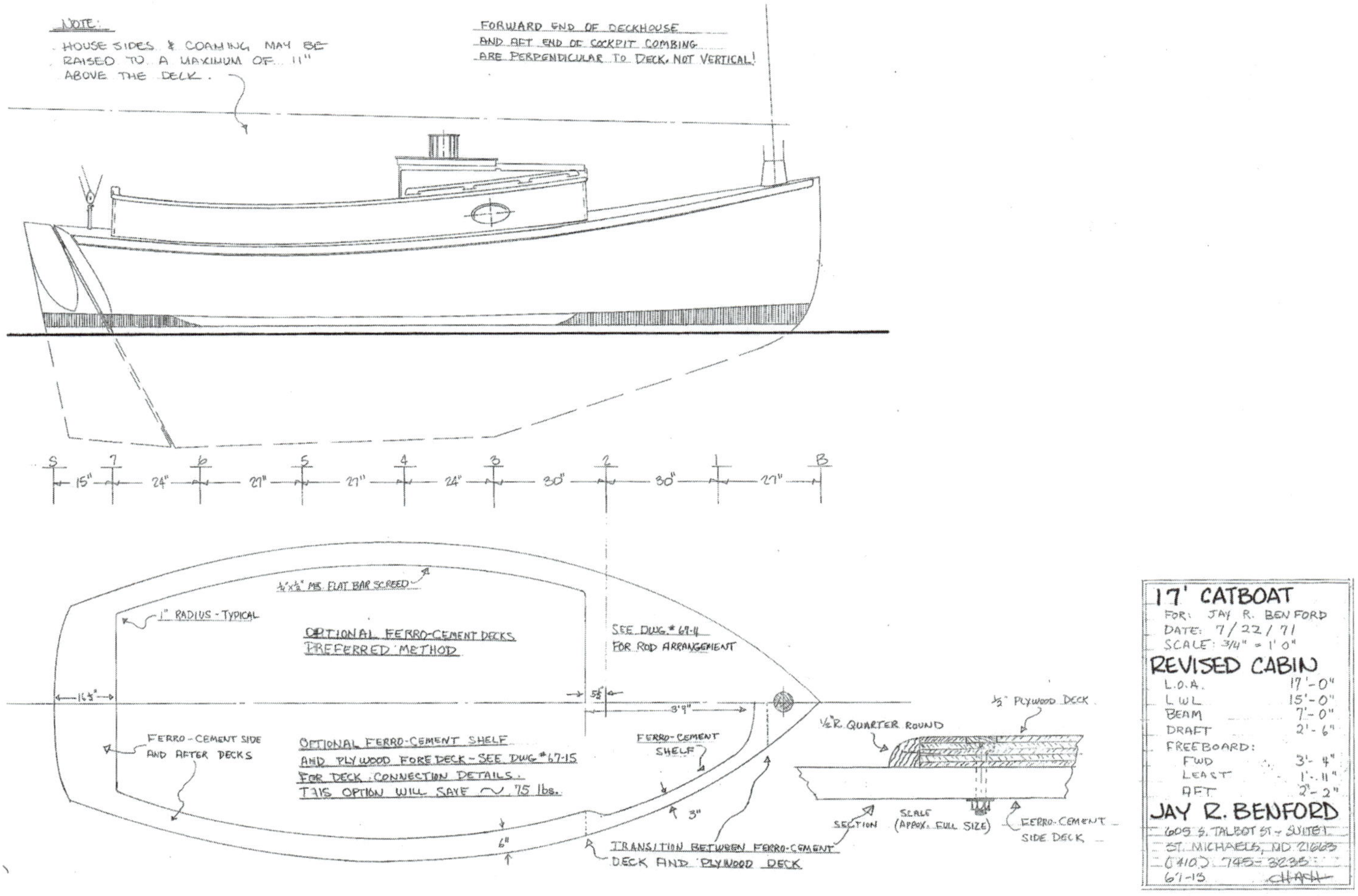
NOTE:
HOUSE SIDES & COAMING MAY BE RAISED TO A MAXIMUM OF 11" ABOVE THE DECK.
FORWARD END OF DECKHOUSE AND AFT END OF COCKPIT COMBING ARE PERPENDICULAR TO DECK, NOT VERTICAL!
S 7 6 5 4 3 2 1 B
15" 24" 27" 27" 24" 30" 30" 27"
¼" x ½" MS. FLAT BAR SCREED
1" RADIUS - TYPICAL
OPTIONAL FERRO-CEMENT DECKS PREFERRED METHOD
SEE DWG. # 67-4 FOR ROD ARRANGEMENT
16½"
5⅝"
3'9"
FERRO-CEMENT SIDE AND AFTER DECKS
OPTIONAL FERRO-CEMENT SHELF AND PLYWOOD FOREDECK - SEE DWG # 67-15 FOR DECK CONNECTION DETAILS. THIS OPTION WILL SAVE ~ 75 lbs.
FERRO-CEMENT SHELF
3"
6"
TRANSITION BETWEEN FERRO-CEMENT DECK AND PLYWOOD DECK
½" R. QUARTER ROUND
½" PLYWOOD DECK
SECTION
SCALE (APPOX. FULL SIZE)
FERRO-CEMENT SIDE DECK
17' CATBOAT
FOR: JAY R. BENFORD
DATE: 7/22/71
SCALE: 3/4" = 1'0"
REVISED CABIN
L.O.A. 17'-0"
LWL 15'-0"
BEAM 7'-0"
DRAFT 2'-6"
FREEBOARD:
FWD 3'-4"
LEAST 1'-11"
AFT 2'-2"
JAY R. BENFORD
605 S. TALBOT ST - SUITE 1
ST. MICHAELS, MD 21663
(410) 745-3235
67-13

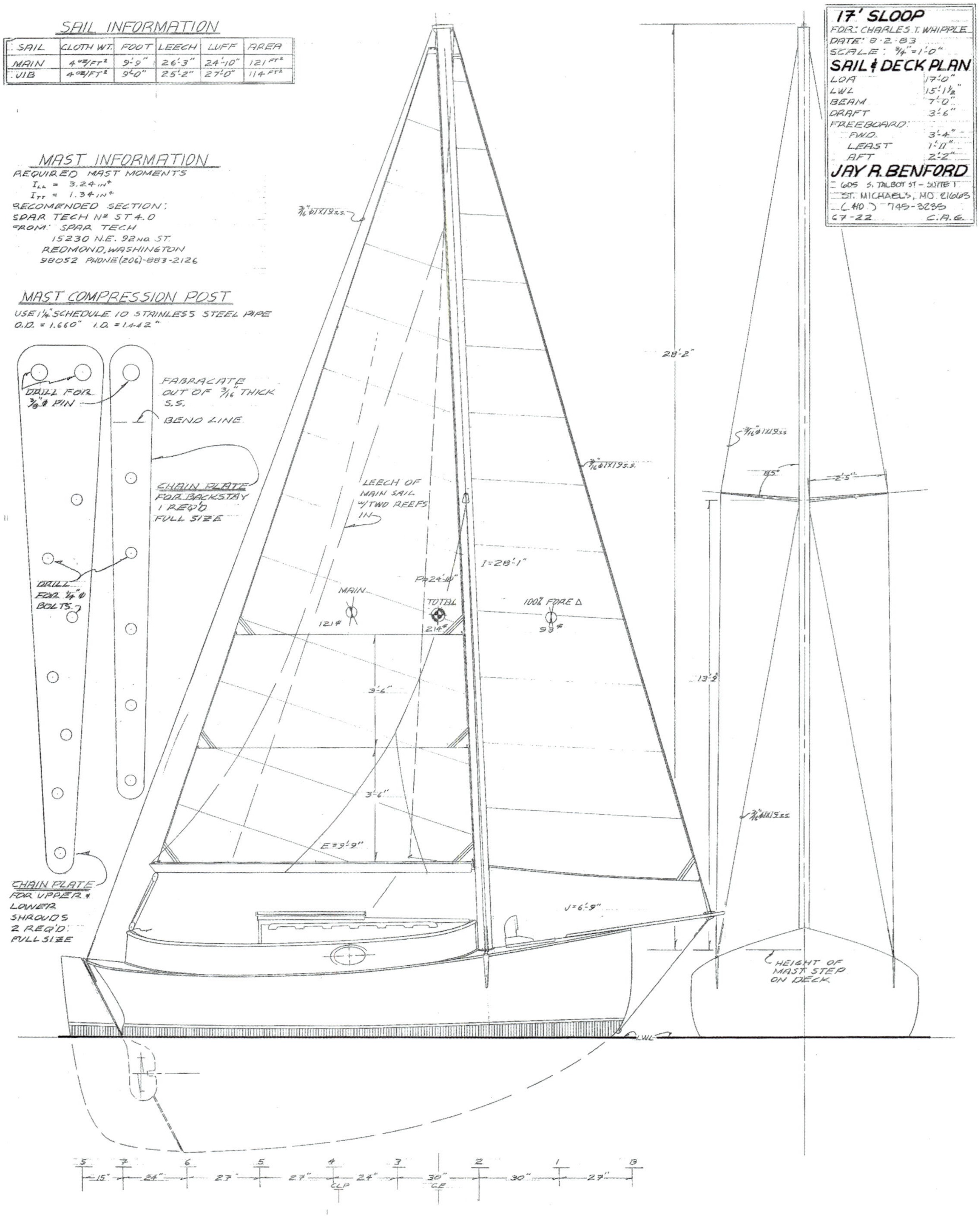

Deck plan for this rig is on the facing page.

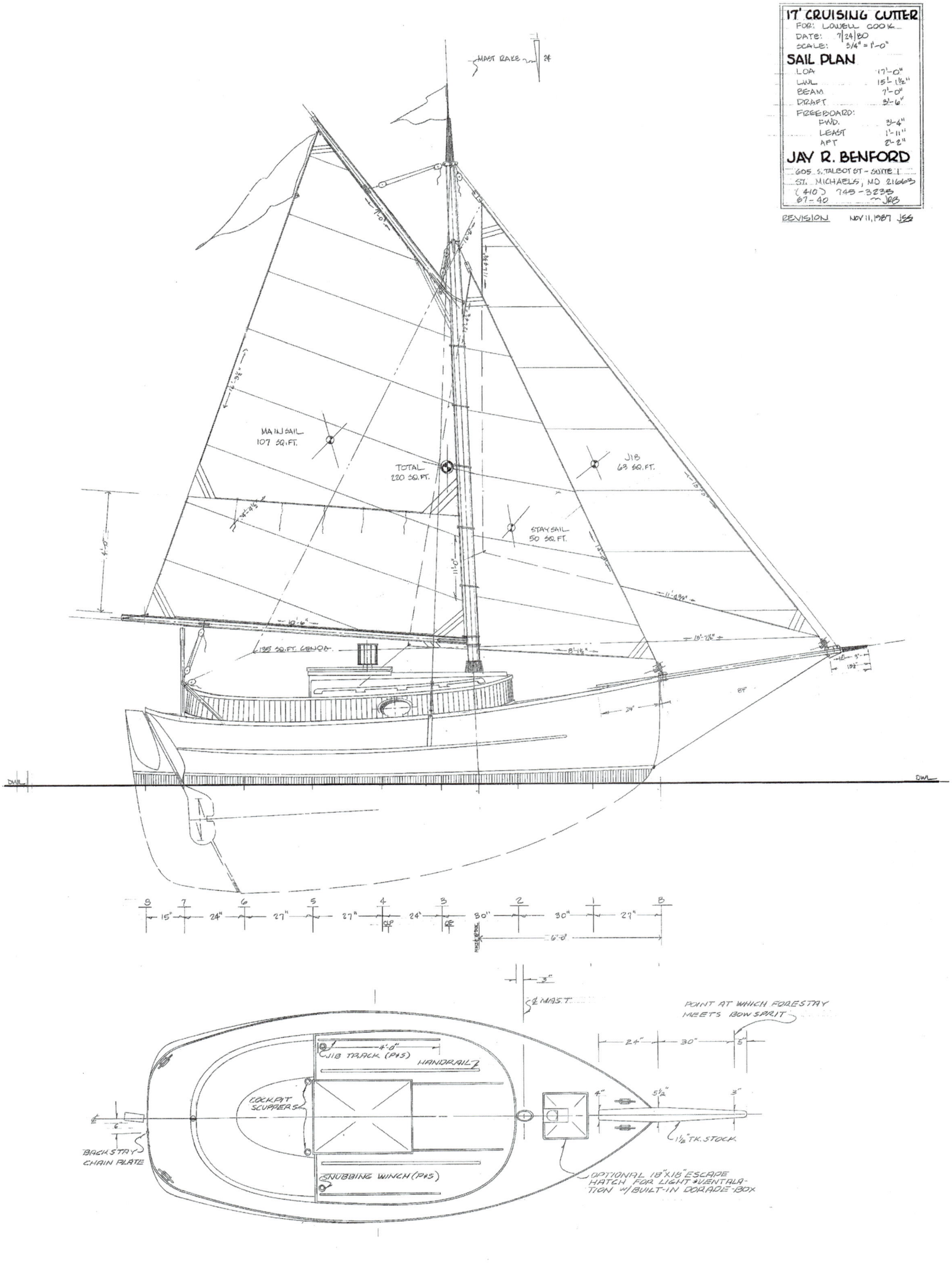
17' CRUISING CUTTER
FOR: LOWELL COOK
DATE: 7/24/80
SCALE: 3/4" = 1'-0"
SAIL PLAN
LOA 17'-0"
LWL 15'-1½"
BEAM 7'-0"
DRAFT 3'-6"
FREEBOARD:
FWD. 3'-4"
LEAST 1'-11"
AFT 2'-2"
JAY R. BENFORD
605 S. TALBOT ST - SUITE 1
ST. MICHAELS, MD 21663
(410) 745-3235
67-40
JRB
REVISION NOV 11, 1987 JSS
MAST RAKE
MAINSAIL 107 SQ. FT.
TOTAL 220 SQ. FT.
JIB 63 SQ. FT.
STAY SAIL 50 SQ. FT.
135 SQ. FT. GENOA
DWL
15" 24" 27" 27" 24" 30" 30" 27"
6'-0"
℄ MAST
POINT AT WHICH FORESTAY MEETS BOW SPRIT
24" 30" 5"
4'-0"
JIB TRACK (P&S)
HANDRAIL
COCKPIT SCUPPERS
BACK STAY CHAIN PLATE
SNUBBING WINCH (P&S)
1½" TK. STOCK
OPTIONAL 18"X18" ESCAPE HATCH FOR LIGHT & VENTALATION W/ BUILT-IN DORADE-BOX

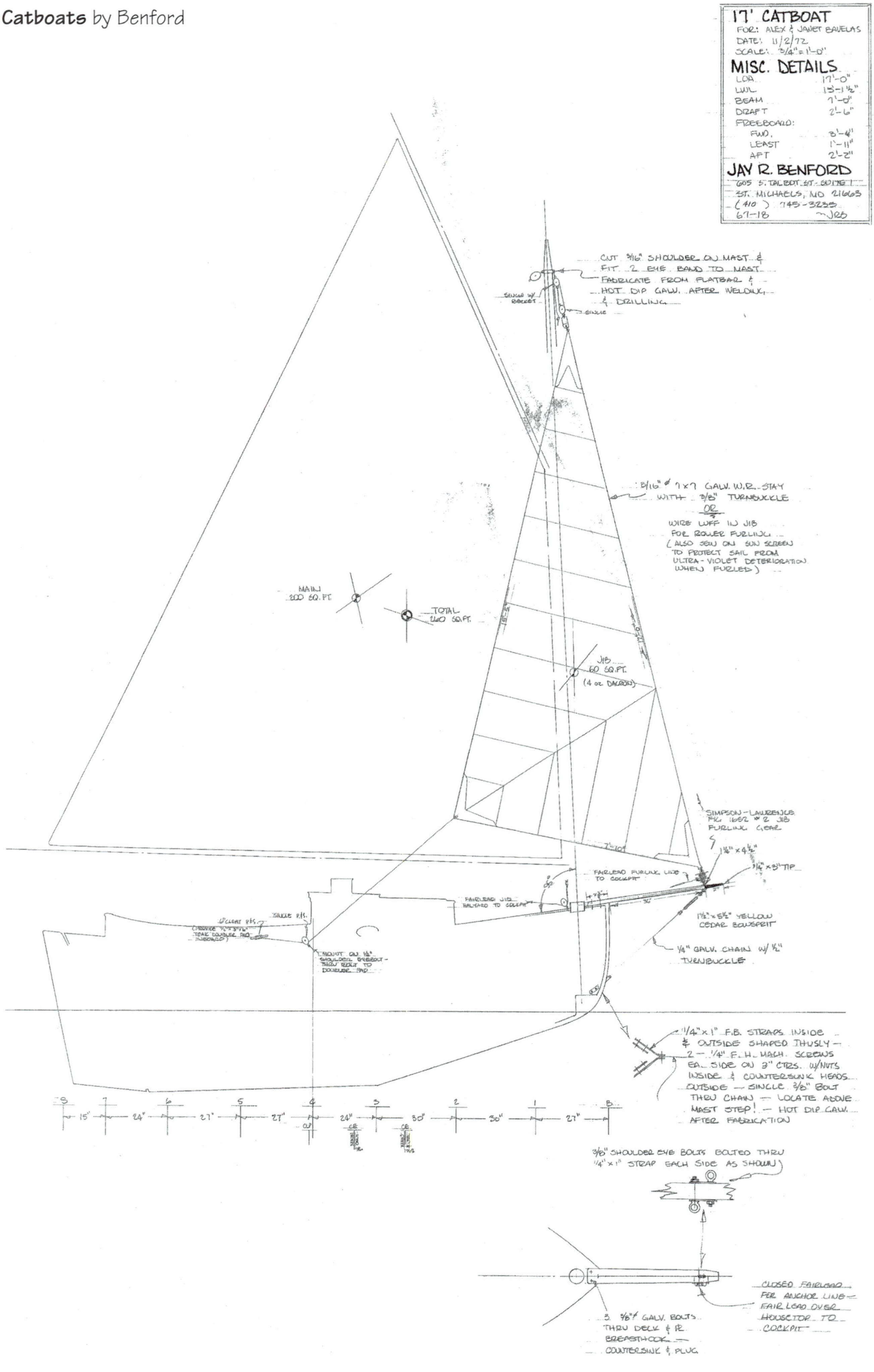
17' CATBOAT
FOR: ALEX & JANET BAVELAS
DATE: 11/2/72
SCALE: 3/4" = 1'-0"
MISC. DETAILS
LOA 17'-0"
LWL 15'-1 1/2"
BEAM 7'-0"
DRAFT 2'-6"
FREEBOARD:
FWD. 3'-4"
LEAST 1'-11"
AFT 2'-2"
JAY R. BENFORD
605 S. TALBOT ST - SUITE 1
ST. MICHAELS, MD 21663
(410) 745-3235
67-18 ~JRB
CUT 3/16" SHOULDER ON MAST & FIT 2 EYE BAND TO MAST FABRICATE FROM FLATBAR & HOT DIP GALV. AFTER WELDING & DRILLING
SINGLE
3/16" ø 7x7 GALV. W.R. STAY WITH 3/8" TURNBUCKLE
OR
WIRE LUFF IN JIB FOR ROLLER FURLING (ALSO SEW ON SUN SCREEN TO PROTECT SAIL FROM ULTRA-VIOLET DETERIORATION WHEN FURLED)
MAIN 200 SQ. FT.
TOTAL 260 SQ. FT.
JIB 60 SQ. FT. (4 oz DACRON)
SIMPSON-LAWRENCE FIG. 1662 #2 JIB FURLING GEAR
7'-10"
1 1/4" x 4 1/2"
1 1/4" x 3" TIP
FAIRLEAD FURLING LINE TO COCKPIT
FAIRLEAD JIB HALYARD TO COCKPIT
36"
1 1/2" x 5 1/2" YELLOW CEDAR BOWSPRIT
1/4" GALV. CHAIN W/ 1/2" TURNBUCKLE
1/4" x 1" F.B. STRAPS INSIDE & OUTSIDE SHAPED THUSLY — 2 — 1/4" F.H. MACH. SCREWS EA. SIDE ON 3" CTRS. W/NUTS INSIDE & COUNTERSUNK HEADS OUTSIDE — SINGLE 3/8" BOLT THRU CHAIN — LOCATE ABOVE MAST STEP! — HOT DIP GALV. AFTER FABRICATION
S 15" 7 24" 6 27" 5 27" 4 24" 3 30" 2 30" 1 27" B
CLR
CE
3/8" SHOULDER EYE BOLTS BOLTED THRU 1/4" x 1" STRAP EACH SIDE AS SHOWN
3 3/8" ø GALV. BOLTS THRU DECK & BREASTHOOK — COUNTERSINK & PLUG
CLOSED FAIRLEAD FOR ANCHOR LINE — FAIRLEAD OVER HOUSETOP TO COCKPIT

The mold framing drawing was created later, after some questions were raised by a Scandinavian builder of the **Liberty** *version. This sort of approach would work for setting up others of the small boats shown in this book.*

Above is a sistership to **Puffin** *in Victoria, BC, and below is a* **Liberty** *sistership in Maine.*

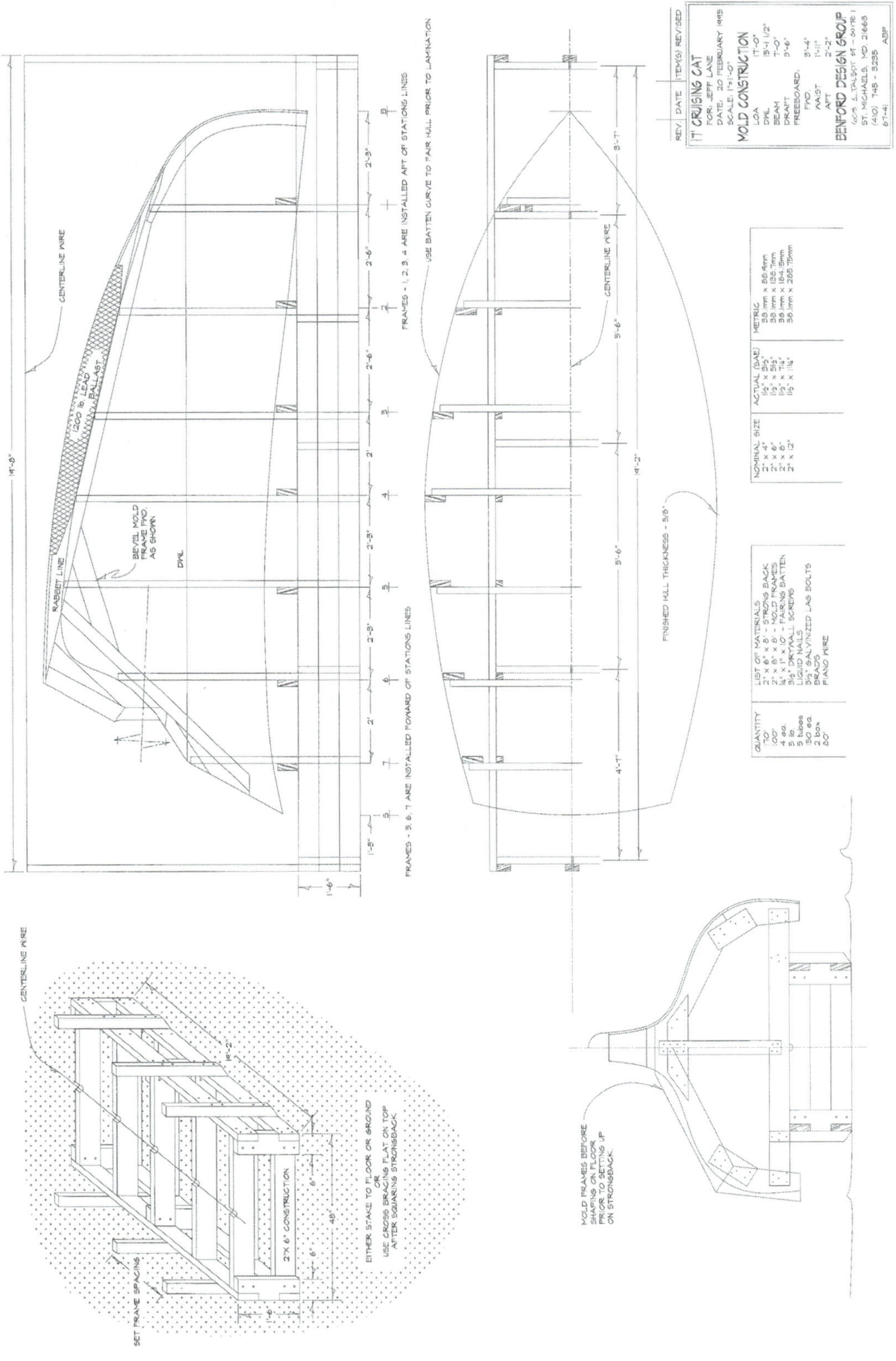

Puffin's *companionway doors had hand craved "P"s on them. The hinge pins were replaced with removable pins, so that the doors could be removed and stowed in the cabin when the boat was in use.*

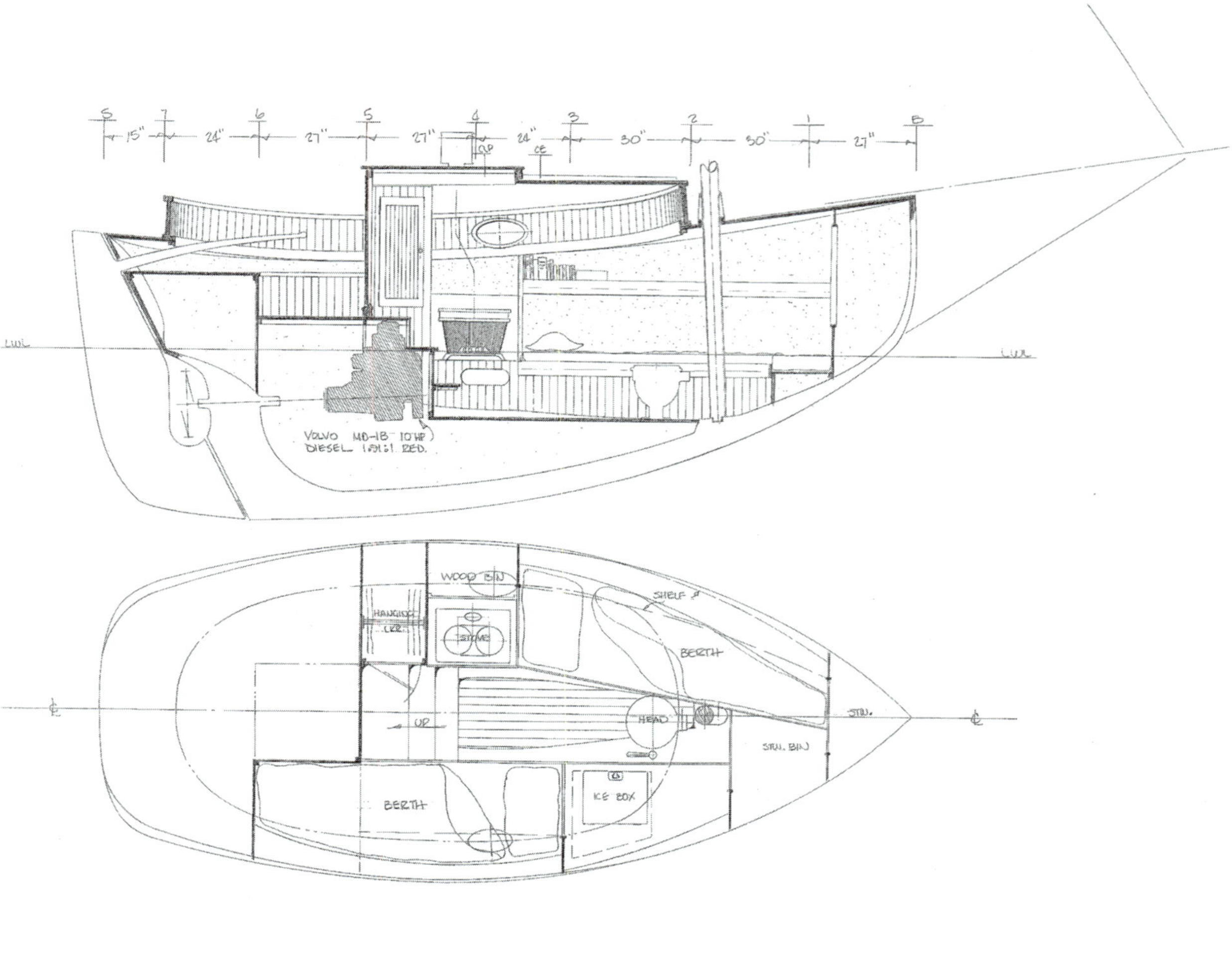

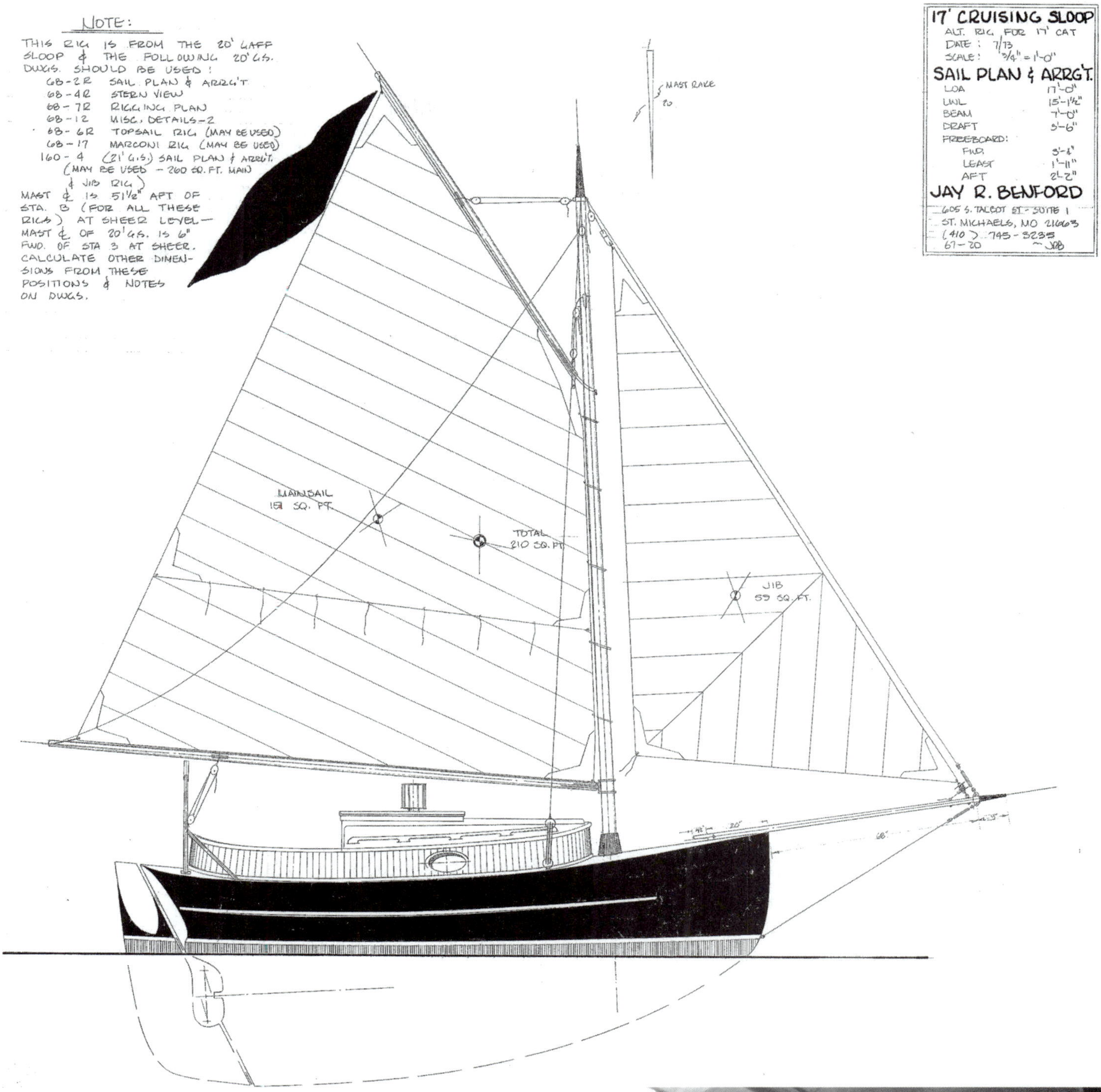

*Looking aft along the port side in **Puffin**'s cabin. The stove pipe joints were reversed later on, so that any condensate inside the stove pipe would run down inside the pipe to the stove instead of dripping out of the joints . . . .*

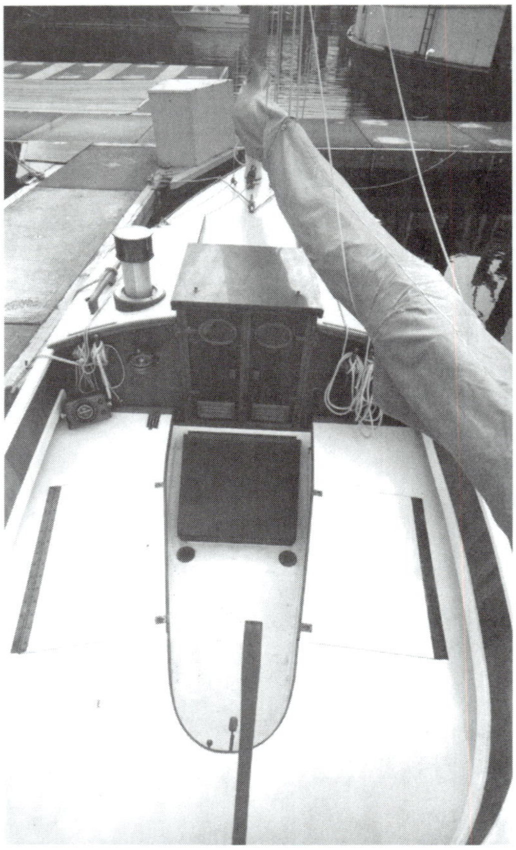

Puffin *has room to daysail with up to six people aboard. The toerail, as can be discerned in the side view photo, has notches at the low point to drain rainwater and spray.*

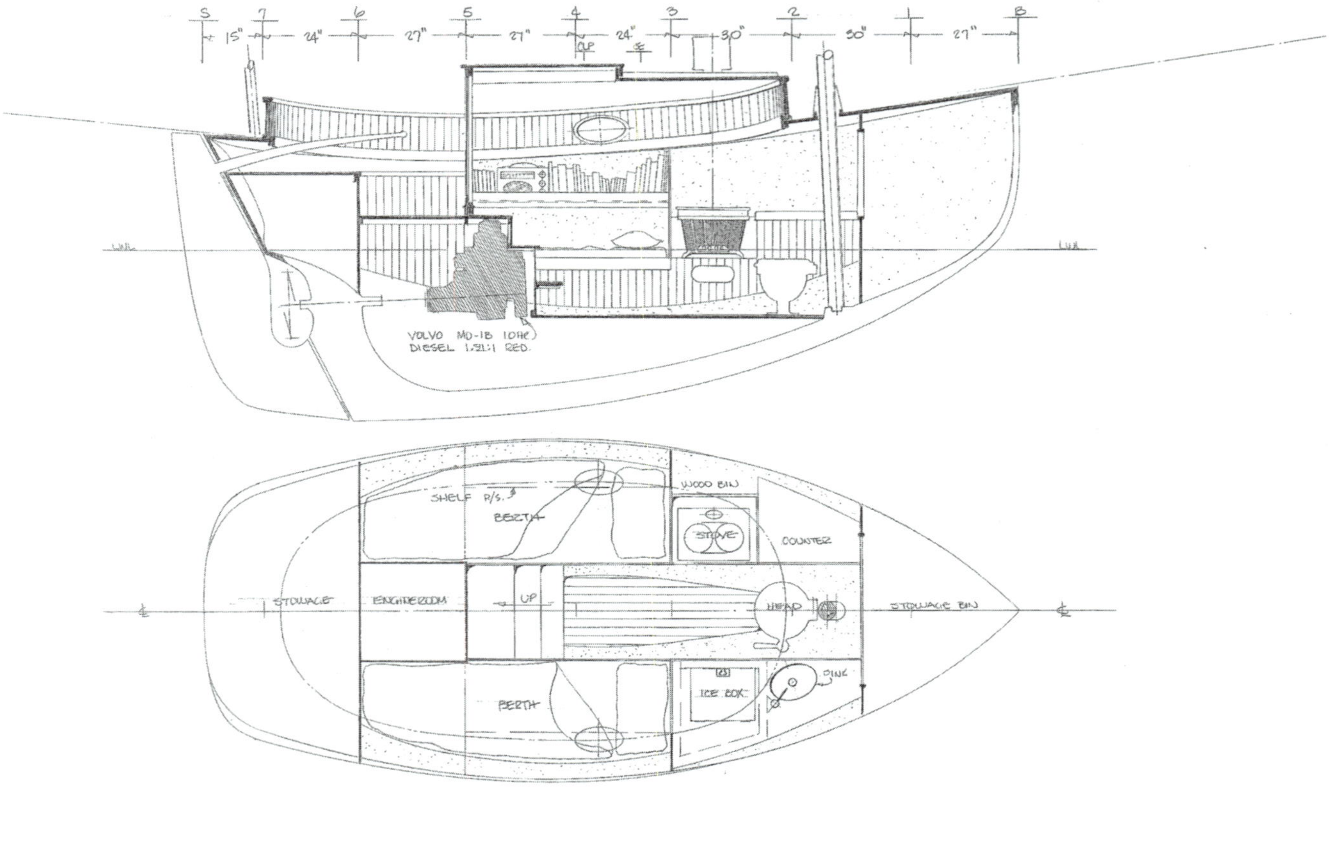

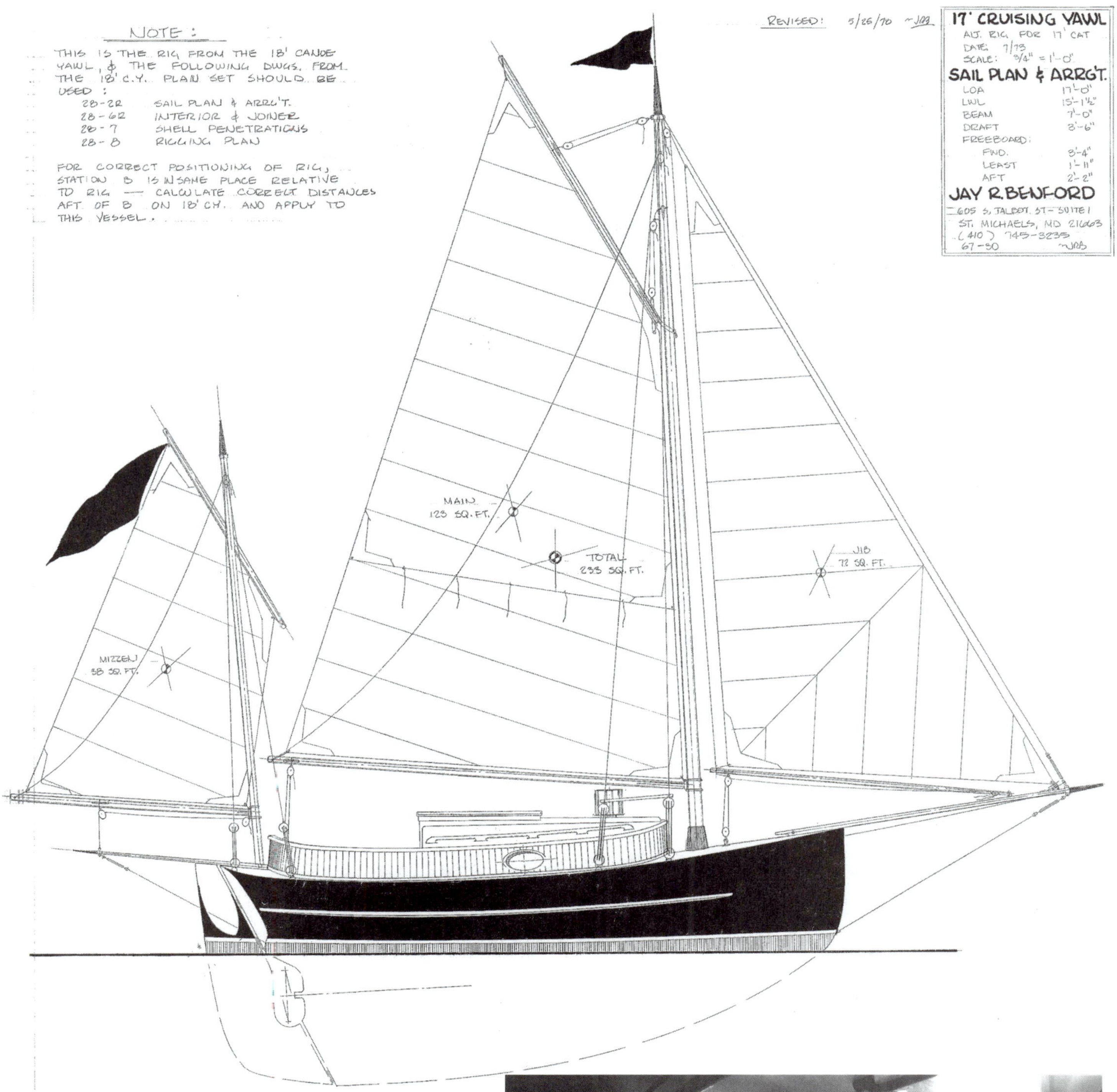

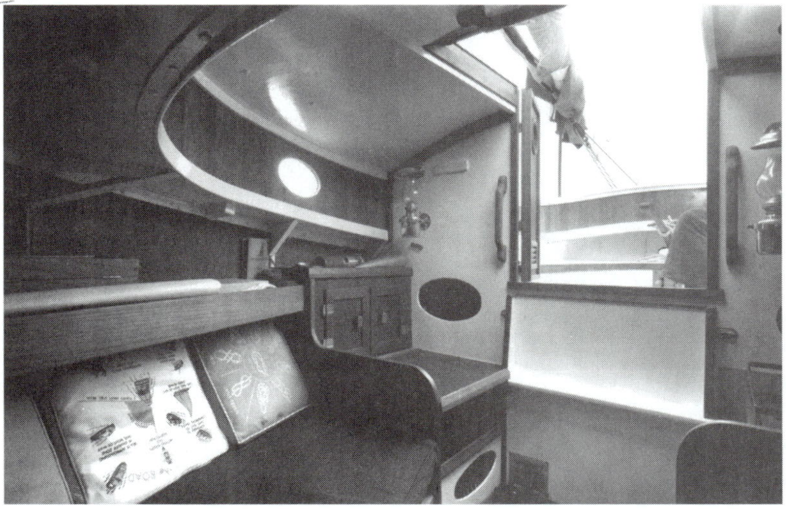

Looking aft along the starboard side in **Puffin***'s cabin. The oval openings into the storage areas make for easy access and work to retain the contents of the lockers. Note the handrails alongside the companionway opening.*

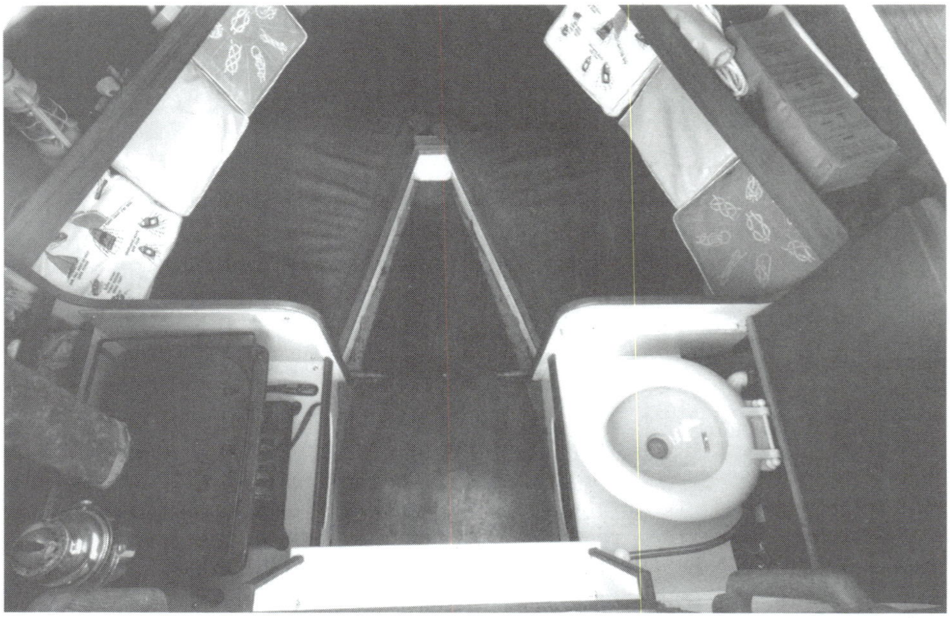

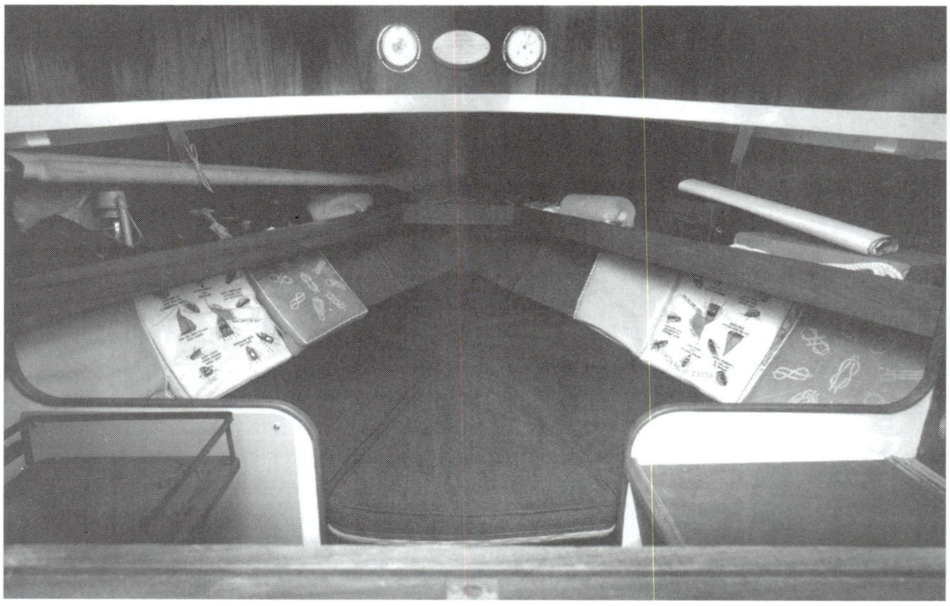

*The view into **Puffin**'s cabin from the companionway, above, and a different angle showing the filler piece in the v-berth, below. The low profile Simpson-Lawrence head allowed us to still have good sitting headroom on it.*

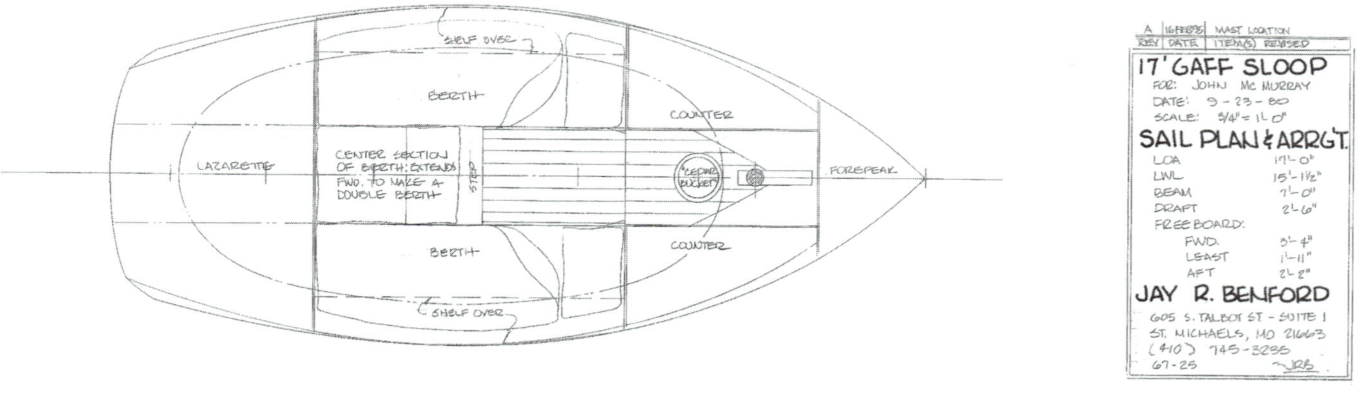

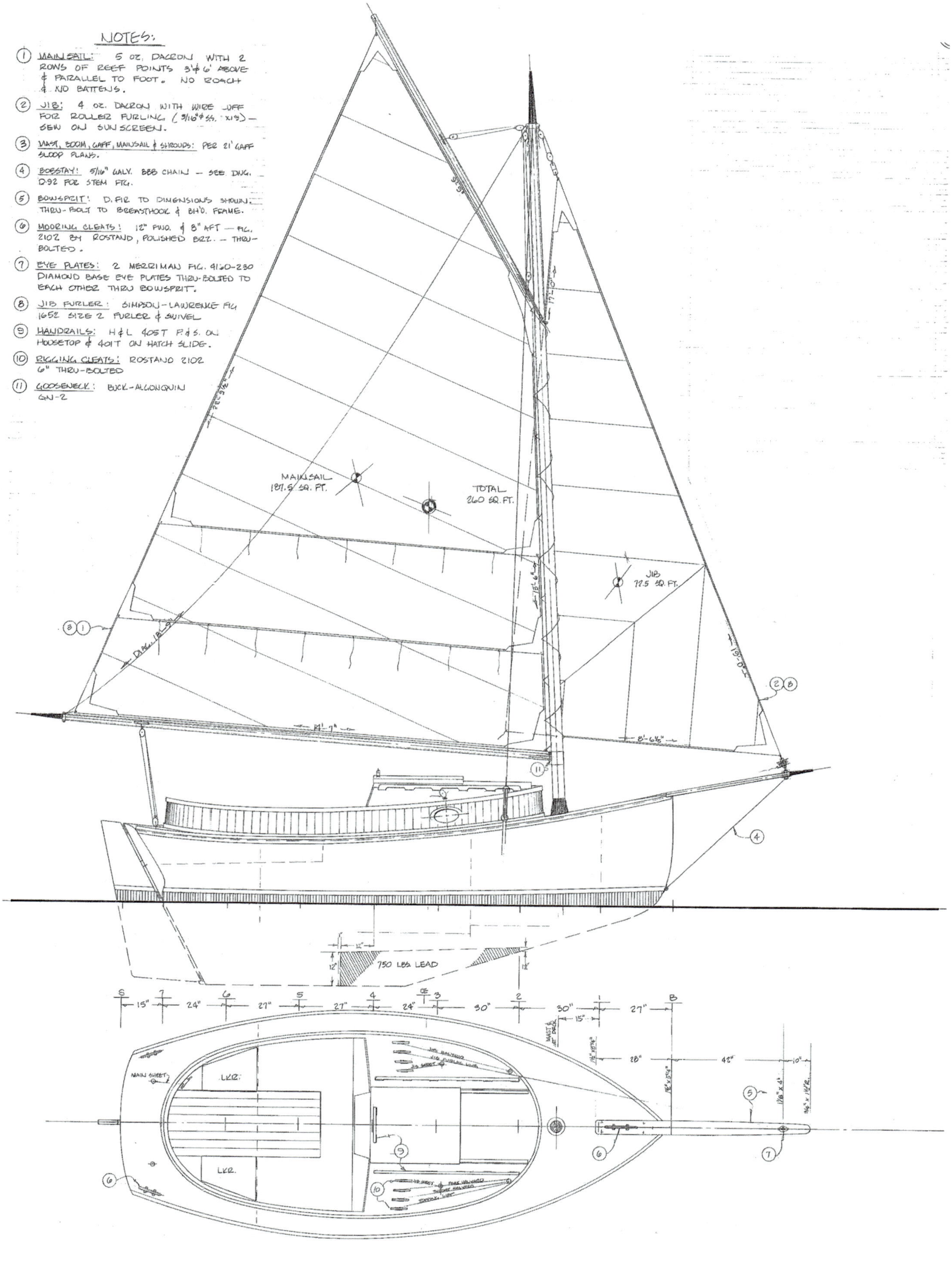
NOTES:
(1) MAINSAIL: 5 OZ. DACRON WITH 2 ROWS OF REEF POINTS 3' & 6' ABOVE & PARALLEL TO FOOT. NO ROACH & NO BATTENS.
(2) JIB: 4 OZ. DACRON WITH WIRE LUFF FOR ROLLER FURLING (3/16"⌀ S.S. 1x19) — SEW ON SUN SCREEN.
(3) MAST, BOOM, GAFF, MAINSAIL & SHROUDS: PER 21' GAFF SLOOP PLANS.
(4) BOBSTAY: 5/16" GALV. BBB CHAIN — SEE DWG. D-92 FOR STEM FTG.
(5) BOWSPRIT: D. FIR TO DIMENSIONS SHOWN. THRU-BOLT TO BREASTHOOK & BH'D. FRAME.
(6) MOORING CLEATS: 12" FWD. & 8" AFT — FIG. 2102 BY ROSTAND, POLISHED BRZ. — THRU-BOLTED.
(7) EYE PLATES: 2 MERRIMAN FIG. 4160-230 DIAMOND BASE EYE PLATES THRU-BOLTED TO EACH OTHER THRU BOWSPRIT.
(8) JIB FURLER: SIMPSON-LAWRENCE FIG 1652 SIZE 2 FURLER & SWIVEL
(9) HANDRAILS: H & L 405T P & S. ON HOUSETOP & 401T ON HATCH SLIDE.
(10) RIGGING CLEATS: ROSTAND 2102 6" THRU-BOLTED
(11) GOOSENECK: BUCK-ALGONQUIN GN-2
MAINSAIL 187.5 SQ. FT.
TOTAL 260 SQ. FT.
JIB 72.5 SQ. FT.
750 LBS LEAD
LKR.
LKR.
MAIN SHEET

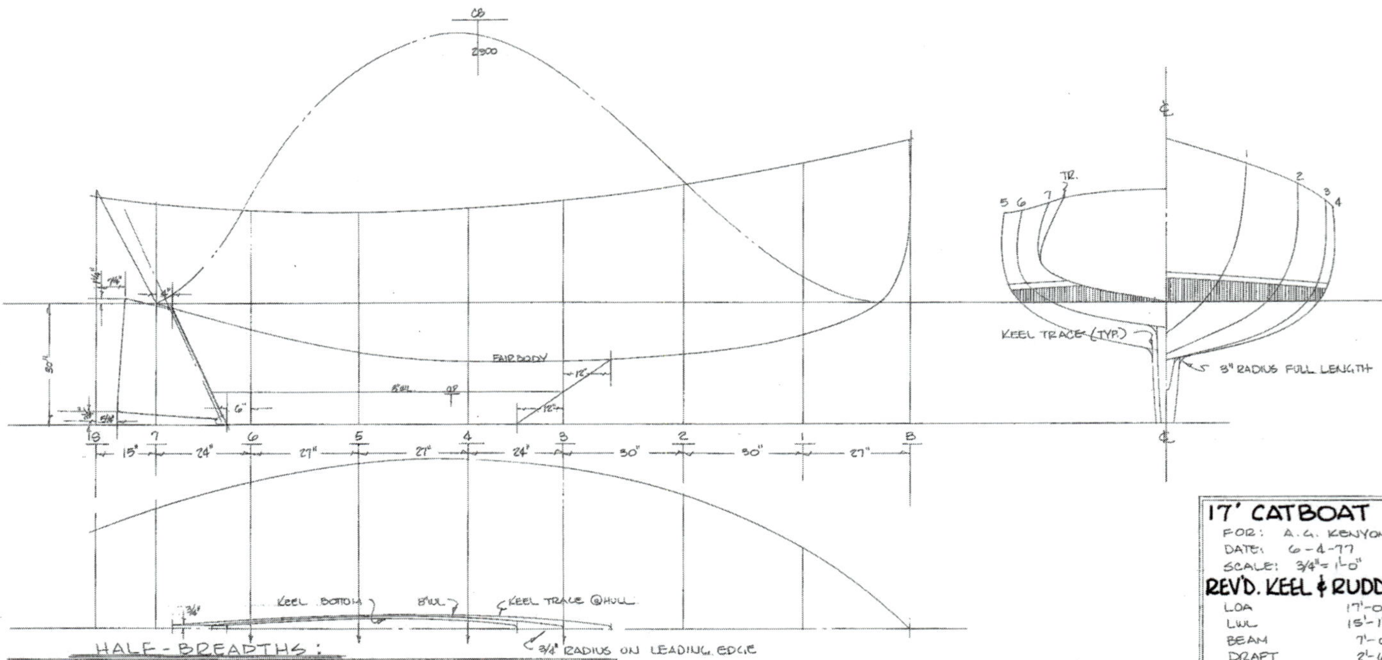

HALF-BREADTHS:

KEEL BOTTOM	0-1-2	0-2-4	0-1-7	—
8" WL	0-1-6	0-3-0	0-2-6	—
KEEL TRACE @ HULL	0-2-2	0-3-3	0-3-3	0-2-1

17' CATBOAT
FOR: A. G. KENYON
DATE: 6-4-77
SCALE: 3/4" = 1'-0"
REV'D. KEEL & RUDDER
LOA 17'-0"
LWL 15'-1½"
BEAM 7'-0"
DRAFT 2'-6"
FREEBOARD:
FWD. 3'-4"
LEAST 1'-11"
AFT 2'-2"
JAY R. BENFORD
605 S. TALBOT ST - SUITE 1
ST. MICHAELS, MD 21663
(410) 745-3235
67-26 ~JRB

LAMINATE SCHEDULE

DEDUCTION	LAYUP "TYPE"	DRY WT. OF GLASS	COMPOSITION (A)	CHANGES
OUTSIDE LAY-UP				
3/32" (B)	"A"	5.5oz/ft²	3/4M + 18W.R. + 18W.R. + 3/4M + AIREX	——
1/8"	"B"	7.5oz/ft²	3/4M + 3 - 18W.R. + 3/4M + AIREX	ADD 18W.R. TO "A"
5/32"	"C"	9.6oz/ft²	3/4M + 4 - 18W.R. + 3/4M + AIREX	ADD 18W.R. TO "B"
3/16"	"D"	11.5oz/ft²	3/4M + 5 - 18W.R. + 3/4M + AIREX	ADD 18W.R. TO "C"
INSIDE LAY-UP				
——	"E"	4.75oz/ft²	18W.R. + 18W.R. + 3/4M + AIREX	——
——	"F"	6.75oz/ft²	3 - 18W.R. + 3/4M + AIREX	ADD 18W.R. TO "E"
——	"G"	10.75oz/ft²	5 - 18W.R. + 3/4M	ADD 2-18W.R. TO "F"

(A) COMPOSITIONS ARE LISTED FROM OUTSIDE TO CORE RESPECTIVELY.

(B) DEDUCTION FOR OUTER SKIN THICKNESS WHEN DETERMINING MOLD SIZES.

(C) 3/4M = 3/4 oz MAT, 18W.R. = 18oz WOVEN ROVING.

NOTES

(1) SEE ALSO SHEET 67-4G FOR ADDITIONAL MOLD INFORMATION & DETAILS.

(2) CORE - AIREX R60.80 (PREVIOUSLY D1/18) PVC FOAM 3/8" THICK.

(3) LIMIT OF FOAM - CUT BACK FOAM 3" BELOW SHEER & AS SHOWN - REDUCE CORE THICKNESS WITH TAPER OF 1 IN 3.

(4) TRANSITION ZONES - CHANGES IN LAMINATE WEIGHT (i.e. DRY WEIGHT OF GLASS) ARE TO OCCUR AT THE RATE OF 2oz. IN 2 INCHES.

(5) KEEL REINFORCEMENT - LAMINATE (G) IS TO RUN THE FULL LENGTH OF THE HULL FROM STEM HEAD TO TOP OF TRANSOM.

(6) REPLACE AIREX CORE WITH 1/2" PLYWOOD IN WAY OF OUTBOARD MOTOR BRACKET. (APPROX. 12" × 12")

(7) DECK - MAY BE CONSTRUCTED OF WOOD OR BALSA SANDWICH - SEE SHEET 67-4W FOR WOOD DECK SCANTLINGS OR SHEET 67-4G FOR BALSA CORE.

(8) BALLAST - 650# FIXED, & 50# LEAD TRIM BALLAST. FIXED BALLAST TO BE 535# SCRAP LEAD & 115# CONC. - PLACE LEAD IN BILGE AS SHOWN & POUR CONCRETE OVER - VIBRATE CONC. TO FILL VOIDS. - 'GLASS OVER AFTER CURE.

(9) FILL SOLID WITH CASTING RESIN & PHENOLIC MICROSPHERES IN WAY OF GUDGEON FITTING.

(10) POUR IN PLACE POLYURETHANE FORM & 'GLASS OVER.

(11) RUDDER TUBE - 1" O.D. × 1/8" WALL × 2'-0" LONG - CUT OPENING IN HULL, & 'GLASS TUBE IN PLACE WITH RUDDER STOCK FITTED TO INSURE ALLIGNMENT.

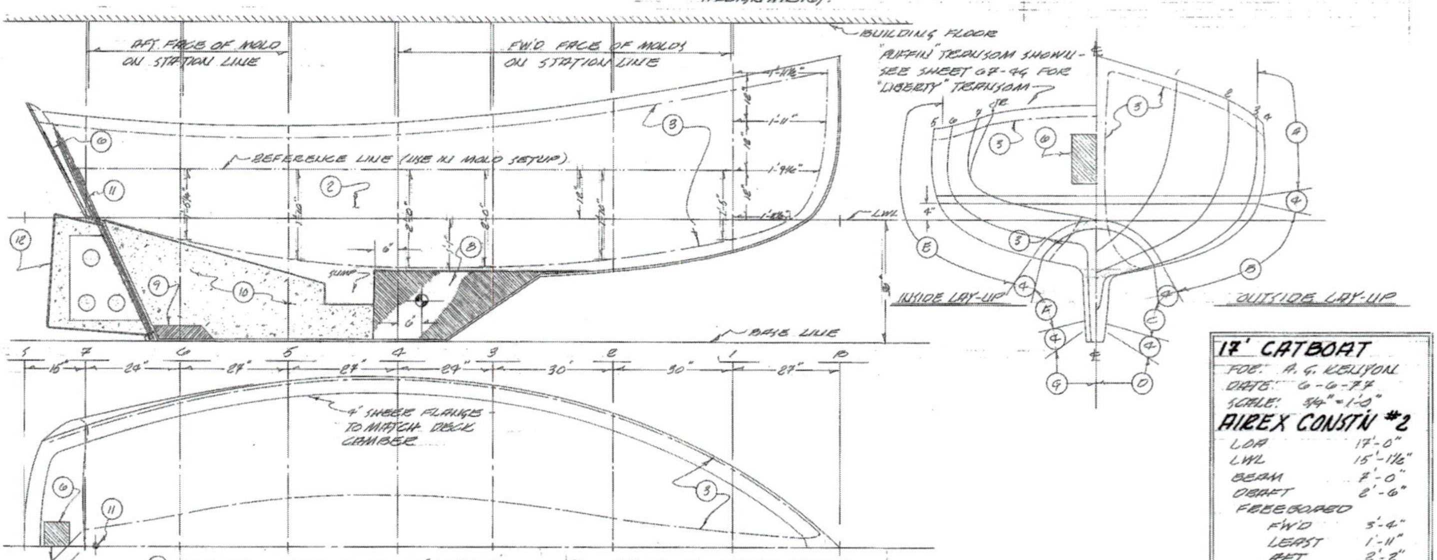

(12) RUDDER - STOCK TO BE 3/4" ⌀ TYPE 316 S.S. 4'-8" LONG - WELD 3/16" BLADE REINFORCEMENT PLATE TO STOCK AS SHOWN - SANDWICH PLATE BETW'N 2 - 3/8" AIREX SHEETS & 'GLASS WITH 3/4oz MAT, & 2 - 18oz WOVEN ROVING - WRAP REINFORCEMENT AROUND EDGES TO DOUBLE THICKNESS AT LEADING & TRAILING EDGES.

17' CATBOAT
FOR: A. G. KENYON
DATE: 6-6-77
SCALE: 3/4" = 1'-0"
AIREX CONST'N #2
LOA 17'-0"
LWL 15'-1½"
BEAM 7'-0"
DRAFT 2'-6"
FREEBOARD
FWD 3'-4"
LEAST 1'-11"
AFT 2'-2"
JAY R. BENFORD
605 S. TALBOT ST - SUITE 1
ST. MICHAELS, MD 21663
(410) 745-3235
67-27 PAC

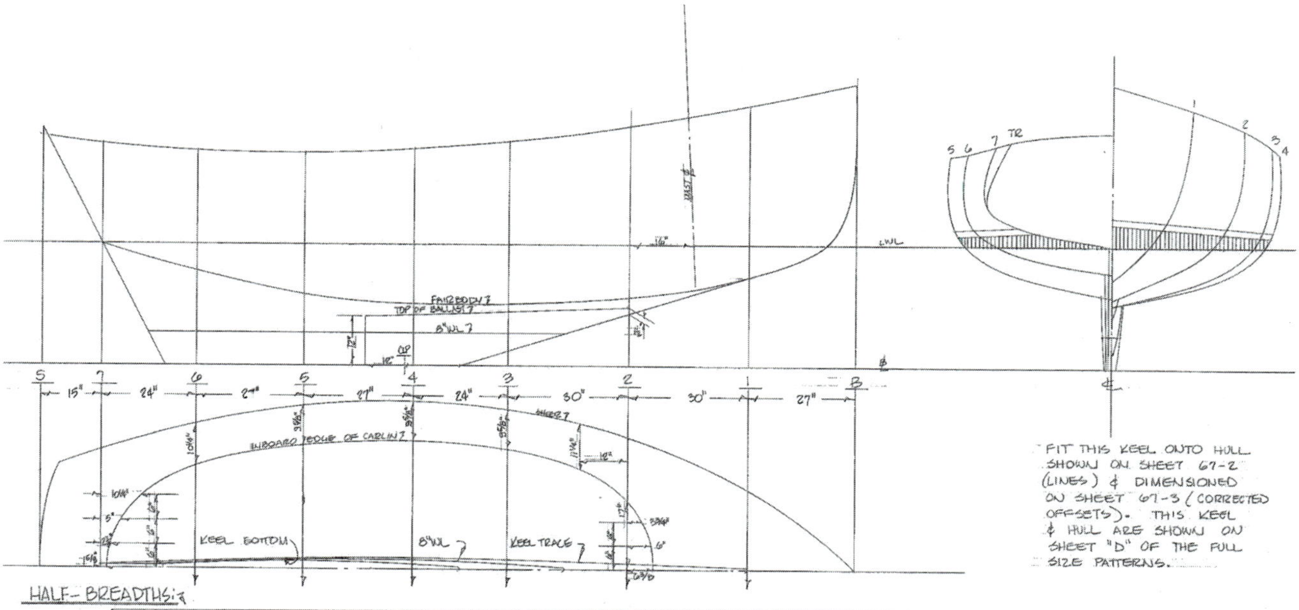

HALF-BREADTHS:

0-1-6	0-2-5	0-2-6	0-2-3	0-1-4	0-0-4	KEEL TRACE @ HULL
0-1-3+	0-2-3	0-2-2	0-1-3	—	—	8" WL
0-1-2	0-1-7	0-1-2	—	—	—	KEEL BOTTOM @ ℄

OFFSETS IN FEET — INCHES — EIGHTHS TO OUTSIDE OF KEEL. LEADING EDGE OF KEEL TO HAVE 1/2" RADIUS. TRAILING EDGE 3/4" HALF-SIDING (1 1/2" TOTAL WIDTH).

JAY R. BENFORD ☆ 17' GAFF SLOOP ☆ WOOD VERSION KEEL	SHEET Nº 28
605 S. TALBOT ST. #1 - ST. MICHAELS, MD 21663 (410) 745-3235 FOR: JOHN Mc MURRAY 9/16/80 ★ 3/4" = 1'-0" ★ ~JRB	DESIGN Nº 67

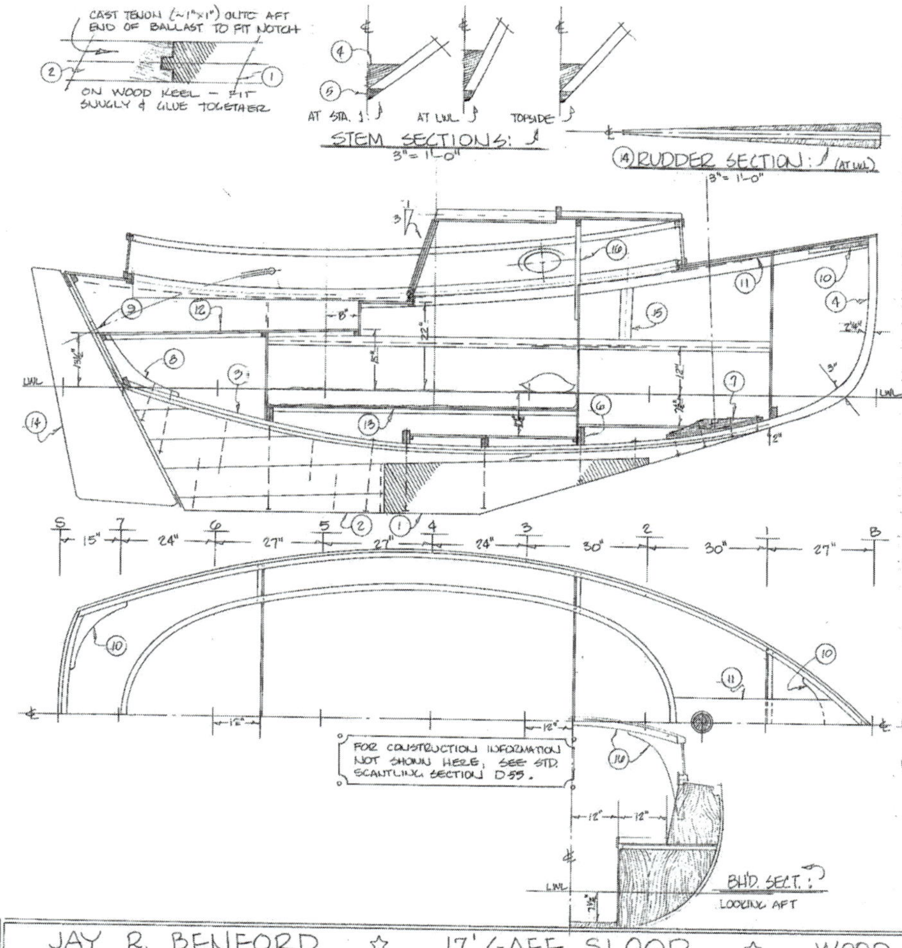

NOTES:

1. BALLAST: 750 LBS. LEAD, CAST WITH 2% ANTIMONY FOR HARDNESS. AFT 2 BOLTS 1/2"⌀; FWD. 2 BOLTS 3/8"⌀; ALL EVERDUR BRZ. OR MONEL
2. KEEL: SHAPED PER SHEET 28 — FROM D. FIR.
3. APRON: LAMINATED TO SHAPE — 1" × 5 1/2" WIDE D. FIR.
4. STEM: LAMINATED TO SHAPE & DEPTHS AS SHOWN — FAIRS INTO & BECOMES APRON BETWEEN STA. 1 & 2.
5. STEMCAP & BAND: 1/2" D. FIR WITH 3/4" BRZ. OR S.S. 1/2 OVAL OVER, EXTENDING FROM STEMHEAD TO JOINING WITH KEEL JUST AFT OF STA 1.
6. FLOORS: D. FIR OR Y. CEDAR SIDED 1 1/2" — MIN. 2" HGT. ABOVE APRON
7. MAST STEP: FROM 3 1/2" × 3 1/2" × 24" D. FIR — CUT 1 1/2" × 1 1/2" × 1" DEEP SOCKET FOR TENON ON END OF MAST. PROVIDE DRAIN HOLE FROM SOCKET.
8. STERN KNEE: SIDED 2 1/2" D. FIR OR Y. CEDAR.
9. SLOT FOR TILLER ON ℄ — PROVIDE HOLES IN OUTBOARD CORNERS OF FOOTWELL TO DRAIN COCKPIT THROUGH TRANSOM.
10. BREASTHOOK & QUARTER KNEES: 1 1/2" D. FIR OR Y. CEDAR
11. DOUBLER/MAST PARTNER: 12" WIDE 1/2" PLY UNDER DECK ON ℄ — MAST HOLE SIZED TO ALLOW FOR 3/8" TO 1/2" WEDGES ALL AROUND.
12. COCKPIT SOLE: 1/2" (12MM) PLY — SLOPED AS SHOWN.
13. BERTH FLAT: 3/8" (10MM) PLY — PROVIDE SECTION ON ℄ TO SLIDE FWD., FILLING SPACE ON ℄, MAKING ONE LARGE BERTH FROM SIDE TO SIDE. FLAT 4" BELOW LWL.
14. RUDDER: 3 LAYERS 1/2" D. FIR OR Y. CEDAR. CENTER LAYER TO RUN FORE & AFT — OUTER LAYERS TO RUN VERTICALLY. CUT WITH STRAIGHT TAPER AS SHOWN FROM BOTTOM TO BOOTTOP LEVEL — TRAILING EDGE 1/4" OR LESS.
15. CHAINPLATE DOUBLER: 3/8" (10MM) PLY GLUED IN PLACE
16. HOUSETOP BEAM & KNEE: LAMINATED BEAM 1 1/8" × 1 1/2" HIGH — KNEE GROWN OR LAM. TO SHAPE — 1 1/8" THICK — HALF-LAP ONTO BULKHEAD

JAY R. BENFORD ☆ 17' GAFF SLOOP ☆ WOOD CONSTRUCTION	SHEET Nº 29
605 S. TALBOT ST. #1 - ST. MICHAELS, MD 21663 (410) 745-3235 FOR: JOHN Mc MURRAY (REVD. 12/5/80 ~JRB) 9/17/80 ★ 3/4" = 1'-0" ★ ~JRB	DESIGN Nº 67

Shortly after doing the **Puffin**, we created the 17' Cruiser version using the same hull and a modified keel and cabin layout. The Alternate 17' Cruiser was done some years later, to give more carrying capacity in the cockpit and let her be driven a bit faster with a somewhat larger engine.

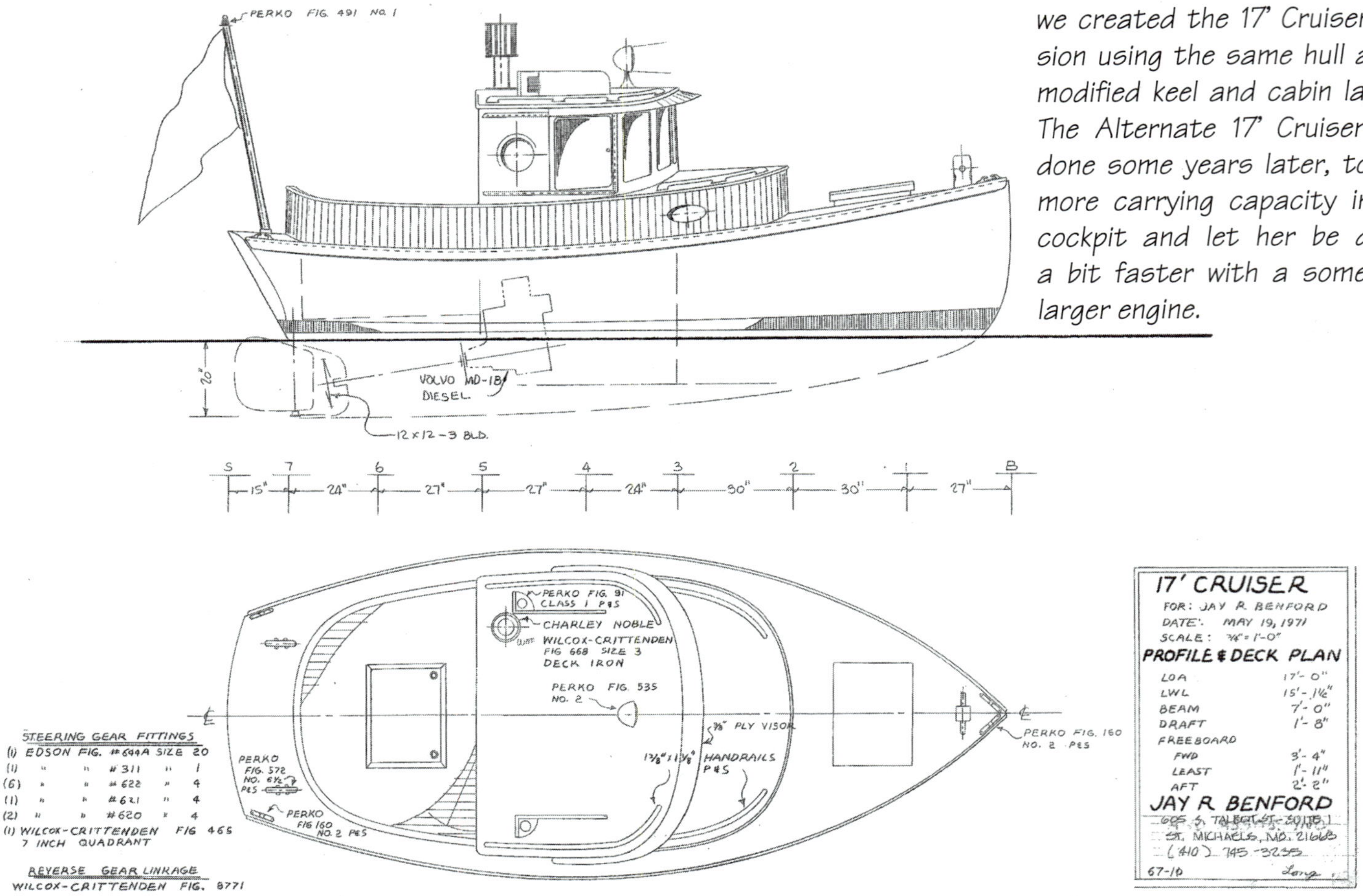

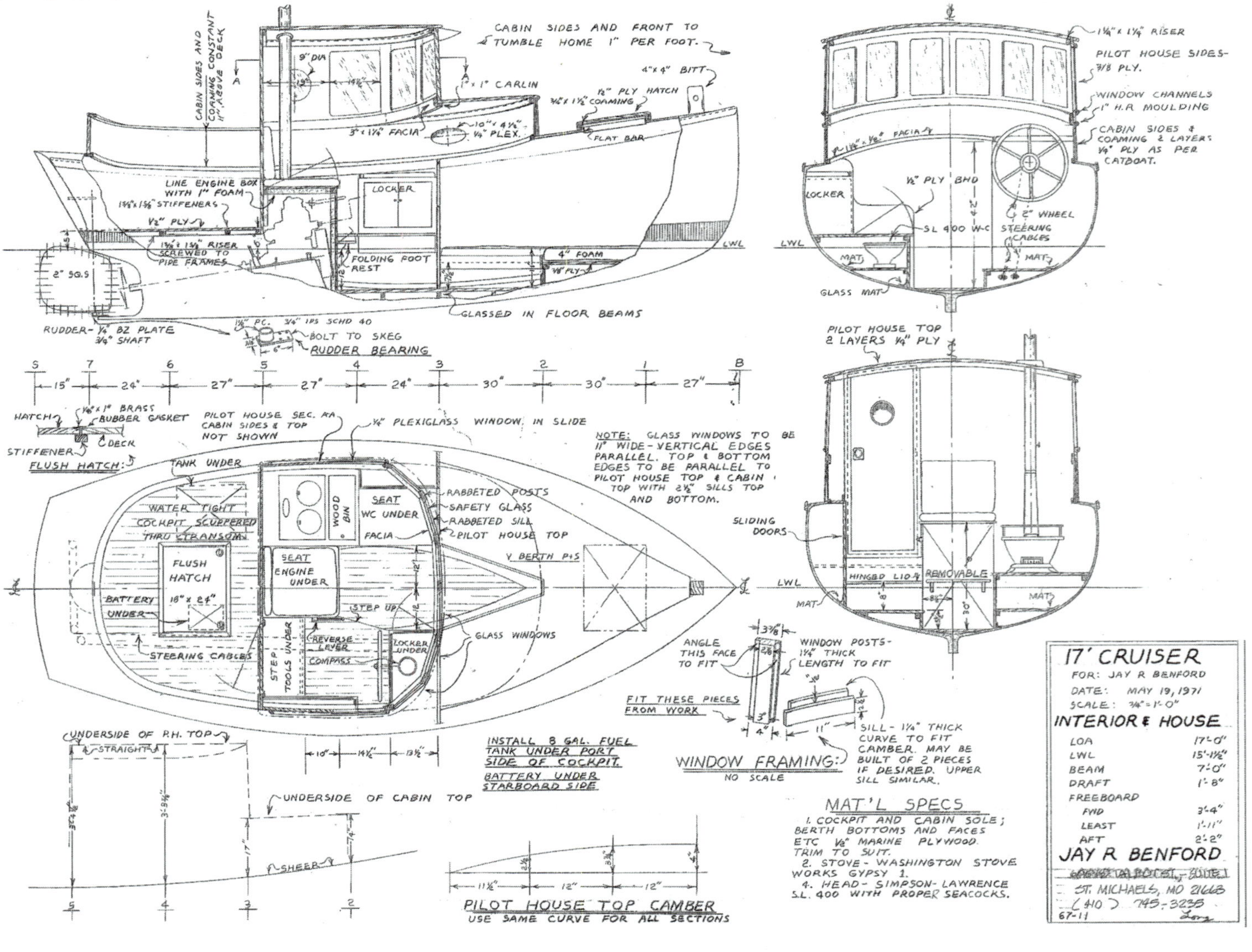

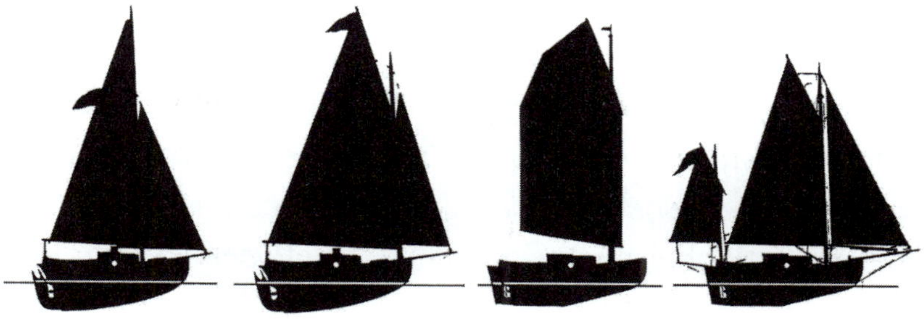

Chapter 2
Scrimshaw
19' Cruising Cat, Sloop or Yawl
Design number 177
1978

This 19-footer is a longer version of our very popular and successful 17' Catboat design, both the ***Puffin*** and ***Liberty*** versions. Like them there is both a deep and an alternate shallow keel version available, along with a sloop rig, lug sloop rig or yawl rig options.

Read what **Cruising World** magazine says in their May 1982 issue, on page 164:

"The prolific naval architect Jay Benford (formerly) of Friday Harbor, Washington, has drawn plans for a new and lovable little pocket cruiser called the ***Scrimshaw* 19**. There's some truth to the saying that smaller boats are harder to design well than larger ones. There is always the temptation to try drawing in more accommodations than the hull can reasonably be expected to hold. Yet, if some modicum of comfort is missing, prospective buyers will be reluctant to plunk down their hard-earned cash. Full-length berths and provisions for a head and galley space are usually deemed minimum requirements.

"The ***Scrimshaw* 19** succeeds admirably in meeting this design challenge, and should exceed many persons' expectations for good looks. Those committed to pocket cruisers sometimes feel neglected because of the relatively small number of boats to choose from, especially under 20 feet. Jay Benford is one designer who seems to enjoy working for the small guy as well as for the client whose tastes and pocketbook demand the "ultimate maxi."

"Jay says when Lin and Larry Pardey stopped in to visit one day, he showed them the rough draft of lines for the ***Scrimshaw* 19**. They both agreed her sections looked much like ***Seraffyn***'s. She has a long keel and outboard rudder, so she should have the feel of a larger boat. This will be especially true considering her designed displacement of 4,120 pounds. Benford cautions readers of his plans that "cruising displacement" means he has included the weight of fuel and water, crew and gear.

"Two rigs are available, a traditional gaff cat rig and a three-quarter sloop rig. With roller-furling gear on the bowsprit, both rigs can be handled from the cockpit. This will be an important feature in rough weather as no one will have to go forward on the small deck.

"Comfort seems to have been the watchword in designing the ***Scrimshaw*19**, and this is clearly evident in the cockpit. The coamings are 13 to 17 inches high and are sloped outboard to prevent sore backs. The footwell width is calculated to be just right for bracing a foot from the other side. A hatch in the cockpit floor provides ready access to the small inboard diesel engine.

"Going down the companionway there is a hanging locker to port that can be used for wet oilskins. Forward of it is space for a wood or oil range with oven. This latter feature is in keeping with the comfort requirements of the blustery Pacific Northwest, where Benford lives. But farther south such an oven would generate too much heat for general use. An alcohol, kerosene or gas range could easily be substituted, but it is nice to know Benford has arranged the space necessary to bake bread.

"Opposite the stove is a lounging seat for the cook, which is angled for comfort. Aft of the seat is a small counter with a sink and locker underneath. Benford says, "The cook has almost everything needed within easy reach, and has the best seat in the house."

"The V-berths forward can be used for seating and converted to a double berth at bedtime. A small head could be fitted in the insert space between the berths, but Benford believes a cedar bucket stowed in a cockpit seat locker is a more reasonable choice."

Since the original versions were done, we've also done a shallower draft version, a junk rigged version and a canoe yawl version. These variations are noted in the table of particulars:

Particulars:		**19' Scrimshaw**	**19' Scrimshaw**	**19' Scrimshaw**	**19' Scrimshaw**
Units		**Imperial**	**Metric**	**Imperial**	**Metric**
Length overall		19'-0"	5.79 m	19'-0"	5.79 m
Length designed waterline		16'-9"	5.11 m	16'-9"	5.11 m
Beam		7'-0"	2.13 m	7'-0"	2.13 m
Draft – *note 2 different drafts*		2'-6"	0.76 m	3'-6"	1.07 m
Freeboard:	Forward	3'-4"	1.02 m	3'-4"	1.02 m
	Least	1'-11"	0.59 m	1'-11"	0.59 m
	Aft	2'-2"	0.66 m	2'-2"	0.66 m
Displacement*, cruising trim.		3,025 Lbs.	1,372 Kg.	4,120 lbs.	1,869 Kg.
Displacement-length ratio		287		391	
Ballast, pounds		900	408 Kg.	1,200	544 Kg.
Ballast ratio		29%		29%	
Sail area – Catboat		342 sq. ft.	31.77 m2	342 sq. ft.	31.77 m2
Sail area-displacement ratio		26.16		21.29	
Sail area – Junk Sloop		272.5 sq. ft.	25.32 m2	272.5 sq. ft.	25.32 m2
Sail area-displacement ratio		20.85		16.97	
Sail area – Marconi Sloop		254 sq. ft.	23.60 m2	254 sq. ft.	23.60 m2
Sail area-displacement ratio		19.43		15.81	
Sail area – Yawl		300 sq. ft.	27.87 m2	300 sq. ft.	27.87 m2
Sail area-displacement ratio		22.95		18.68	
Prismatic coefficient		.54		.54	
Pounds per inch immersion		390		390	
Entrance half-angle		22°		22°	
Water tankage, gallons		13	49 liters	13	49 liters
Fuel tankage, gallons		13	49 liters	13	49 liters
Headroom		4'-4"	1.32 m	4'-10"	1.47 m

***CAUTION:** The figure for displacement quoted here is for the boat in cruising trim. That is, with the fuel and water tanks filled, the crew on board, as well as the crews' gear and stores in the lockers. This should not be confused with the "shipping weight" often quoted as "displacement" by some manufacturers. This should be taken into account when comparing figure and ratios between this and other designs.

STATION	S	7	6	5	4	3	2	1
	HEIGHTS							
℄ TO KEEL		(0'-3¼")	0'-0"	0'-1¼"	0'-4"	0'-8¼"	1'-4¾"	2'-6"
" " B1	(4'-1½")	3'-8½"	3'-1 1/16"	2'-7"	2'-4"	2'-4¼"	2'-9¼"	3'-10"
" " B2	(4'-5⅞")	4'-0⅛"	3'-4¼"	2'-10⅝"	2'-8¾"	2'-10 5/16"	3'-4 13/16"	—
LWL TO BOOTTOP	0'-5⅝"	0'-5⅛"	0'-4⅝"	0'-4⅜"	0'-4¼"	0'-4½"	0'-5¼"	0'-6¼"
" " SHEER	2'-2⅝"	2'-0¾"	1'-10 15/16"	1'-10½"	1'-11⅝"	2'-1⅝"	2'-5⅜"	2'-10⅝"
	HALF-BREADTHS							
℄ TO KEEL HALF-SIDING	—	—	0'-¾"	0'-3¼"	0'-4"	0'-3¾"	0'-2¼"	0'-0½"
" " 34" WL	—	—	0'-3¾"	1'-10"	2'-2¾"	1'-11 5/16"	1'-1 3/16"	0'-2¼"
" " 42" WL	—	0	2'-4 9/16"	3'-1 5/16"	3'-2⅛"	2'-11⅜"	2'-1"	0'-8 1/16"
" " 50" WL	2'-0¾"	2'-5 15/16"	3'-0⅝"	3'-4⅞"	3'-5⅞"	3'-3 1/16"	2'-6⅜"	1'-2⅜"
" " 58" WL	2'-3¾"	2'-7⅞"	3'-1¼"	3'-5"	3'-6⅜"	3'-4"	2'-8¾"	1'-5 1/16"
" " SHEER	2'-0¼"	2'-5⅝"	3'-0⅛"	3'-4 7/16"	3'-5¾"	3'-3 13/16"	2'-9¼"	1'-8½"
	DIAGONALS							
DIAG A	—	—	0'-3½"	0'-9"	1'-0⅝"	1'-0 9/16"	0'-8⅛"	0'-2 7/16"
DIAG B	—	—	0'-1⅜"	0'-5 3/16"	0'-7⅛"	0'-6¾"	0'-3⅝"	—

NOTES:

1. LINES MUST BE LOFTED FULL SIZE- DO NOT SCALE PRINTS -- LINES & OFFSETS TO THE OUTSIDE OF THE SHELL. DEDUCT APPROPRIATE SHELL THICKNESS FOR HOLDS, WEB FRAMES, BHDS. ETC.
2. BOOTTOP STRIPE IS 1½" HIGH IN PROFILE FULL LENGTH. OFFSETS TO BE TOP EDGE OF STRIPE.
3. THESE PLANS ARE THE PROPERTY OF THE DESIGNER & MAY BE USED ONLY AS AUTHORIZED. NO MORE THAN ONE BOAT WILL BE BUILT FROM THESE PLANS WITHOUT PERMISSION OF THE DESIGNERS.
4. ANY ALTERATION FROM THESE PLANS RELIEVES THE DESIGNER OF FURTHER RESPONSIBILITY.
5. DECK CAMBER IS 4" IN 7'-0"
6. OFFSETS AT STA.7 FOR FAIRING PURPOSES ONLY -- DO NOT PUT MOLD FRAME AT STA.7 -- SPAN FROM STA. 6 TO TRANSOM WITH PLANKING.

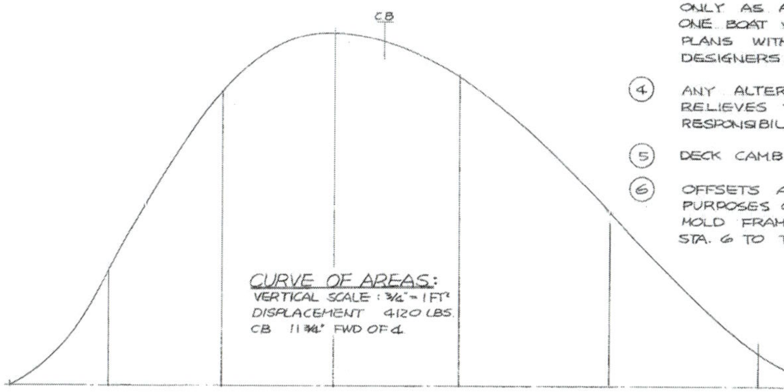

CURVE OF AREAS:
VERTICAL SCALE: 3/16" = 1 FT²
DISPLACEMENT 4120 LBS.
CB 11¾" FWD OF 4

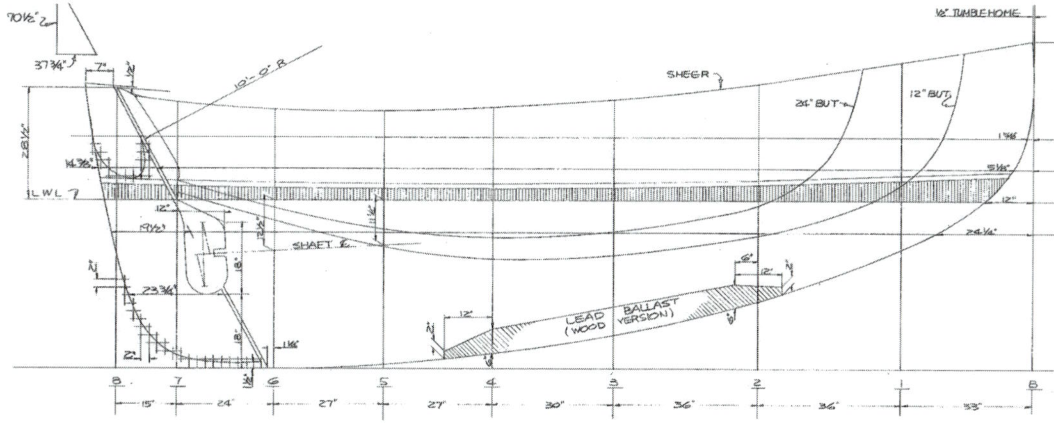

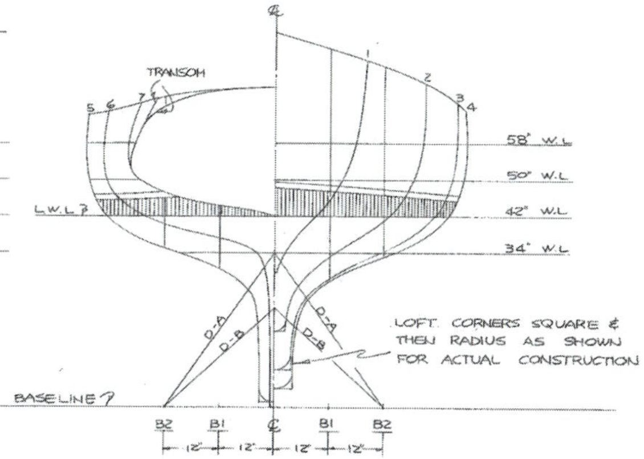

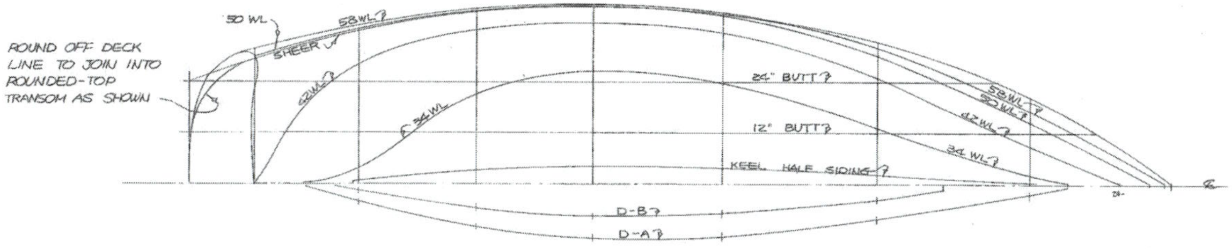

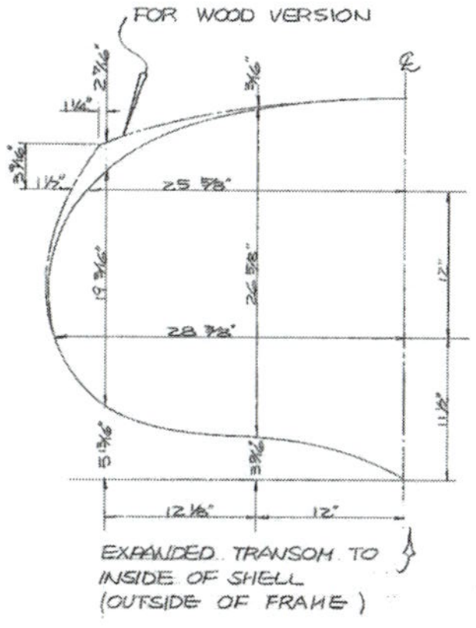

19' CRUISING CAT
FOR: MacRAE YACHTS
DATE: NOV. 6 1978
SCALE: ¾" = 1'-0"

LINES & OFFSETS

L.O.A.	19'-0"
L.W.L.	16'-9"
BEAM	7'-0"
DRAFT	3'-6"
FREEBOARD	
FWD	3'-4"
LEAST	1'-11"
AFT	2'-2"

JAY R. BENFORD
605 S. TALBOT ST., SUITE 1
ST. MICHAELS, MD 21663
(410) 745-3235
177-3

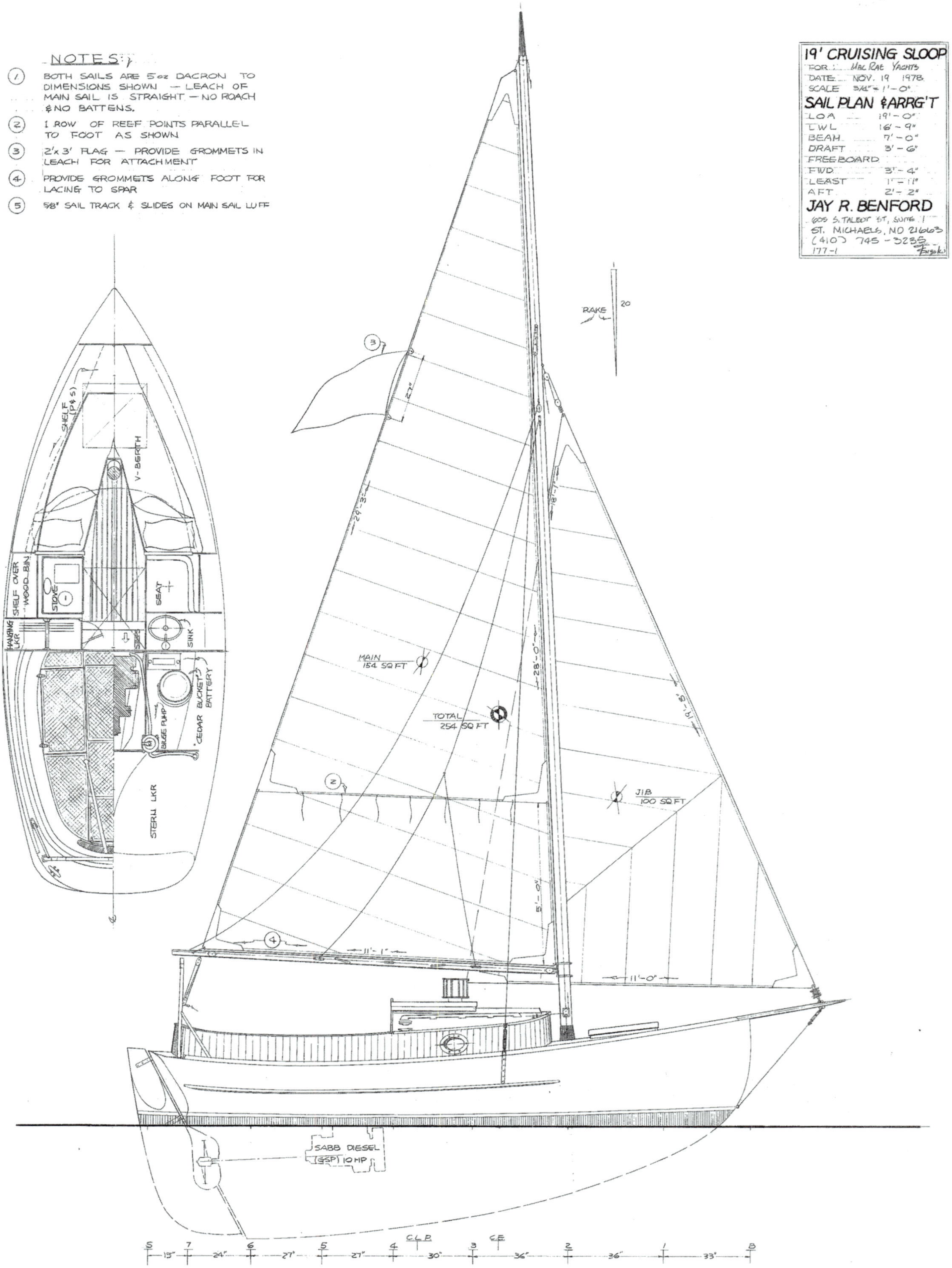
NOTES:
1 BOTH SAILS ARE 5 oz DACRON TO DIMENSIONS SHOWN — LEACH OF MAIN SAIL IS STRAIGHT — NO ROACH & NO BATTENS.
2 1 ROW OF REEF POINTS PARALLEL TO FOOT AS SHOWN
3 2'x 3' FLAG — PROVIDE GROMMETS IN LEACH FOR ATTACHMENT
4 PROVIDE GROMMETS ALONG FOOT FOR LACING TO SPAR
5 5/8" SAIL TRACK & SLIDES ON MAIN SAIL LUFF
19' CRUISING SLOOP
FOR MAC RAE YACHTS
DATE NOV. 19 1978
SCALE 3/4" = 1'-0"
SAIL PLAN & ARRG'T
LOA 19'-0"
LWL 16'-9"
BEAM 7'-0"
DRAFT 3'-6"
FREEBOARD
FWD 3'-4"
LEAST 1'-11"
AFT 2'-2"
JAY R. BENFORD
605 S. TALBOT ST, SUITE 1
ST. MICHAELS, MD 21663
(410) 745-3235
177-1
RAKE
20
MAIN
154 SQ FT
TOTAL
254 SQ FT
JIB
100 SQ FT
SHELF (P&S)
V-BERTH
SHELF OVER WOOD BIN
STOVE
SEAT
HANGING LKR
STEPS
SINK
CEDAR BUCKET
BATTERY
BILGE PUMP
STERN LKR
SABB DIESEL (GSP) 10 HP
CLP
CE

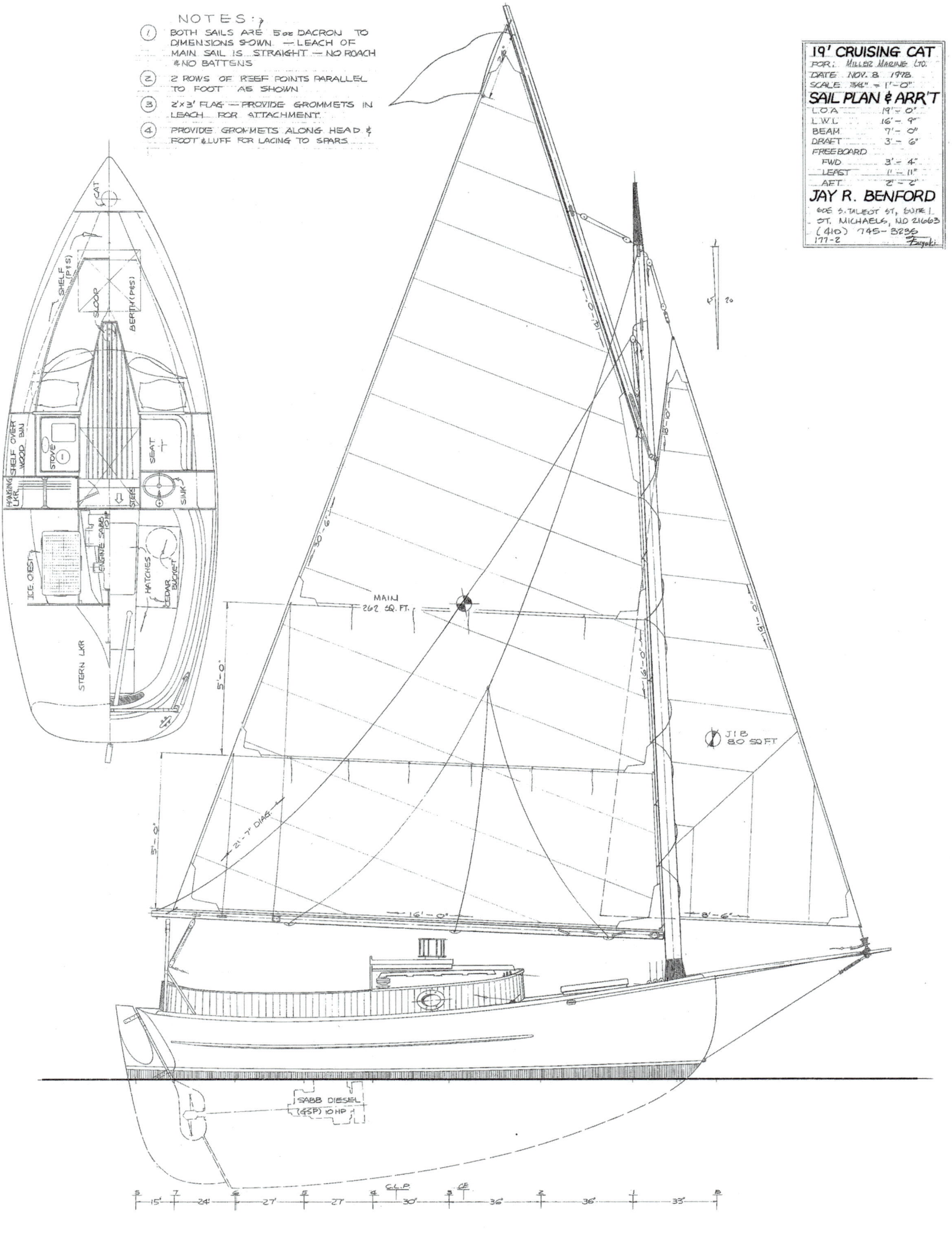
NOTES:
1. BOTH SAILS ARE 5oz DACRON TO DIMENSIONS SHOWN. — LEACH OF MAIN SAIL IS STRAIGHT — NO ROACH & NO BATTENS
2. 2 ROWS OF REEF POINTS PARALLEL TO FOOT AS SHOWN
3. 2'x3' FLAG — PROVIDE GROMMETS IN LEACH FOR ATTACHMENT.
4. PROVIDE GROMMETS ALONG HEAD & FOOT & LUFF FOR LACING TO SPARS
19' CRUISING CAT
FOR: MILLER MARINE LTD.
DATE NOV. 8 1978
SCALE 3/4" = 1'-0"
SAIL PLAN & ARR'T
L.O.A 19' - 0"
L.W.L 16' - 9"
BEAM 7' - 0"
DRAFT 3' - 6"
FREEBOARD
FWD 3' - 4"
LEAST 1' - 11"
AFT 2' - 2"
JAY R. BENFORD
605 S. TALBOT ST, SUITE 1
ST. MICHAELS, MD 21663
(410) 745-3235
177-2
CAT
SHELF (P&S)
SLOOP
BERTH (P&S)
SHELF OVER WOOD BIN
STOVE
SEAT
HANGING LKR
STEPS
SINK
ICE CHEST
ENGINE SABB 10 HP
HATCHES
CEDAR BUCKET
STERN LKR
MAIN 262 SQ. FT.
JIB 80 SQ FT
30'-6"
19'-0"
18'-0"
15'-0"
16'-0"
5'-0"
2'-7" DIAG.
16'-0"
8'-6"
SABB DIESEL (GSP) 10 HP
C.L.P
C.E
15' 24' 27' 27' 30' 36' 36' 33'

Tardis *under construction in Auckland, New Zealand. She had a strip-planked core with cold-molded veneers over. The strip-planking is trimmed off about the turn of the bilge and resumed, making the last planks run parallel to the sheer.*

NOTES:

1. AIREX CORE: CORE MATERIAL TO BE 3/8" AIREX 01-18 PVC FOAM 5 LBS. DENSITY
2. TRANSITION ZONES: CHANGES IN LAMINATE WEIGHT (DRY WEIGHT OF GLASS) ARE TO OCCUR AT A RATE OF 2oz. IN 2 INCHES.
3. KEEL REINFORCEMENT: LAMINATE (G) IS RUN THE FULL LENGTH OF THE HULL FROM THE STEM HEAD TO TOP OF TRANSOM.
4. BULKHEAD ATTACHMENT: MOUNT BULKHEAD ON 3/8" THICK AIREX PADS & GLASS WITH 3 - 6" WIDE 1½ oz MAT STRIPS EACH SIDE.
5. FILL SOLID WITH CASTING RESIN WITH PHENOLIC MICROSPHERES IN WAY OF GUDGEON.
6. POUR IN PLACE POLYURETHANE FORM & GLASS OVER
7. DEAD WOOD FOR PROPELLER SHAFT TO BE A FIR BLOCK 'GLASSED OVER
8. REPLACE FOAM IN WAY OF CHAINPLATE WITH 3/8" PLYWOOD - SEAL WITH RESIN BEFORE LAMINATING OVER. (ONLY SLOOP RIG)
9. MAST STEP: BUILD AS SHOWN FROM 1½" FIR MOUNT ON FOAM PADS IN WAY OF SOLID GLASS AREA & GLASS TO SHELL WITH 2 PAIRS OF 1½ oz MAT & 24 oz WR
10. BOBSTAY FITTING: USE ½" ø "U" BOLT, REINFORCE INSIDE OF HULL IN WAY OF FITTING WITH 1 PAIR OF 1½" oz MAT & 24 WR
11. RUDDER FITTINGS: SEE SHEET D-102
12. COMPANION HATCH: SEE SHEET D-119
13. REINFORCE AROUND MAST: APPLY EXTRA 1 PAIR OF 1½ oz MAT & 24 oz WR.
14. MORE DETAILS OF THE DECK: SEE SHEET 177-6
15. BALLAST: ~1200 LBS LEAD — TAKE SHAPE FROM LINES & LOFTING. — GLASS OVER AFTER SETTING IN PLACE.

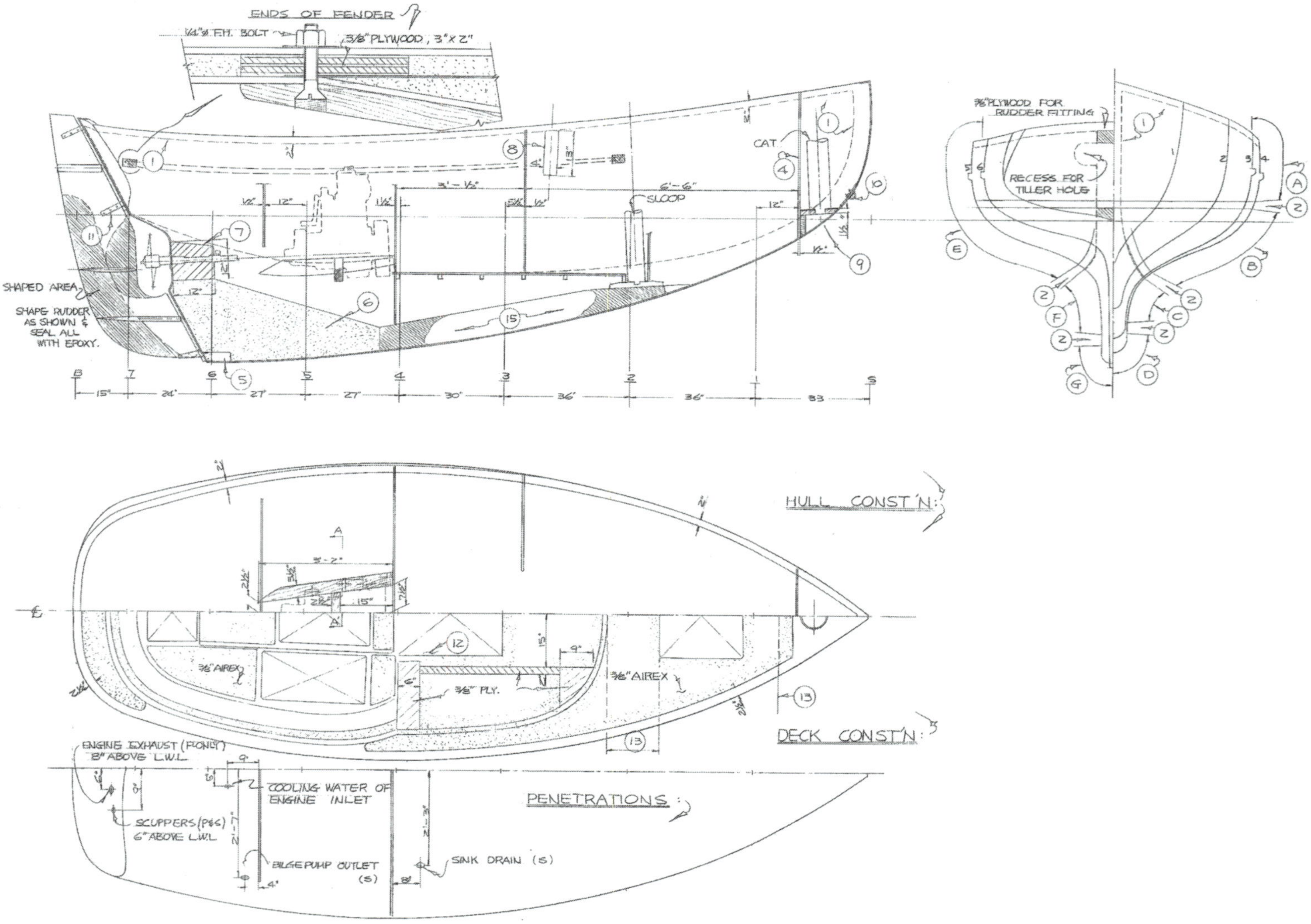

Tardis *under construction in Auckland, New Zealand. She has a skin of 'glass over the wood. And the resulting fine finish is visible.*

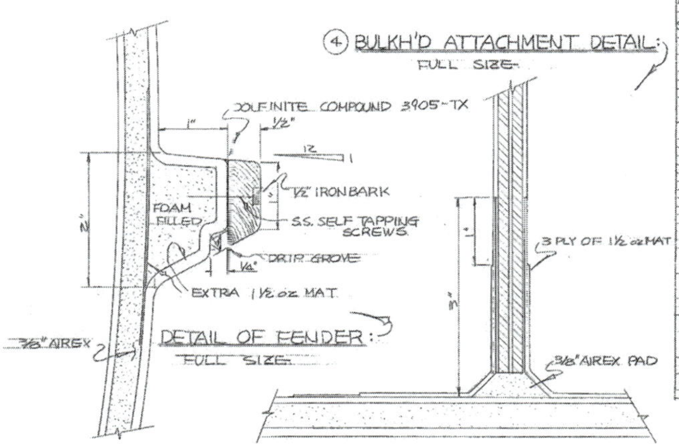

LAMINATE SCHEDULE

DEDUCTION SKIN THICKNS	LAMINATE "TYPE"	WT OF GLASS DRY (OZ/FT)	COMPOSITION	CHANGES
	OUTSIDE LAMINATE			
.133"	"A"	5.58	6oz "C" + ¾oz "M" + 24oz "R" + 1½oz "M" + AIREX	—
.181"	"B"	7.08	6oz "C" + ¾oz "M" + 24oz "R" + 1½oz "M" + 1½oz "M" + AIREX	ADD 1½ oz M
.232"	"C"	9.75	6oz "C" + ¾oz "M" + 24oz "R" + 1½oz "M" + 24oz "R" + 1½oz "M"	ADD 24oz R
.314"	"D"	15.08	6oz "C" + ¾oz "M" + 24oz "R" + 1½oz "M" + 24oz "R" + 1½oz "M" + 24oz "R"	ADD 2-24oz R
	INSIDE LAMINATE			
.099"	"E"	3.67	AIREX + 1½oz "M" + 24oz "R"	—
.147"	"F"	5.17	AIREX + 1½oz "M" + 24oz "R" + 1½oz "M"	ADD 1½oz M
.245"	"G"	9.33	AIREX + 1½oz "M" + 24oz "R" + 1½oz "M" + 24oz "R" + 1½oz "M"	ADD 24oz R + 1½oz M
.280" (+.375")	"H" (DECK)	10.75	6oz "C" + ¾oz "M" + 24oz "R" + 1½oz "M" + AIREX + 1½oz "M" + 24oz "R" + 1½oz "M"	—

"C" = CLOTH
"M" = MAT
"R" = WOVEN ROVING

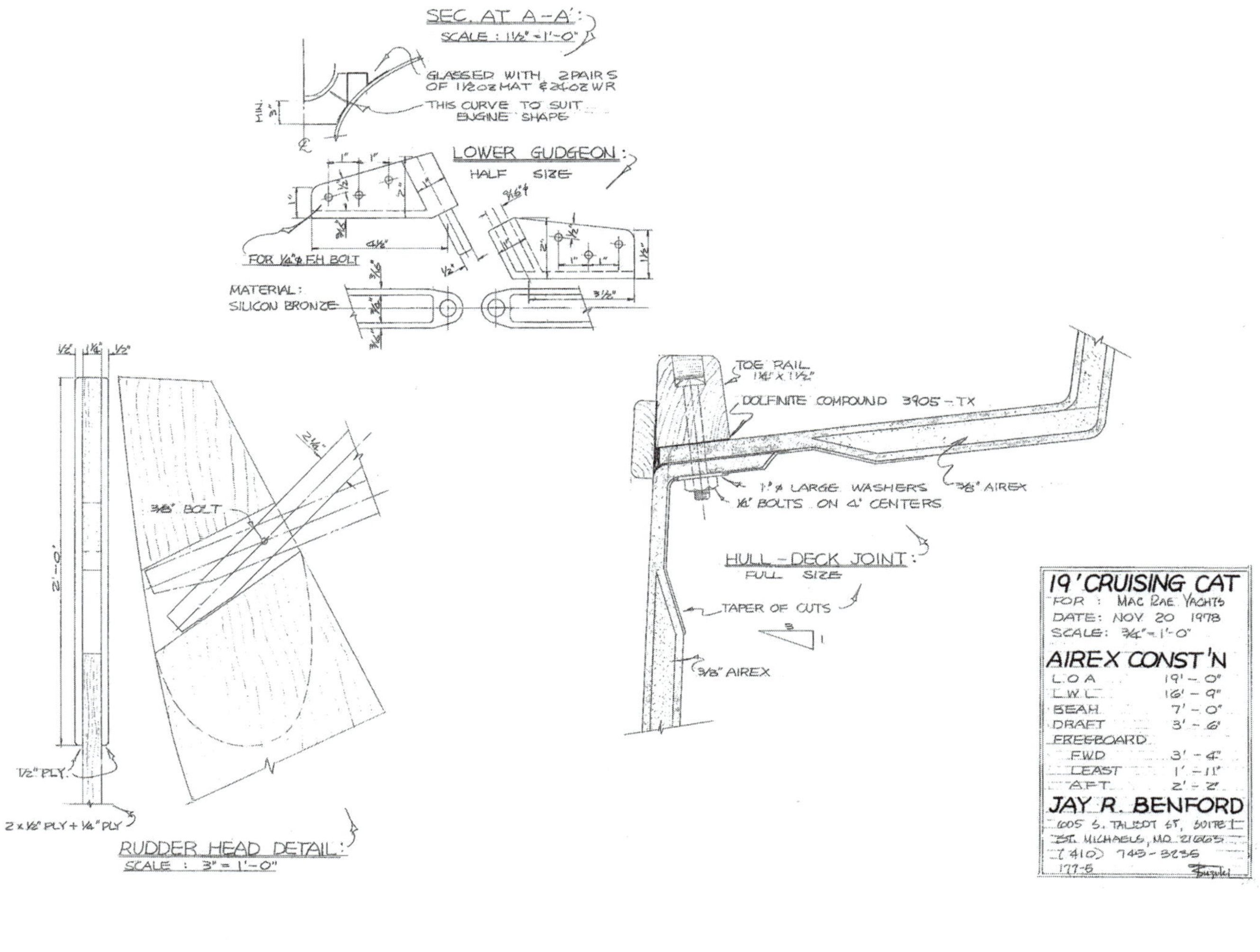

Chickadee *and her proud builder rolling her out of the building shop, left. The cockpit slope is readily visible in this pre-launch photo, right. She has the shallower draft version of this design.*

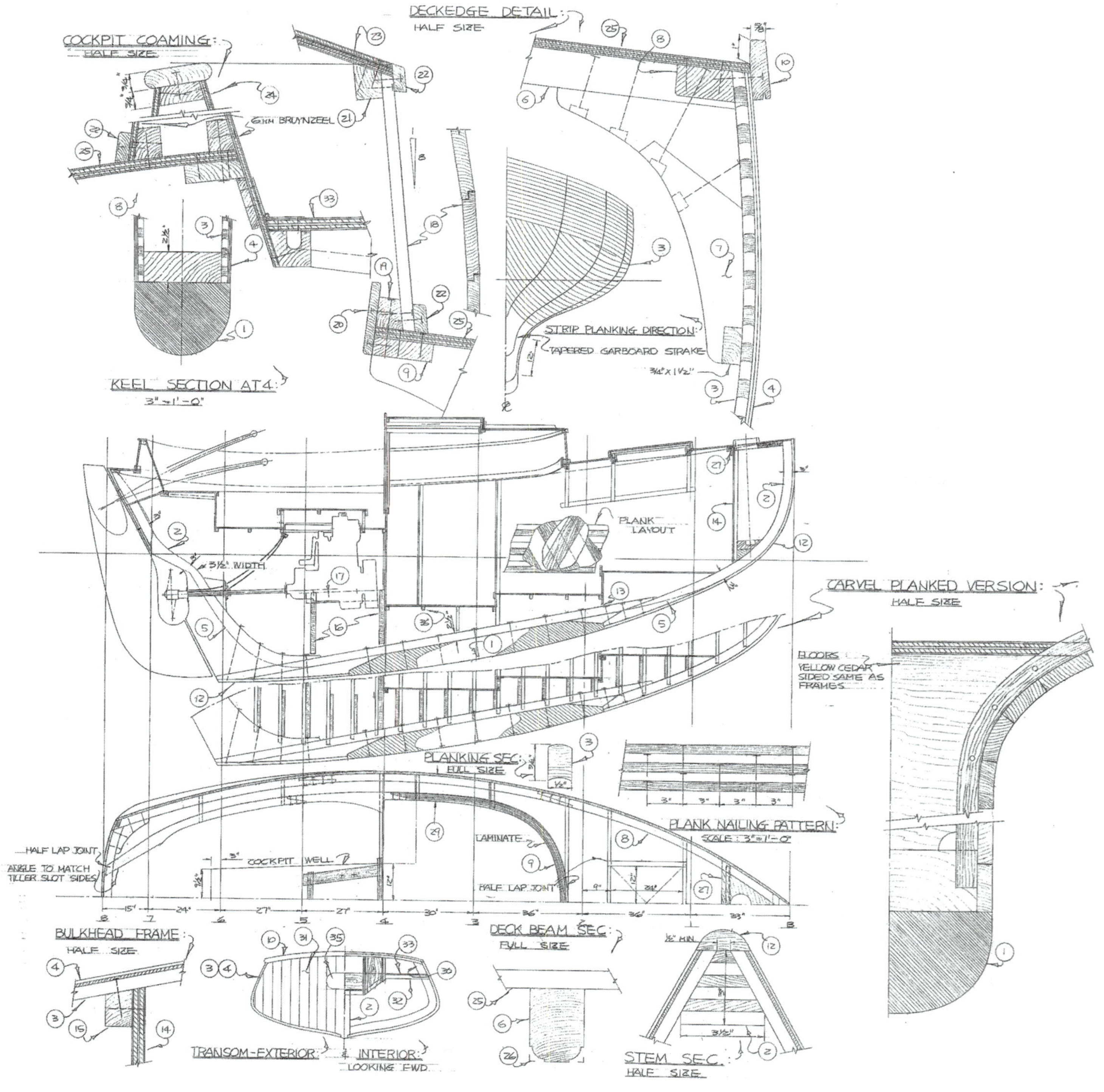

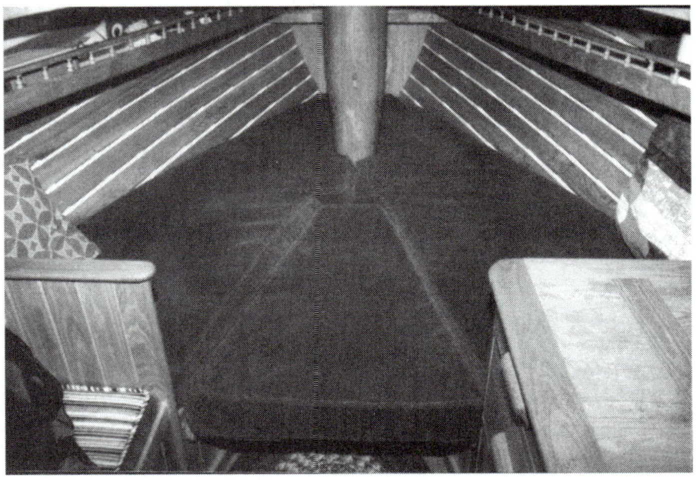

Chickadee *has a well detailed interior, left. The filler for the v-berth, visible in this photo, right, makes the forward berth into a double.*

SCANTLINGS:

(1) LEAD BALLAST: — 1200 LBS — TAKE SHAPE FROM LINES & LOFTING — 7- ½"ø BRONZE KEEL BOLTS ON 12" CENTERS LOCATED AS SHOWN.

(2) STEM & STERNPOST:-LAMINATED TO SHAPE AS SHOWN & TRIMMED TO FIT. ALL TO BE CLEAR FIR OR YELLOW CEDAR.

(3) INNER PLANKING: — ½" x ¾" MILLED TO SHAPE SHOWN — AT MID SECTION STRIPS TO BE FIR FOR 12" ON EITHER SIDE OF RABBET & TO 6" BELOW SHEER — ALL OTHER STRIPS TO BE RED CEDAR.

(4) OUTER PLANKING: — DOUBLE DIAGONAL ⅛" VERT. GRAIN RED CEDAR — ORIENT INNER LAYER 30° AFT & OUTER 30° FWD — EPOXY GLUE ALL PLANKING

(5) TOP OF LEAD KEEL TO BE RABBET LINE — RABBET FWD. & AFT OF KEEL AS SHOWN FAIRS INTO STEM & STERNPOST.

(6) 1⅛" x 1½" ALASKA CEDAR DECK BEAMS — NOTCH FOR HARPIN & CARLIN & BOLT TOGETHER.

(7) KNEES:- SHAPED AS SHOWN SAWN FROM 2 PCS. OF 1⅛" x 3½" ALASKA CEDAR

(8) 1¼" x 2½" ALASKA CEDAR HARPIN SAWN TO SHAPE AS SHOWN — SCARF & BOLT AS SHOWN.

(9) ¾" x 2½" ALASKA CEDAR CARLIN SAWN TO SHAPE AS SHOWN — SCARF & BOLT AS SHOWN.

(10) SHEER GUARD & TOE RAIL FROM ¾" x 2½" IRONBARK — SHAPE & DRIP GROOVE AS SHOWN — SCREW THRU INTO HARPIN.

(11) NOTE SQUARE CORNER ON TRANSOM - SHEER INTERSECTION — USE EXPANDED TRANSOM SHAPE ON SHEET 177 - 3 FOR TRANSOM FRAME — NOTE THIS SHAPE IS FOR AFT FACE OF FRAME & FRAME BEVELS OUT FOR GREATER WIDTH AS IT GOES FORWARD — TAKE BEVELS FROM LOFT FLOOR.

(12) ¾" IRONBARK STEM CAP & WORM GUARD ALONG BOTTOM EDGE OF KEEL — ROUND OFF AS SHOWN & TO FAIR INTO KEEL SECTION

(13) NOTE ALL PLYWOOD BULKHEADS TO STOP 1" CLEAR OF KEEL TO KEEP BILGE WATER OUT OF END GRAIN & TO PROVIDE LIMBERS.

(14) BULKHEADS ½" MARINE PLYWOOD (FIR) OR 10 MM BRUYNZEEL PLY.

(15) BULKHEAD FRAME TO BE 1⅛" x 1½" YELLOW (ALASKA) CEDAR SAWN TO SHAPE — ON FWD. SIDE OF BHD'S. FWD OF STA. 4 & ON AFT SIDE OF BHD'S AFT OF STA 4

(16) 1½" ALASKA CEDAR FLOORS AT ENGINE BEDS — AFT FLOOR CENTERED AT AFT ENG. MOUNT & FWD FLOOR CENTERED 8½" FWD. OF FWD. ENG MOUNT — CARRY FLOORS UP TO TOP OF BED & NOTCH DOWN TO PROVIDE ½" CLEARANCE TO ENGINE.

(17) ENGINE BEDS 2½" x 3½" W. OAK (OR FIR) FITTED TO HULL AS SHOWN — NOTCH FWD. FLOOR FOR BED & LAG. FASTEN

(18) CABIN & COAMING 11" HIGH OFF DECK ALL AROUND — HEIGHT CONSTANT & 1" IN 8" TUMBLE HOME ALL AROUND 7/16" x 3½" TEAK OR ½" x 3½" FIR STAVES RUN VERTICALLY — HALF-LAP SEAMS & FILL WITH KNIFING GRADE OF THIOKOL PER DETAILED SECTION — NOTE OUTBOARD SEAM TO BE WIDER WITH MORE THIOKOL & INBOARD SEAM SET TIGHT.

(19) CABIN FRAME ON DECK FROM 1⅛" x 1¼" ALASKA CEDAR SAWN TO SHAPE AT ENDS & BEVELED OUTBOARD FOR TUMBLE HOME OF HOUSE & COAMING — THRU-BOLT (3/16"ø GALV.) TO CARLIN ON 8" CTRS. & BED IN THIOKOL.

(20) FACIA OF ⅜" x 3" TEAK — TO PROJECT ABOVE CABIN FRAME AS SHOWN TO FORM HAND - GRAB

(21) HOUSE TOP FRAME FROM 1⅛" x 1½" ALASKA CEDAR BEVEL FOR HOUSE TOP

(22) ½" x 1¼" TEAK MOLDINGS — CUT BEVELS AS SHOWN & BED IN THIOKOL.

(23) HOUSE TOP 2 LAYERS (6 MM + 4 MM) BRUYNZEEL PLYWOOD CLOUT NAILED & GLUED TOGETHER STAGGER BUTTS FROM ONE LAYER TO NEXT COVER WITH 2 LAYERS OF ¾ OZ FIBERGLASS MAT & EPOXY RESIN — RUN MAT OVER EDGES & END UNDER MOLDING.

(24) COCKPIT COAMING TO FORM CONTINUATION OF LINE OF HOUSE TOP & BUILD UP AS SHOWN.

(25) DECK 10 MM BRUYNZEEL OR ½" FIR PLYWOOD — ANCHOR FAST NAIL TO DECK FRAMING ON 6" TO 8" CENTERS.

(26) BOTTOM EDGES OF DECK BEAMS MAY BE CUT AS SHOWN IF DESIRED — KNEES ALSO.

(27) 1½" x 11" + BREASTHOOK & MAST BLOCK — HALF-LAP ONTO HARPIN & SAW CAMBER IN TOP — THRU-BOLT TO HARPIN — ALASKA CEDAR

(28) 1¼" ALASKA CEDAR KNEES — THRU-BOLT TO HARPIN.

(29) INCREASE TO 8⅝" FROM STA. 3 TO STA. 4 & FAIR INTO CURVES AT ENDS.

(30) 1⅛" x 2½" ALASKA CEDAR TRANSOM FRAME SAWN TO SHAPE & CAMBER — SEE NOTE (11)

(31) ¾" x 3½" TEAK TRANSOM PLANKING — HALF-LAP ETC. PER NOTE (18)

(32) 1⅛" x 1½" ALASKA CEDAR FRAMING FOR COCKPIT SEAT.

(33) COCKPIT SEAT FLATS 10 MM BRUYNZEEL OR ½" FIR PLYWOOD — HATCHES BY OWNER TO SUIT

(34) FIR BLOCK FOR MAST STEP BUILT UP FROM 4" STOCK TO 6½" ABOVE LWL AS SHOWN 3" x 3" x 2" DEEP SOCKET FOR MAST HEEL (NOTE - REVISE LENGTH OF MAST HEEL TO FIT)

(35) CUT SLOT IN TRANSOM FOR TILLER

(36) CABIN SOLE BEAMS 1⅛" x 1½" ALASKA CEDAR — SOLE ½" PLYWOOD — PROVIDE LIFT OUT SECTIONS FOR BILGE ACCESS TO SUIT

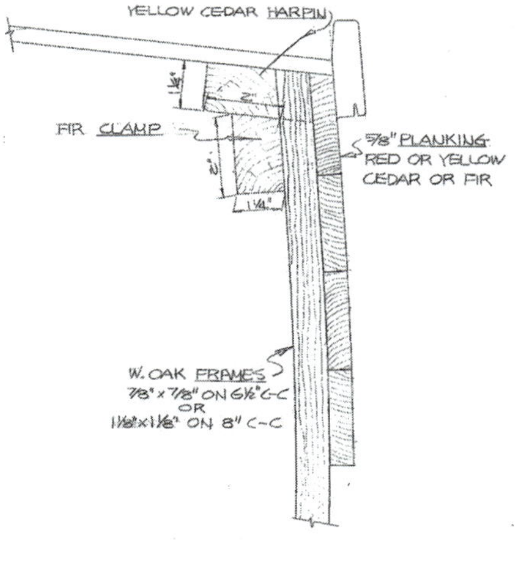

19' CRUISING CAT
FOR: MILLER MARINE SERVICES
DATE: DEC. 14 1978
SCALE: ¾" = 1'-0" & AS NOTED

WOOD CONS'T.

LOA	19'-0"
LWL	16'-9"
BEAM	7'-0"
DRAFT	3'-6"
FREEBOARD	
FWD	3'-4"
LEAST	1'-11"
AFT	2'-2"

JAY R. BENFORD
605 S. TALBOT ST., SUITE 1
ST. MICHAELS, MD 21663
(410) 745-3235
177-8

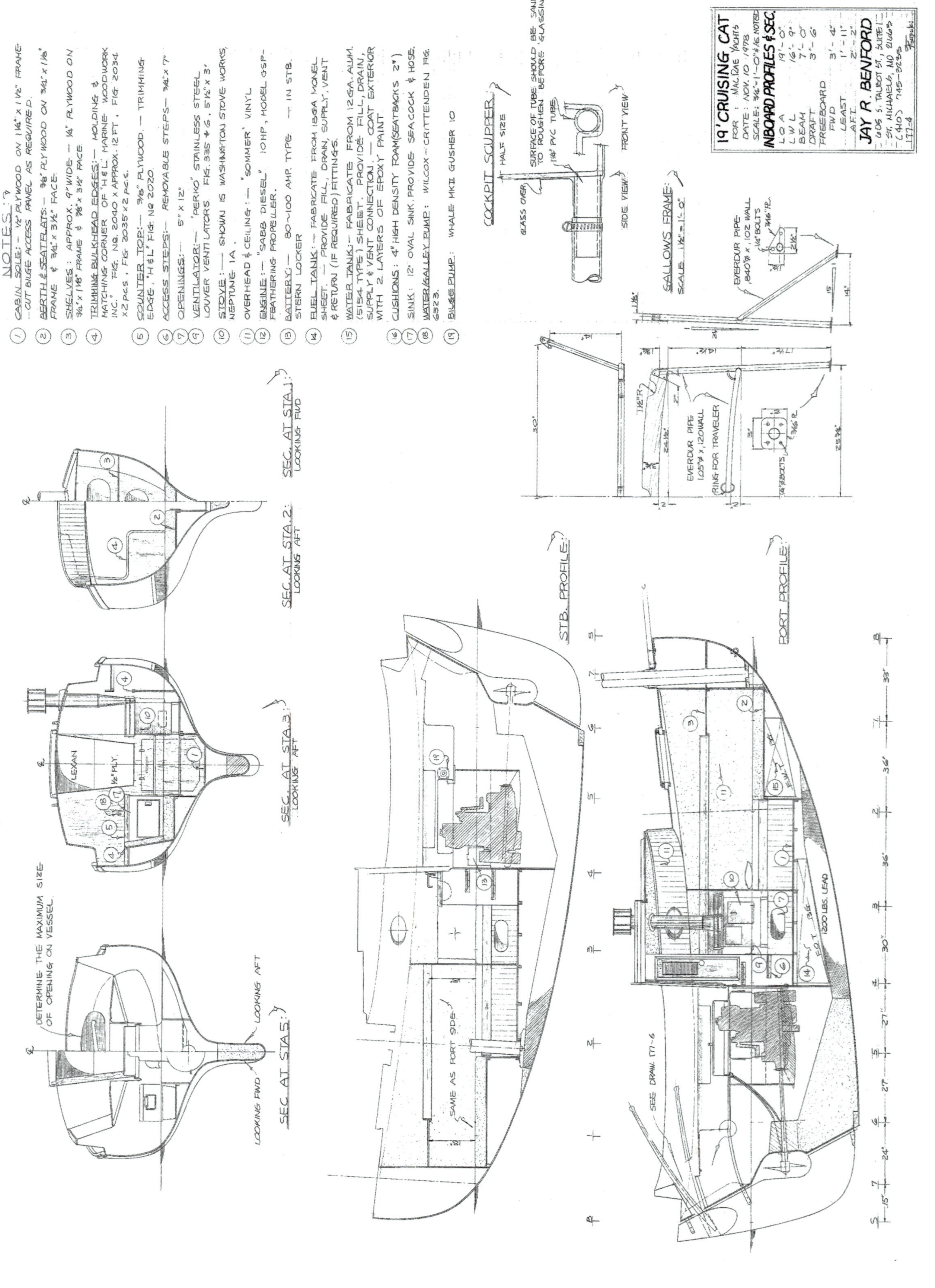
NOTES:
(1) CABIN SOLE: — ½" PLYWOOD ON 1¼" x 1½" FRAME — CUT BILGE ACCESS PANEL AS REQUIRED.
(2) BERTH & SEATFLATS: — ⅜" PLYWOOD ON ¾" x 1⅛" FRAME & ¾" x 3½" FACE.
(3) SHELVES: APPROX. 9" WIDE — ¼" PLYWOOD ON ¾" x 1⅛" FRAME & ⅜" x 3½" FACE.
(4) TRIMMING BULKHEAD EDGES: — MOLDING & MATCHING CORNER OF "H&L MARINE WOODWORK INC." FIG. NO. 2040 x APPROX. 12 FT, FIG 2034 x 2 pcs, FIG 2035 x 2 pcs.
(5) COUNTER TOP: — ⅜" PLYWOOD. — TRIMMING EDGE, "H&L" FIG. NO 2020.
(6) ACCESS STEPS: — REMOVABLE STEPS — ¾" x 7"
(7) OPENINGS: — 5" x 12"
(9) VENTILATOR: — "PERKO" STAINLESS STEEL LOUVER VENTILATORS FIG. 335 #6, 5½" x 3"
(10) STOVE: — SHOWN IS WASHINGTON STOVE WORKS, NEPTUNE 1A.
(11) OVERHEAD & CEILING: — "SOMMER" VINYL
(12) ENGINE: — "SABB DIESEL" 10HP, MODEL GSP - FEATHERING PROPELLER.
(13) BATTERY: — 80~100 AMP. TYPE — IN STB. STERN LOCKER
(14) FUEL TANK: — FABRICATE FROM 18GA MONEL SHEET. — PROVIDE FILL, DRAIN, SUPPLY, VENT & RETURN (IF REQUIRED) FITTINGS.
(15) WATER TANK: — FABRICATE FROM 12GA. ALUM. (5154 TYPE) SHEET. PROVIDE FILL, DRAIN, SUPPLY & VENT CONNECTION. — COAT EXTERIOR WITH 2 LAYERS OF EPOXY PAINT.
(16) CUSHIONS: 4" HIGH DENSITY FOAM (SEATBACKS 2")
(17) SINK: 12" OVAL SINK. PROVIDE SEACOCK & HOSE.
(18) WATER/GALLEY PUMP: WILCOX-CRITTENDEN FIG. 6523.
(19) BILGE PUMP: WHALE MKII GUSHER 10
DETERMINE THE MAXIMUM SIZE OF OPENING ON VESSEL
LOOKING FWD
LOOKING AFT
SEC AT STA 5
LEXAN
½" PLY.
SEC. AT STA. 3
LOOKING AFT
SEC. AT STA. 2
LOOKING AFT
SEC. AT STA. 1
LOOKING FWD
SAME AS PORT SIDE
STB. PROFILE
SEE DRAW. 177-6
F.O.T. 13G
F.W.T.
1200 LBS. LEAD
PORT PROFILE
15" 24" 27" 27" 30" 36" 36" 33"
COCKPIT SCUPPER
HALF SIZE
GLASS OVER
SURFACE OF TUBE SHOULD BE SANDED TO ROUGHEN BEFORE GLASSING
1¼" PVC TUBE
SIDE VIEW
FRONT VIEW
EVERDUR PIPE 1.05"⌀ x .120 WALL
RING FOR TRAVELER
EVERDUR PIPE .840"⌀ x .102 WALL
¼" BOLTS
GALLOWS FRAME
SCALE 1½" = 1'-0"
19' CRUISING CAT
FOR: MACRAE YACHTS
DATE: NOV. 10 1978
SCALE: ¾"=1'-0" & AS NOTED
INBOARD PROFILES & SEC.
LOA 19'-0"
LWL 16'-9"
BEAM 7'-0"
DRAFT 3'-6"
FREEBOARD
FWD 3'-4"
LEAST 1'-11"
AFT 2'-2"
JAY R. BENFORD
605 S. TALBOT ST., SUITE 1
ST. MICHAELS, MD 21663
(410) 745-3235
177-4

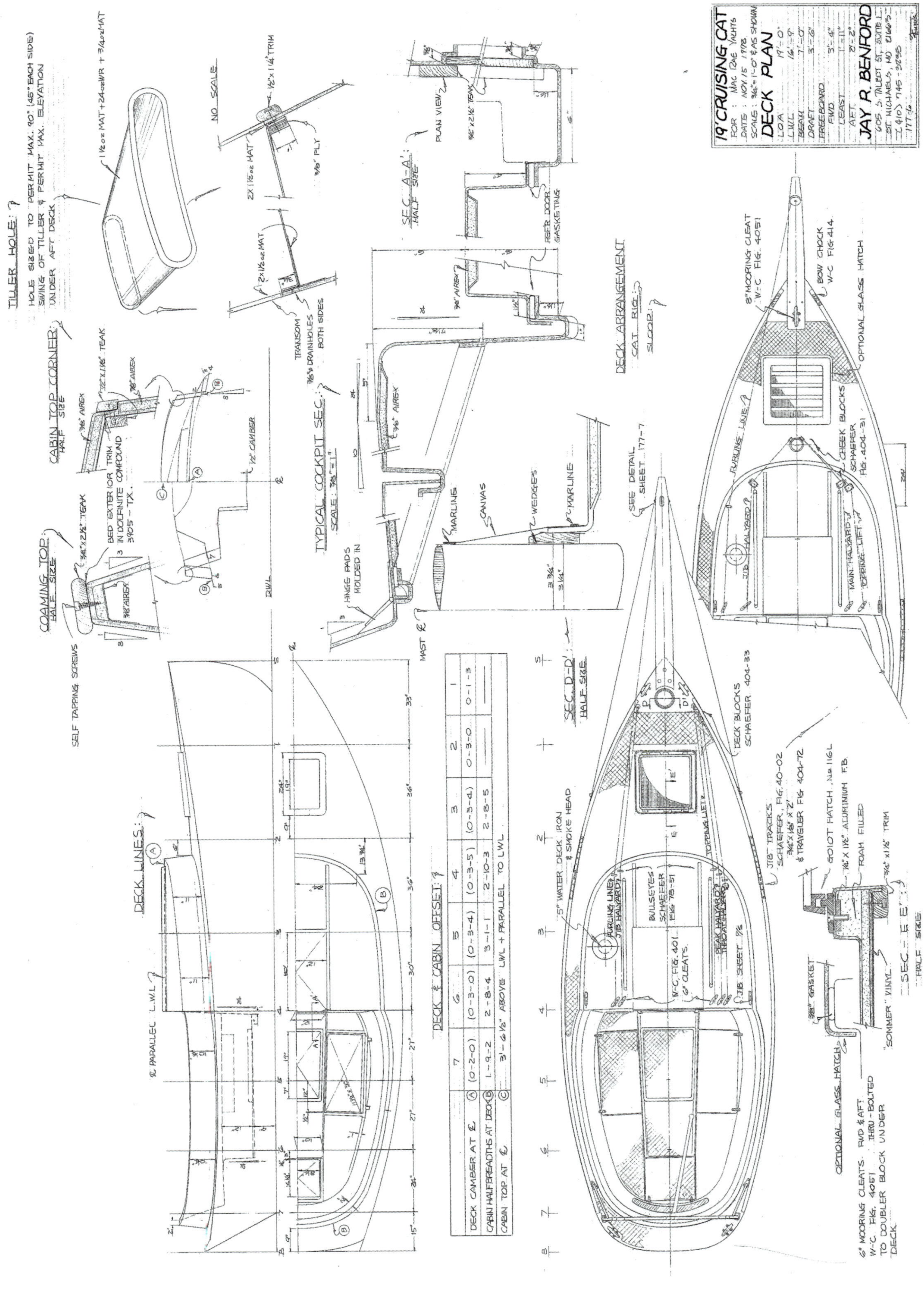
19' CRUISING CAT
FOR: MAC RAE YACHTS
DATE: NOV. 15 1998
SCALE: 3/4"=1'-0" & AS SHOWN
DECK PLAN
LOA 19'-0"
LWL 16'-9"
BEAM 7'-0"
DRAFT 3'-6"
FREEBOARD FWD 3'-4"
LEAST 1'-11"
AFT 2'-2"
JAY R. BENFORD
605 S. TALBOT ST. SUITE 1
ST. MICHAELS, MD 21663
T (410) 745-3235
177-6
DECK ARRANGEMENT
CAT RIG
SLOOP
DECK LINES
DECK & CABIN OFFSET
	7	6	5	4	3	2	1
DECK CAMBER AT ℄ (A)	(0-2-0)	(0-3-0)	(0-3-4)	(0-3-5)	(0-3-4)	0-3-0	0-1-3
CABIN HALFBREADTHS AT DECK (B)	1-9-2	2-8-4	3-1-1	2-10-3	2-8-5	—	
CABIN TOP AT ℄	3'-6 1/2" ABOVE LWL + PARALLEL TO LWL						
TILLER HOLE
HOLE SIZED TO PERMIT MAX. 90° (45° EACH SIDE) SWING OF TILLER & PERMIT MAX. ELEVATION UNDER AFT DECK
1 1/2oz MAT + 24oz WR + 3/4oz MAT
NO SCALE
CABIN TOP CORNER HALF SIZE
COAMING TOP HALF SIZE
SELF TAPPING SCREWS
BED EXTERIOR TRIM IN DOLFINITE COMPOUND 3905-TX.
TYPICAL COCKPIT SEC. SCALE: 3/8"=1"
TRANSOM
3/8"Ø DRAINHOLES BOTH SIDES
HINGE PADS MOLDED IN
SEC. A-A' HALF SIZE
PLAN VIEW
REEFER DOOR GASKETING
SEC. D-D' HALF SIZE
MARLINE
CANVAS
WEDGES
SEE DETAIL SHEET 177-7
MAST ℄
DWL
℄ PARALLEL L.W.L.
5" WATER DECK IRON & SMOKE HEAD
FURLING LINE / JIB HALYARD
BULLSEYES SCHAEFER FIG 78-51
W-C FIG. 401 6" CLEATS
PEAK HALYARD / THROAT HALYARD
TOPPING LIFT
JIB SHEET
DECK BLOCKS SCHAEFER 404-33
JIB TRACKS SCHAEFER FIG. 40-02 3/4" x 1/8" x 2' & TRAVELER FIG 404-72
GOIOT HATCH No 116L ALUMINIUM F.B.
FOAM FILLED
3/4" x 1 1/8" TRIM
SEC. E-E' HALF SIZE
"SOMMER" VINYL
GASKET
OPTIONAL GLASS HATCH
6" MOORING CLEATS FWD & AFT. W-C FIG. 4061 THRU-BOLTED TO DOUBLER BLOCK UNDER DECK.
8" MOORING CLEAT W-C FIG. 4051
BOW CHOCK W-C FIG 414
CHEEK BLOCKS SCHAEFER FIG. 404-31
MAIN HALYARD
TOPPING LIFT
JIB HALYARD
FURLING LINE

	NAME	LINE	BLOCK	SCHAEFER FIGURE Nº's
1	MAIN SHEET	3/8" DACRON	1-FIDDLE , 1-FIDDLE WITH CAMCLEAT & BECKET	404-43 , 26-53
2	JIB SHEET	3/8" "	SHEET LEAD BLOCK WITH SLIDER (2 REQ'D)	404-72
3	PEAK HALYARD	5/16" "	1-SINGLE BLOCK , 1-SINGLE BLOCK WITH BECKET	404-03 , 404-13
4	THROAT HALYARD	5/16" "	1-SINGLE BLOCK , 1 x DOUBLE BLOCK WITH BECKET	404-03 , 404-33
5	TOPPING LIFT	5/16" "	(THROUGH EXTRA SHEAVE ON THROAT HALYARD DOUBLE WITH BECKET)	
6	LAZY JACKS	1/4" "		
7	FURLING LINE	1/4" "		
8	REEFING LINE	5/16" "	4-CHEEK BLOCKS	404-30
9	JIB HALYARD	5/16" "	1-SINGLE BLOCK , 1-SINGLE BLOCK WITH BECKET	404-03 , 404-13

19' CRUISING CAT
FOR: MAC RAE YACHTS
DATE NOV. 28 1978
SCALE 3/4"=1'-0" & AS NOTED

RIGGING PLAN

LOA	19'-0"
LWL	16'-9"
BEAM	7'-0"
DRAFT	3'-6"
FREEBOARD	
FWD	3'-4"
LEAST	1'-11"
AFT	2'-2"

JAY R. BENFORD
605 S. TALBOT ST., SUITE 1
ST. MICHAELS, MD 21663
(410) 745-3235
177-7

(14) ALL SPARS TO BE MADE OF SITKA SPRUCE, AND SOLID, CIRCULAR IN SECTION, TO DIAMETERS SHOWN.

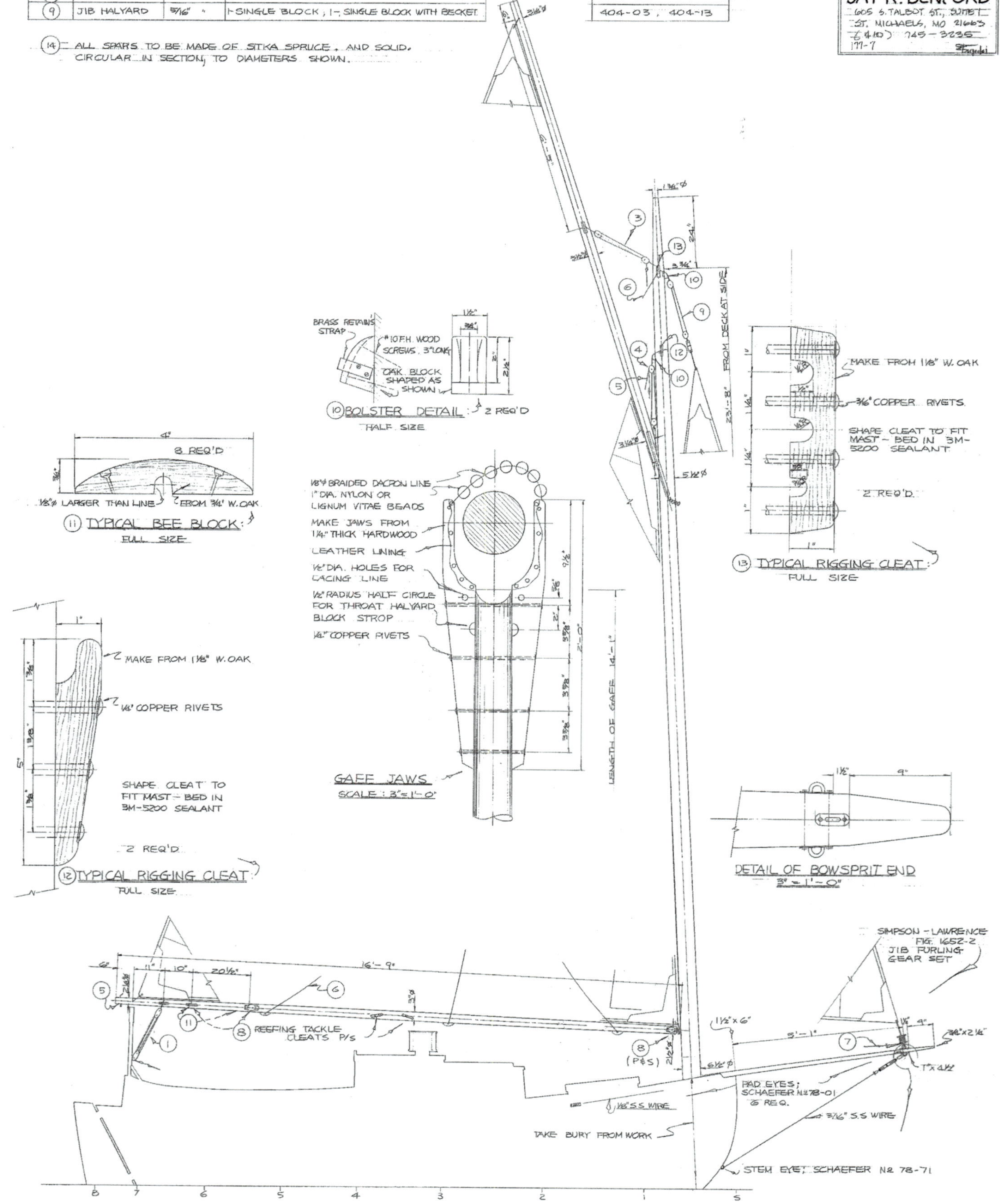

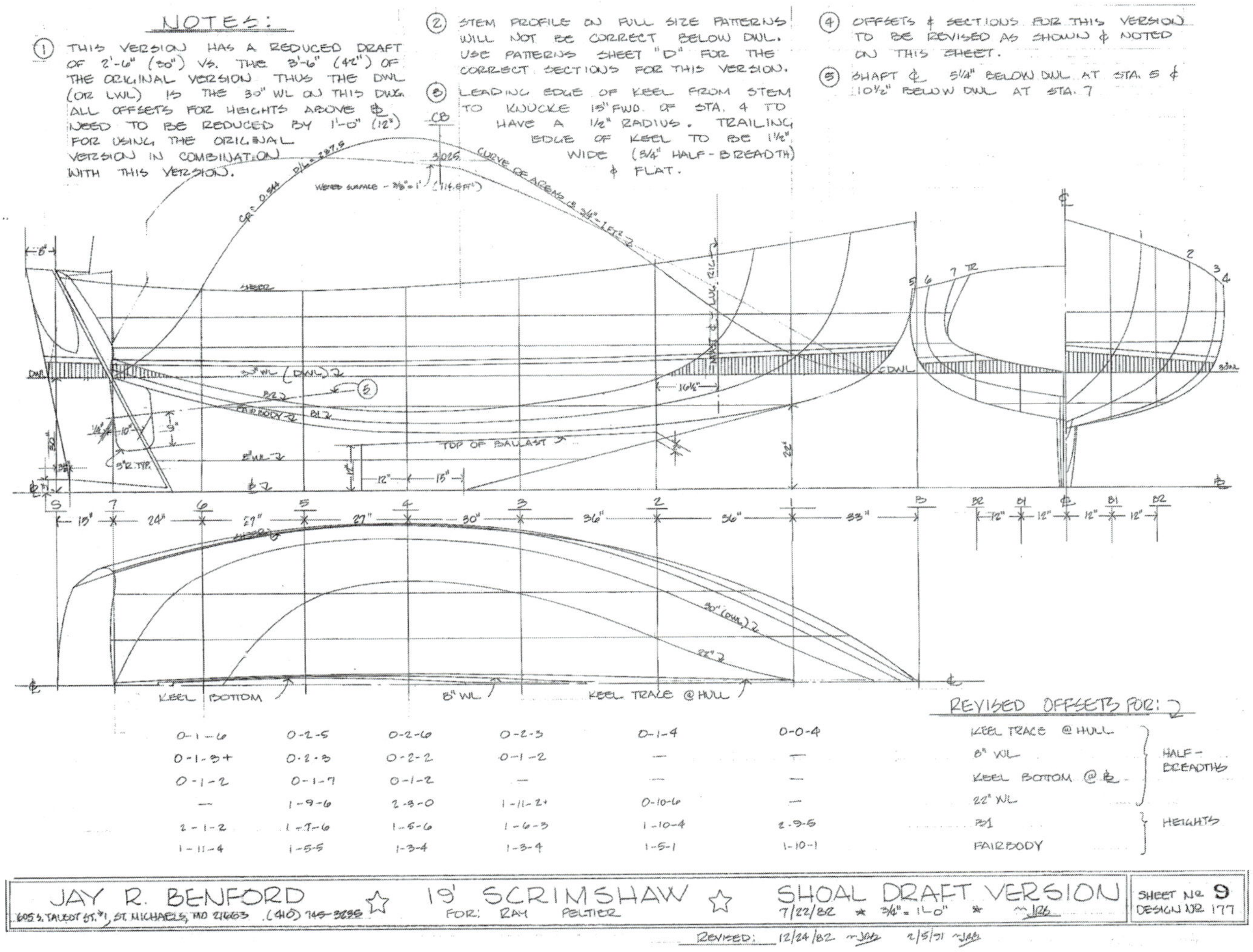
NOTES:
① THIS VERSION HAS A REDUCED DRAFT OF 2'-6" (30") VS. THE 3'-6" (42") OF THE ORIGINAL VERSION. THUS THE DWL (OR LWL) IS THE 30" WL ON THIS DWG. ALL OFFSETS FOR HEIGHTS ABOVE ℄ NEED TO BE REDUCED BY 1'-0" (12") FOR USING THE ORIGINAL VERSION IN COMBINATION WITH THIS VERSION.
② STEM PROFILE ON FULL SIZE PATTERNS WILL NOT BE CORRECT BELOW DWL. USE PATTERNS SHEET "D" FOR THE CORRECT SECTIONS FOR THIS VERSION.
③ LEADING EDGE OF KEEL FROM STEM TO KNUCKLE 15" FWD. OF STA. 4 TO HAVE A 1/2" RADIUS. TRAILING EDGE OF KEEL TO BE 1 1/2" WIDE (3/4" HALF-BREADTH) & FLAT.
④ OFFSETS & SECTIONS FOR THIS VERSION TO BE REVISED AS SHOWN & NOTED ON THIS SHEET.
⑤ SHAFT ℄ 5 1/4" BELOW DWL AT STA. 5 & 10 1/2" BELOW DWL AT STA. 7
CB
CURVE OF AREAS @ 3/4" = 1 FT²
SHEER
DWL
30" WL (DWL)
TOP OF BALLAST
KEEL BOTTOM
8" WL
KEEL TRACE @ HULL
REVISED OFFSETS FOR: 2
0-1-6 | 0-2-5 | 0-2-6 | 0-2-3 | 0-1-4 | 0-0-4 | KEEL TRACE @ HULL
0-1-3+ | 0-2-3 | 0-2-2 | 0-1-2 | — | — | 8" WL
0-1-2 | 0-1-7 | 0-1-2 | — | — | — | KEEL BOTTOM @ ℄
— | 1-9-6 | 2-3-0 | 1-11-2+ | 0-10-6 | — | 22" WL
2-1-2 | 1-7-6 | 1-5-6 | 1-6-3 | 1-10-4 | 2-9-5 | B1
1-11-4 | 1-5-5 | 1-3-4 | 1-3-4 | 1-5-1 | 1-10-1 | FAIRBODY
HALF-BREADTHS
HEIGHTS
JAY R. BENFORD ☆ 19' SCRIMSHAW ☆ SHOAL DRAFT VERSION
605 S. TALBOT ST. #1, ST. MICHAELS, MD 21663 (410) 745-3235
FOR: RAY PELTIER
7/22/82 ★ 3/4" = 1'-0" ★ ~JRB
SHEET NR 9
DESIGN NR 177
REVISED: 12/24/82 ~JRB 2/5/91 ~JRB

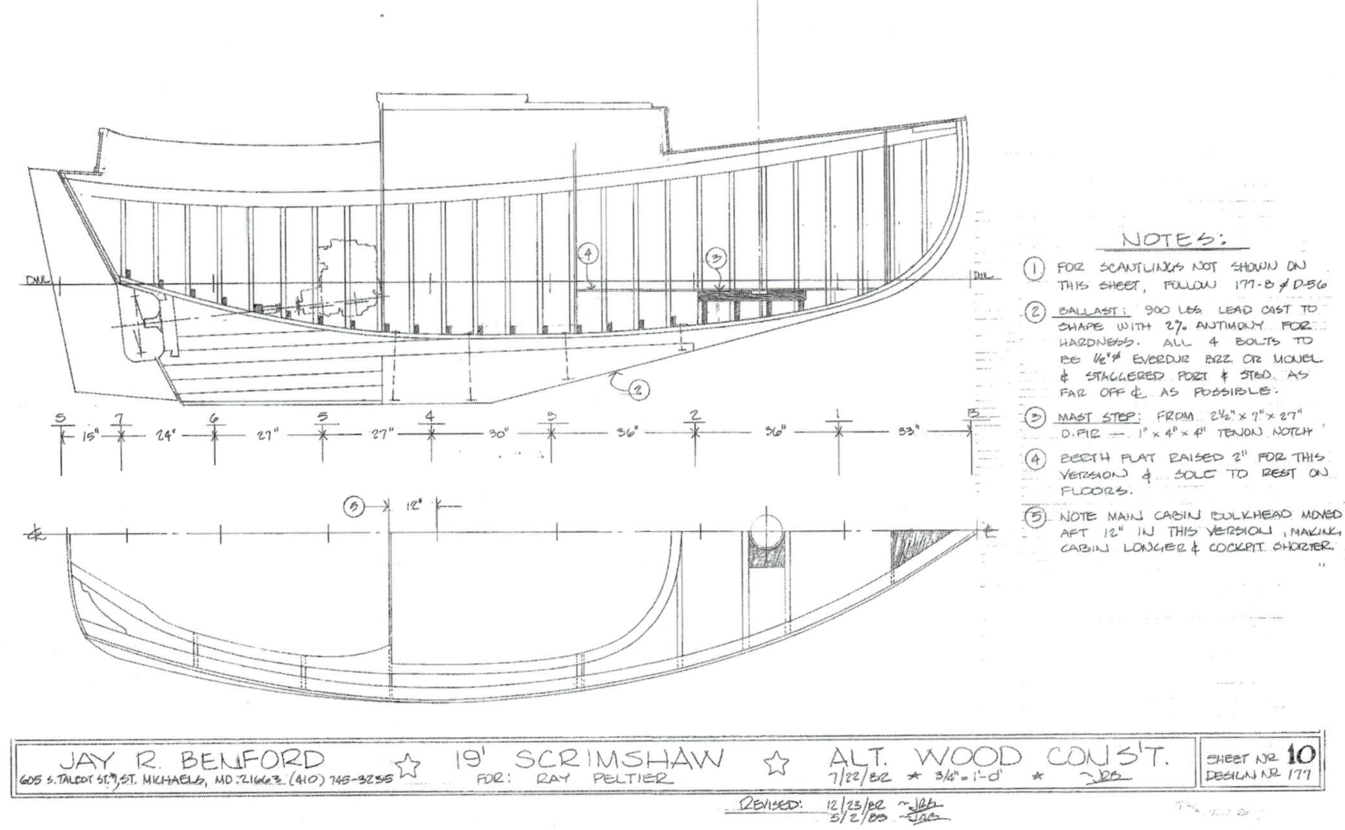
DWL
NOTES:
① FOR SCANTLINGS NOT SHOWN ON THIS SHEET, FOLLOW 177-8 & D-56
② BALLAST: 900 LBS. LEAD CAST TO SHAPE WITH 2% ANTIMONY FOR HARDNESS. ALL 4 BOLTS TO BE 1/2"⌀ EVERDUR BRZ. OR MONEL & STAGGERED PORT & STBD. AS FAR OFF ℄ AS POSSIBLE.
③ MAST STEP: FROM 2 1/2" × 7" × 27" D. FIR — 1" × 4" × 4" TENON NOTCH
④ BERTH FLAT RAISED 2" FOR THIS VERSION & SOLE TO REST ON FLOORS.
⑤ NOTE MAIN CABIN BULKHEAD MOVED AFT 12" IN THIS VERSION, MAKING CABIN LONGER & COCKPIT SHORTER.
JAY R. BENFORD ☆ 19' SCRIMSHAW ☆ ALT. WOOD CONS'T.
605 S. TALBOT ST. #1, ST. MICHAELS, MD 21663 (410) 745-3235
FOR: RAY PELTIER
7/22/82 ★ 3/4" = 1'-0" ★ ~JRB
SHEET NR 10
DESIGN NR 177
REVISED: 12/23/82 ~JRB 5/2/83 ~JRB

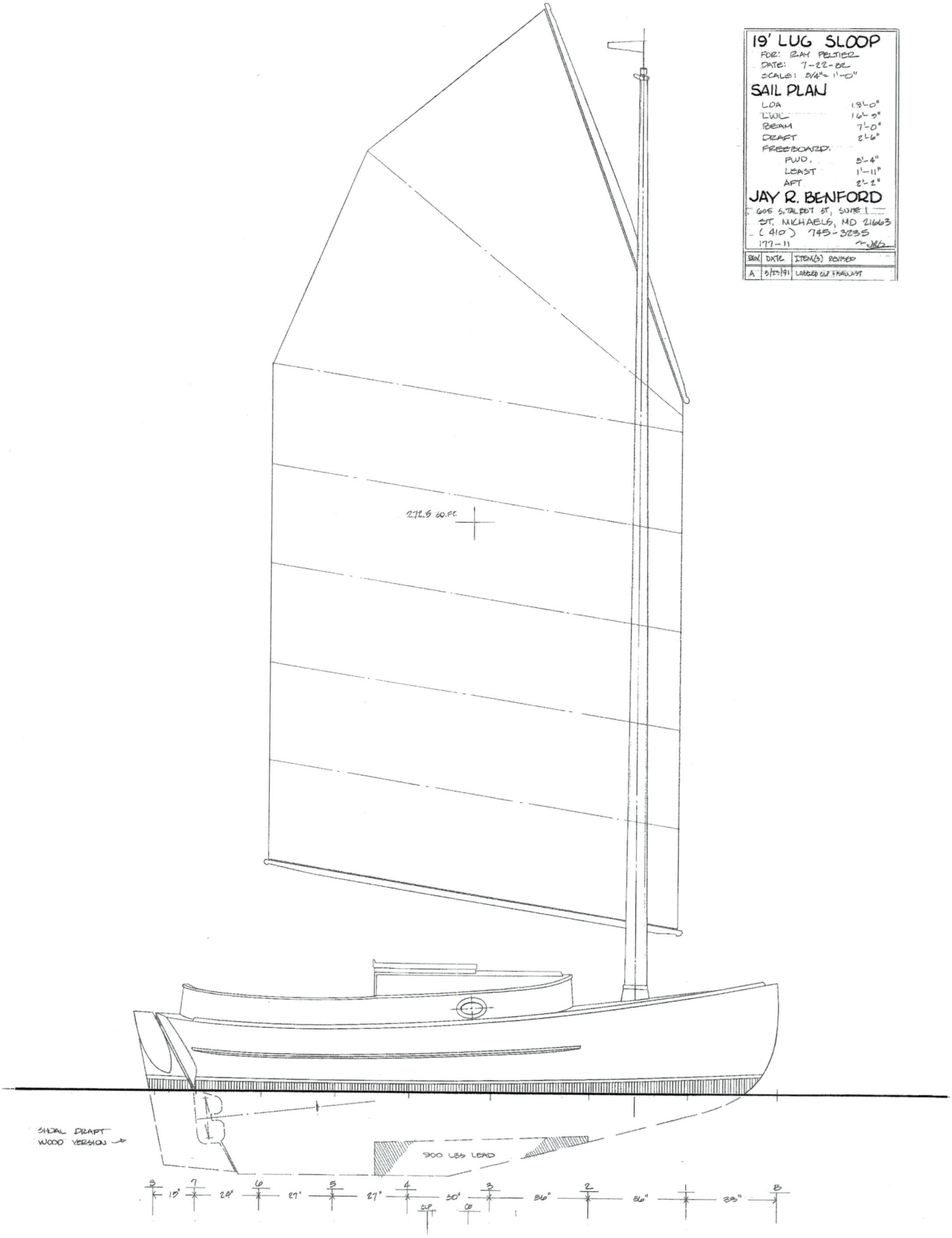
19' LUG SLOOP
FOR: RAY PELTIER
DATE: 7-22-82
SCALE: 3/4"=1'-0"
SAIL PLAN
LOA 19'-0"
LWL 16'-5"
BEAM 7'-0"
DRAFT 2'-6"
FREEBOARD:
FWD. 3'-4"
LEAST 1'-11"
AFT 2'-2"
JAY R. BENFORD
605 S. TALBOT ST, SUITE 1
ST. MICHAELS, MD 21663
(410) 745-3235
177-11
REV. DATE ITEM(S) REVISED
A 5/23/91 LABELED CLP + BALLAST
272.5 SQ.FT.
SHOAL DRAFT WOOD VERSION
900 LBS LEAD
CLP
CE
15"
24"
27"
27"
30"
36"
36"
33"

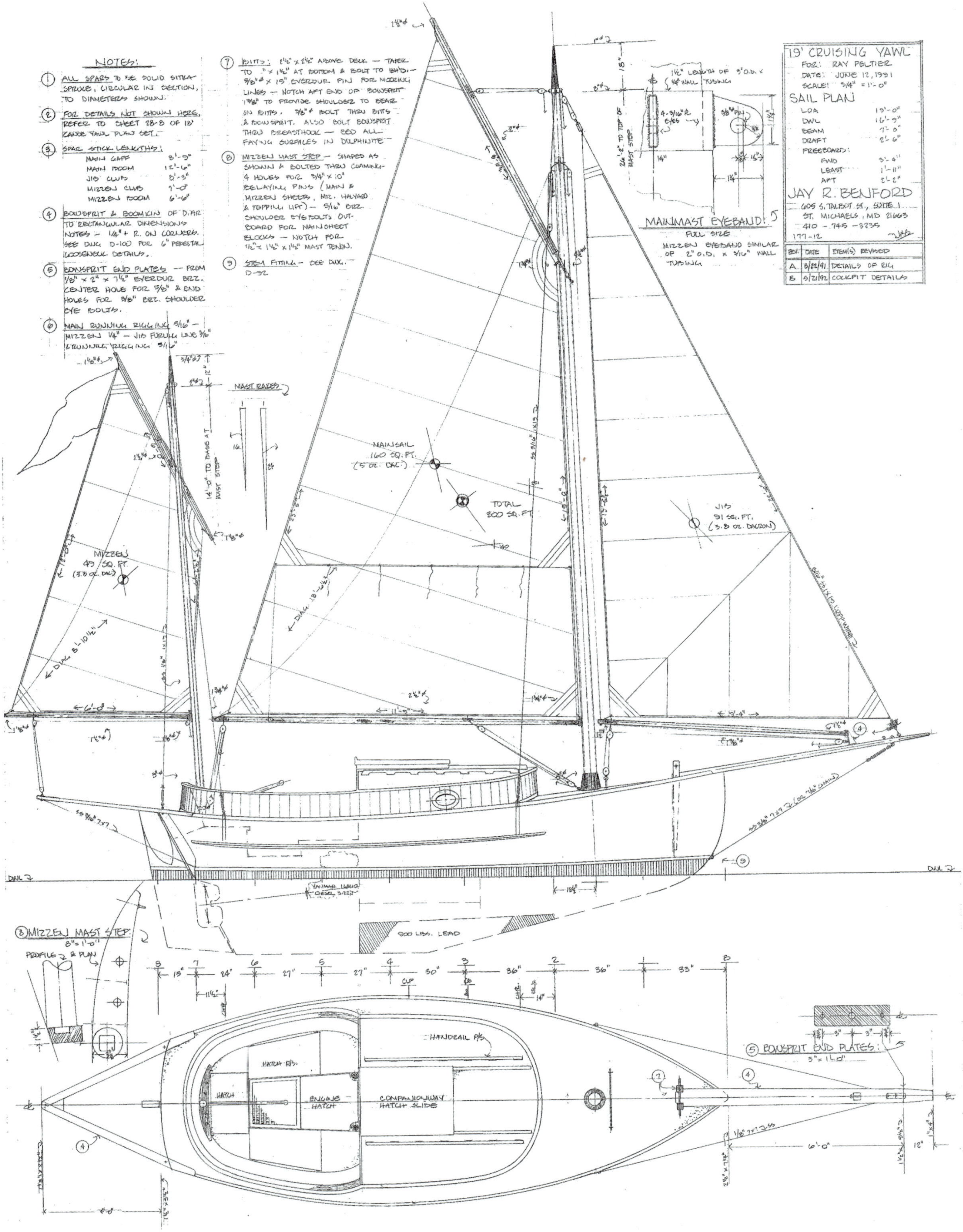

NOTES:
1 ALL SPARS TO BE SOLID SITKA SPRUCE, CIRCULAR IN SECTION, TO DIAMETERS SHOWN.
2 FOR DETAILS NOT SHOWN HERE, REFER TO SHEET 28-8 OF 13' CANOE YAWL PLAN SET.
3 SPAR STICK LENGTHS: MAIN GAFF 8'-9" MAIN BOOM 12'-6" JIB CLUB 8'-3" MIZZEN CLUB 7'-0" MIZZEN BOOM 6'-6"
4 BOWSPRIT & BOOMKIN OF D.FIR TO RECTANGULAR DIMENSIONS NOTED — 1/4" R. ON CORNERS. SEE DWG D-100 FOR 6" PEDESTAL GOOSENECK DETAILS.
5 BOWSPRIT END PLATES — FROM 1/8" x 2" x 7 1/2" EVERDUR BRZ. CENTER HOLE FOR 5/8" & END HOLES FOR 3/8" BRZ. SHOULDER EYE BOLTS.
6 MAIN RUNNING RIGGING 5/16" — MIZZEN 1/4" — JIB FURLING LINE 3/16" & RUNNING RIGGING 5/16"
7 BITTS: 1 1/2" x 2 1/2" ABOVE DECK — TAPER TO " x 1 1/2" AT BOTTOM & BOLT TO BH'D. — 5/8" x 15" EVERDUR PIN FOR MOORING LINES — NOTCH AFT END OF BOWSPRIT 1 3/8" TO PROVIDE SHOULDER TO BEAR ON BITTS. 3/8" BOLT THRU BITTS & BOWSPRIT. ALSO BOLT BOWSPRIT THRU BREASTHOOK — BED ALL FAYING SURFACES IN DOLPHINITE
8 MIZZEN MAST STEP — SHAPED AS SHOWN & BOLTED THRU COAMING — 4 HOLES FOR 3/4" x 10" BELAYING PINS (MAIN & MIZZEN SHEETS, MIZ. HALYARD & TOPPING LIFT) — 5/16" BRZ. SHOULDER EYEBOLTS OUTBOARD FOR MAINSHEET BLOCKS — NOTCH FOR 1/2" x 1 1/2" x 1 1/2" MAST TENON.
9 STEM FITTING — SEE DWG. D-32
19' CRUISING YAWL
FOR: RAY PELTIER
DATE: JUNE 12, 1991
SCALE: 3/4" = 1'-0"
SAIL PLAN
LOA 19'-0"
DWL 16'-9"
BEAM 7'-0"
DRAFT 2'-6"
FREEBOARD:
FWD 3'-4"
LEAST 1'-11"
AFT 2'-2"
JAY R. BENFORD
605 S. TALBOT ST., SUITE 1
ST. MICHAELS, MD 21663
410-745-3235
177-12
REV. DATE ITEM(S) REVISED
A 8/22/91 DETAILS OF RIG
B 5/21/92 COCKPIT DETAILS
MAINMAST EYEBAND 5
FULL SIZE
MIZZEN EYEBAND SIMILAR OF 2" O.D. x 3/16" WALL TUBING
MAST RAKES
MAINSAIL 160 SQ. FT. (5 OZ. DAC.)
TOTAL 300 SQ. FT.
JIB 91 SQ. FT. (3.8 OZ. DACRON)
MIZZEN 49 SQ. FT. (3.8 OZ. DAC.)
DWL
YANMAR 1GM10 DIESEL
900 LBS. LEAD
8 MIZZEN MAST STEP
8" = 1'-0"
PROFILE & PLAN
5 BOWSPRIT END PLATES
3" = 1'-0"
HANDRAIL P/S
HATCH P/S
HATCH
ENGINE HATCH
COMPANIONWAY HATCH SLIDE

Tardis *built by Johnson Yachts, left, under construction. The bulkheads were set up along with the mold frames. You can see the notch for the ballast and laminated backbone. Just launched, right.*

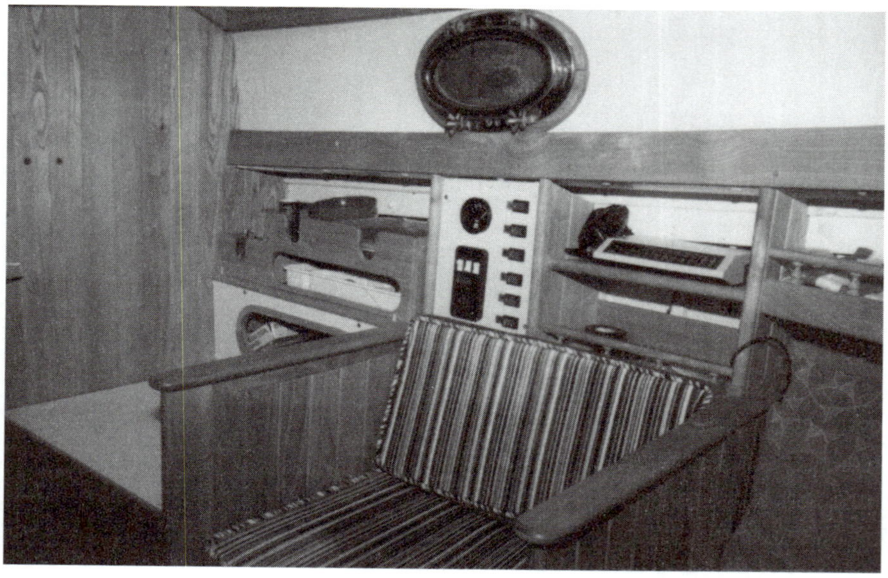

Chickadee *has a well detailed interior, with a clever built-in drawer set, the top one of which houses a fold-out table. Aft of the drawers is the lounging seat, across from the other lounging seat and cover over the head.*

Chapter 3
19' *Gunkholer*

Design Number 19
1969 & 1975

The common definition of a "gunkhole" is a very shallow body of water, usually with a soft and sticky bottom. Thus the name for these shallow draft cruisers – gunkholers are boats that can cruise in shallow water.

The concept for this design was done as a preliminary study in 1969, for friend Dick Wagner when he was getting his Old Boathouse operation underway. Six years later, we had a client who asked us to finish off the plans and these are what you will see on the following pages.

The first 19' Gunkholer was built by Paul Miller while he had his shop at the north end of Vancouver Island. He called her ***Catspaw*** and we adopted this as the name for the design. Paul later built another one for a client and it was named ***Sylvester***. Many thanks to Paul for supplying the photos of their building and sailing that are in this book.

The sturdy little flat bottom 19-foot centerboard catboat, with her shallow draft is ideal for gunkholing where shoals are a concern in one's cruising grounds. Her accommodations are generous for her 19 feet of length. She has two berths that make into a double, two lounging seats (one over the head) with good sitting headroom, sink, stove and hanging locker. Her catboat rig and plywood construction spell simplicity and practicality. Paul Miller's ***Catspaw*** was the first of this series and has proven to be very practical and salty little cruiser. With the very shallow draft, there's hardly any need for a dinghy, for the boat can be beached.

Particulars:	**19'**	**19'**
Units	**Imperial**	**Metric**
Length overall	19'-6"	5.94 m
Length designed waterline	18'-0"	5.49 m
Beam	7'-9"	2.36 m
Draft (cb up/down)	1'-4"/4'-8"	0.41/1.42 m
Freeboard: Forward	3'-8"	1.12 m
Least	2'-0"	0.61 m
Aft	2'-7"	0.79 m
Displacement*, cruising trim.	3,475 lbs.	1,576 kg.
Displacement-length ratio	266	
Ballast, pounds	500	227 kg.
Ballast ratio	14%	
Sail area	256 ft.2	23.78 m2
Sail area-displacement ratio	17.85	
Prismatic coefficient	.613	
Water tankage, gallons	20	76 liters
Fuel tankage, gallons	10	38 liters
Headroom	4-8½"	

*__CAUTION__: The figure for displacement quoted here is for the boat in cruising trim. That is, with the fuel and water tanks filled, the crew on board, as well as the crews' gear and stores in the lockers. This should not be confused with the "shipping weight" often quoted as "displacement" by some manufacturers. This should be taken into account when comparing figure and ratios between this and other designs.

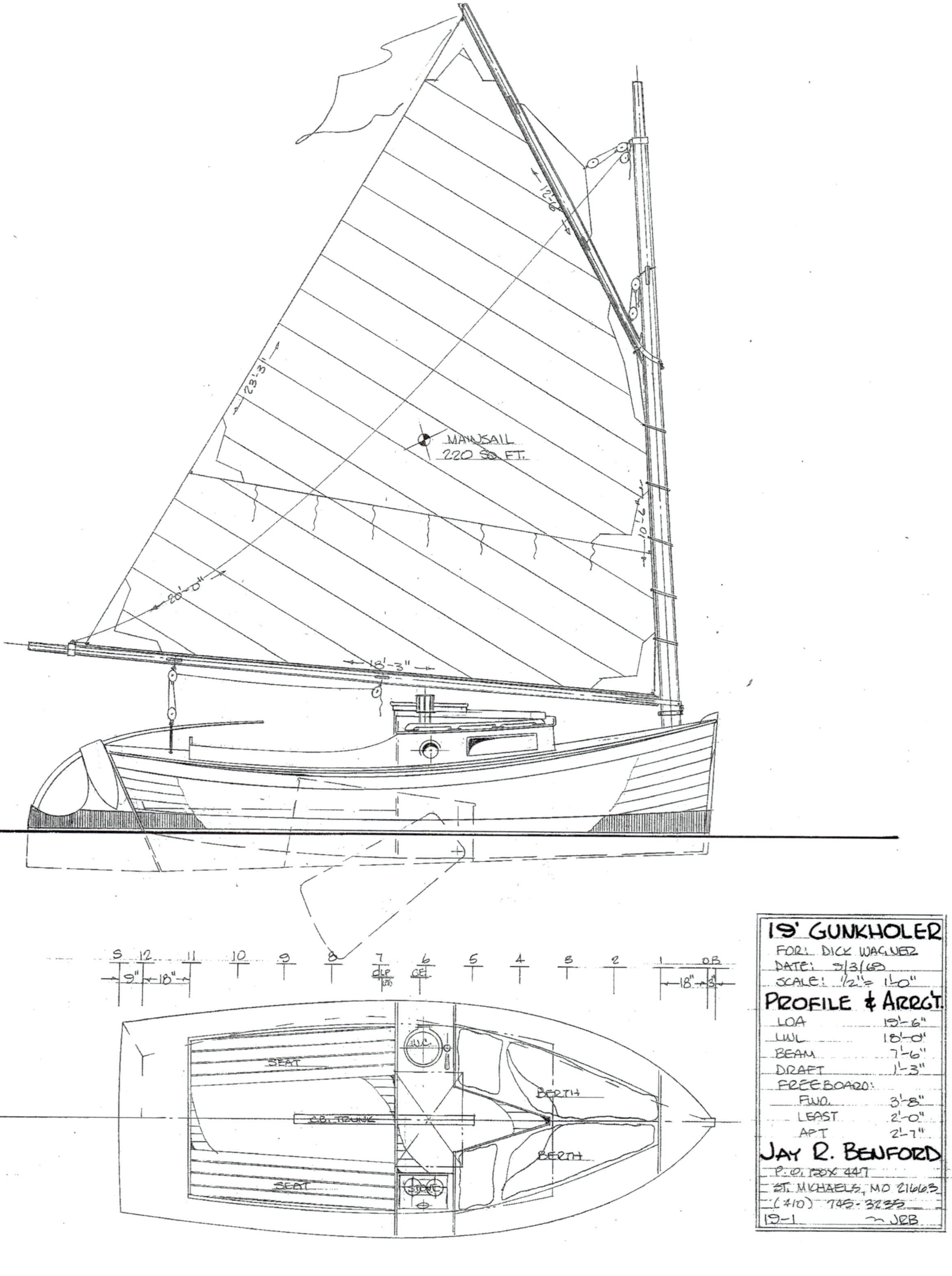
MAINSAIL
220 SQ. FT.
23'-3"
12'-6"
10'-6"
20'-0"
18'-3"
W.C.
SEAT
C.B. TRUNK
BERTH
BERTH
SEAT
STOVE
19' GUNKHOLER
FOR: DICK WAGNER
DATE: 5/3/69
SCALE: 1/2" = 1'-0"
PROFILE & ARRG'T.
LOA 19'-6"
LWL 18'-0"
BEAM 7'-6"
DRAFT 1'-3"
FREEBOARD:
FWD. 3'-8"
LEAST 2'-0"
AFT 2'-7"
JAY R. BENFORD
P.O. BOX 447
ST. MICHAELS, MD 21663
(410) 745-3235
19-1
JRB

Catspaw *beach cruising near the north end of Vancouver Island.*

The concept shown on the preceding page was done as an idea for a small cruising boat that might be easily built and provide a good small cruising boat. In looking at how the final version evolved, you can see that the cabin was extended to add useful interior accommodation space and the centerboard shifted aft to try to help manage the weather helm.

The Lines plan shows her simple, skiff-like hull form. The chine shape is kept fine forward for easing into the seas and broader aft to support the weight of the crew in the cockpit. Note that the mast is located here too, so that its position can be accurately located as the boat is built. The shape of the deckhouse and cockpit are also dimensioned.

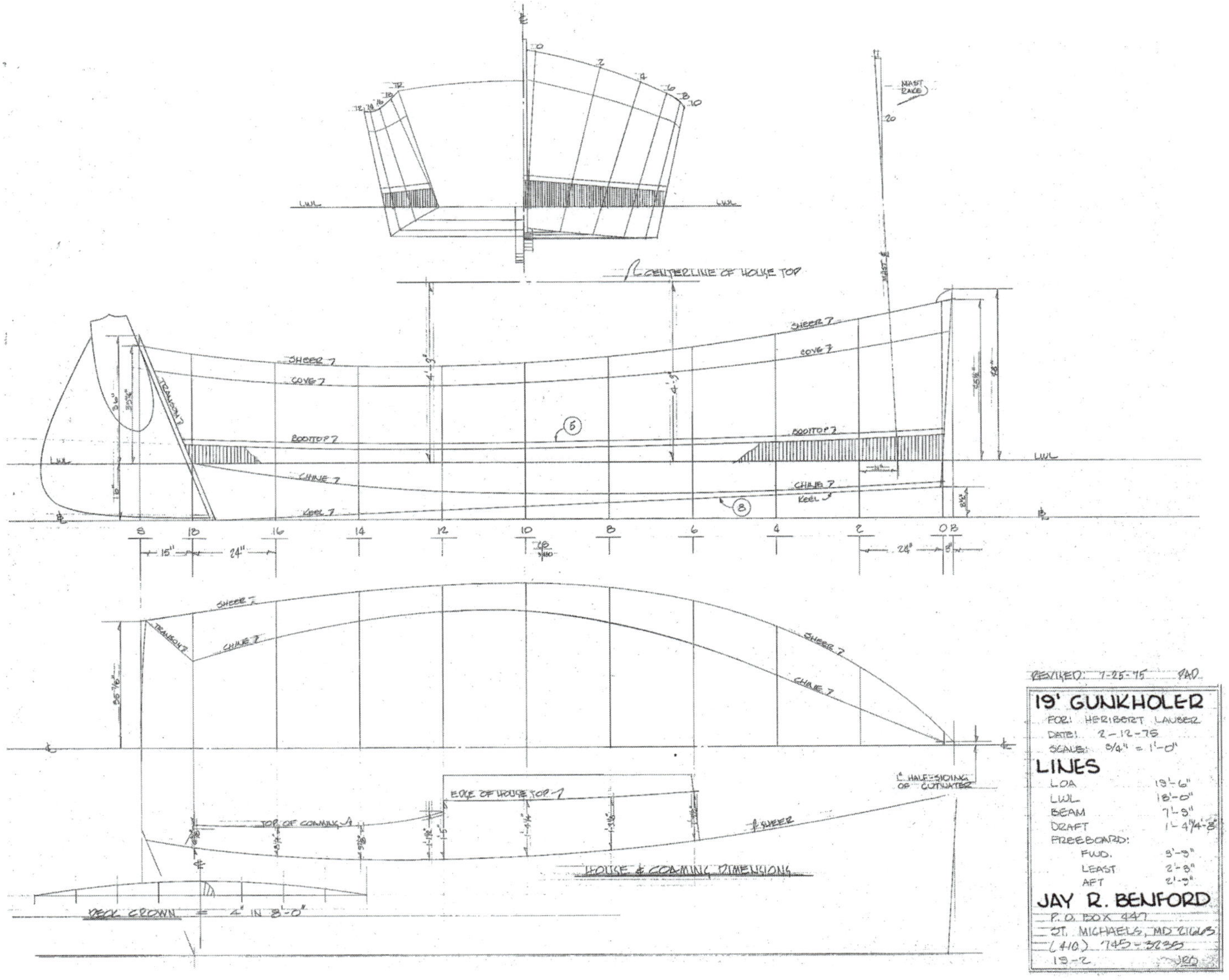

Catspaw *sailing easily on a close reach. Note outboard engine is on a bracket mounted on the side of the rudder.*

The offsets and beginning of the notes on this sheet are the norm for those to be found on a lines and offsets drawing. They supplement the dimensioning and are useful cautions. The notes on this design have a numbered sequence that starts on one page and continues on to the next. Many designs, including most of ours, have notes where the numbering sequence starts anew on each page of the drawing. The rigging schedule table is a good summary of these parts and the equipment used aboard. I would suggest that having more than one anchor – preferably of different styles – is the cheapest of insurances. For instance, on this boat, I'd also have a 40 pound 3-piece Luke-Herreshoff as a storm anchor.

STATION	18	16	14	12	10	8	6	4	2	0
	HALF – BREADTHS									
℄ TO CHINE	2-0-4	2-7-1	2-11-6	3-2-3	3-2-4	2-11-2	2-4-7	1-8-0	0-10-3	—
℄ TO CARLIN	2-7-5	2-11-1	3-1-5	3-2-0 3-3-4	3-2-4	3-1-1	2-9-0	—	—	—
℄ TO SHEER	3-2-1	3-5-4	3-8-2	3-9-7	3-10-6	3-9-2	3-5-2	2-9-7	1-10-2	0-3-6
	HEIGHTS									
BL TO CHINE	1-4-0	1-0-3	0-9-4	0-7-6	0-7-2	0-7-1	0-7-2	0-8-0	0-9-0	(0-10-0)
LWL TO BOOTTOP	0-6-6	0-6-0	0-5-4	0-5-3	0-5-5	0-6-0	0-6-3	0-7-0	0-7-6	0-9-0
LWL TO COVE	2-0-6	1-10-6	1-9-6	1-9-6	1-10-4	1-11-7	2-1-6	2-4-4	2-8-0	3-0-0
LWL TO SHEER	2-6-7	2-4-3	2-3-2	2-3-2	2-4-1	2-5-6	2-8-3	2-11-6	3-4-0	3-8-6

NOTES:

(1) LINES & OFFSETS TO OUTSIDE OF PLANKING IN FEET – INCHES – EIGHTHS – DEDUCT FOR PLANKING THICKNESS AS REQUIRED.

(2) ANY ALTERATION FROM THESE PLANS RELIEVES THE DESIGNER FROM ANY FURTHER RESPONSIBILITY.

(3) IT IS UNDERSTOOD THAT NO MORE THAN ONE BOAT WILL BE BUILT FROM THESE PLANS WITHOUT THE PERMISSION OF THE DESIGNER.

(4) LINES MUST BE LOFTED & FAIRED FULL SIZE — DO NOT SCALE PRINTS & OFFSETS!

(5) BOOTTOP STRIPE IS 1½" HIGH IN PROFILE FULL LENGTH. OFFSETS TO TOP EDGE OF STRIPE

(6) THESE PLANS ARE THE PROPERTY OF THE DESIGNER & MAY BE USED ONLY AS AUTHORIZED IN WRITING BY THE DESIGNER.

(7) ℄ TO CARLIN OFFSETS ARE TO INBOARD EDGE OF CARLIN.

(8) KEEL IS STRAIGHT LINE IN PROFILE FROM 8½" ABOVE BL AT 0 TO TOUCHING BL AT INTERSECTION OF TRANSOM LINE AT BL.

(9) RIGGING CLEATS – MAKE AS SHOWN FROM 1⅛" W. OAK OR SIMILAR HARDWOOD

(10) TOPPING LIFT – GIBB CHEEK BLOCK CAT. #430 TO PORT & PERKO EYE STRAP FIG 881 #1 TO STARB'D. RIG AS SHOWN ON SAIL PLAN ON BOTH SIDES OF SAIL – LAZY JACKS ARE LOOPED THRU THIMBLES & KNOTTED AT FWD BEE BLOCKS ON BOOM.

(11) BOLSTER – FROM 1½" THICK W. OAK CUT ¼" DEEP GROOVE IN TOP EDGE – SHAPE HEEL TO FIT MAST – BED IN 3M-5200 SEALANT & SCREW FASTEN.

SCHEDULE OF RUNNING RIGGING — * STROPS OF ⅜" DACRON 3 STRAND – WORMED, PARCELED, & SERVED.

NAME	NO.	— DIAGRAM —	LINE	FITTINGS	LENGTH	BLOCKS
* PEAK HALYARD	1	STROP AROUND MAST; DOUBLE BLOCK; STROP AROUND GAFF; SHOULDER EYE BOLT	⅜" DACRON (10mm)	⅜" THIMBLE; ¼" SHOULDER EYE BOLT	80 FEET	1-SINGLE w/BECKET; 1-SINGLE; 1-DOUBLE
* THROAT HALYARD	1	STROP AROUND MAST; EMPTY SHEAVE IN DOUBLE; STROP AROUND GAFF	⅜" DACRON (10mm)	⅜" THIMBLE	65 FEET	1-SINGLE w/BECKET; 1-SINGLE
TOPPING LIFT	1	SEE SAIL PLAN	¼" DACRON (6mm)	EYE STRAP – PERKO FIG 881 #1; 2-¼" THIMBLES	120 FEET	1-SINGLE CHEEK BLOCK
* MAIN SHEET	1	STROP AROUND BOOM; DOUBLE; SINGLE w/BECKET; SINGLE; SHOULDER EYE BOLT	⅜" DACRON (10mm)	1-⅜" THIMBLE; 2-¼" SHOULDER EYE BOLTS.	100 FEET	1-SINGLE w/BECKET; 1-SINGLE; 1-DOUBLE
DOCK LINES	4	SPLICE INTO LOOP; 24"	⅜" NYLON (10mm)		25 FEET	
GROUND TACKLE	1	6'; ¼" CHAIN	⅜" NYLON (10mm)	22S DANFORTH ANCHOR; 2-¼" SHACKLES; 1-⅜" THIMBLE	100 FEET	

(12) CHEEK BLOCKS – GIBB CAT. #490 OR SIMILAR.

(13) BEE BLOCKS – FROM ¾" W. OAK – SEE DETAIL (13).

(14) GOOSENECK – MERRIMAN HOLBROOK FIG. 4138 SIZE 2 OR SIMILAR. (BRONZE) THRU BOLT – 4 BOLTS

(15) BOOM STRAPS – FROM 3/16" BRONZE PLATE – BEND JAWS TO FIT GOOSENECK – LET INTO SIDES OF BOOM & THRU-BOLT

(16) CLEATS – WILCOX-CRITTENDEN FIG. 401 BURN. BRONZE 5" CLEATS OR SIMILAR.

(17) BULKHEADS – ½" PLYWOOD ON 1"x1" FIR FRAME.

(18) BERTH & SEAT FLATS – ½" PLYWOOD ON 1"x1" FIR FRAME – CUT ACCESS PANELS FOR STOWAGE AS REQUIRED.

(19) SHELVES – P&S – ⅜" PLYW'D ON 1"x1" FIR FRAME – FIT WITH 3½" x ½" TEAK FACIA.

(20) KNEES – P&S AT FORE END OF HOUSE – LAMINATE – SIDED 1½" THRU-BOLT TO DECK BEAM, CLAMP & SHELF CLAMP.

(21) PORTLIGHTS – PERKO FIG. 796 SIZE #1 (12"x5") OR SIMILAR – LOCATE ℄ 6" FWD STA. 8 & 2" AFT STA. 10. FIT WITH INSECT SCREEN & FINISH FRAME.

(22) DEAD LIGHT – PERKO FIG. 838 6" LIGHT – LOCATE TO STB'D 1'-6" OFF ℄.

(23) CABIN SOLE – ¾" x 3" FIR – SCREW FASTEN TO FLOORS SIDED 1" PROVIDE FOR BILGE ACCESS PANELS AS REQUIRED.

(24) HEAD – WILCOX-CRITTENDEN "JUNIOR 51" OR SIMILAR. FIT WITH VENTED LOOP & PROPER SEACOCKS – OPTIONAL SELF CONTAINED HEAD TO BE MARINE MONOMATIC SANITATION SYSTEM MODEL #1121 (MANHATTAN MARINE FIG. M1561)

REVISED: 7-25-75 PAD

19' GUNKHOLER
FOR: HERIBERT LAUBER
DATE: 2-13-75
SCALE: AS NOTED

TABLE OF OFFSETS

LOA	19'-6"
LWL	18'-0"
BEAM	7'-9"
DRAFT	1'-4"/4'-8"
FREEBOARD:	
FWD.	3'-9"
LEAST	2'-3"
AFT	2'-9"

JAY R. BENFORD
P.O. BOX 447
ST. MICHAELS, MD 21663
(410) 745-3235
19-3 ~JRB

Sylvester *close hauled, above, with a reef tucked in. Photo below shows the boat with the deck on and house sides up, awaiting completion of the interior, prior to putting on the housetop.*

The sail area was increased over the preliminary, for we felt that the preliminary study version was under-canvassed. Even this sail is not over-rigging her, but is a nice balance between light and heavier air needs. The two rows of reef points will let her adapt to varying wind conditions.

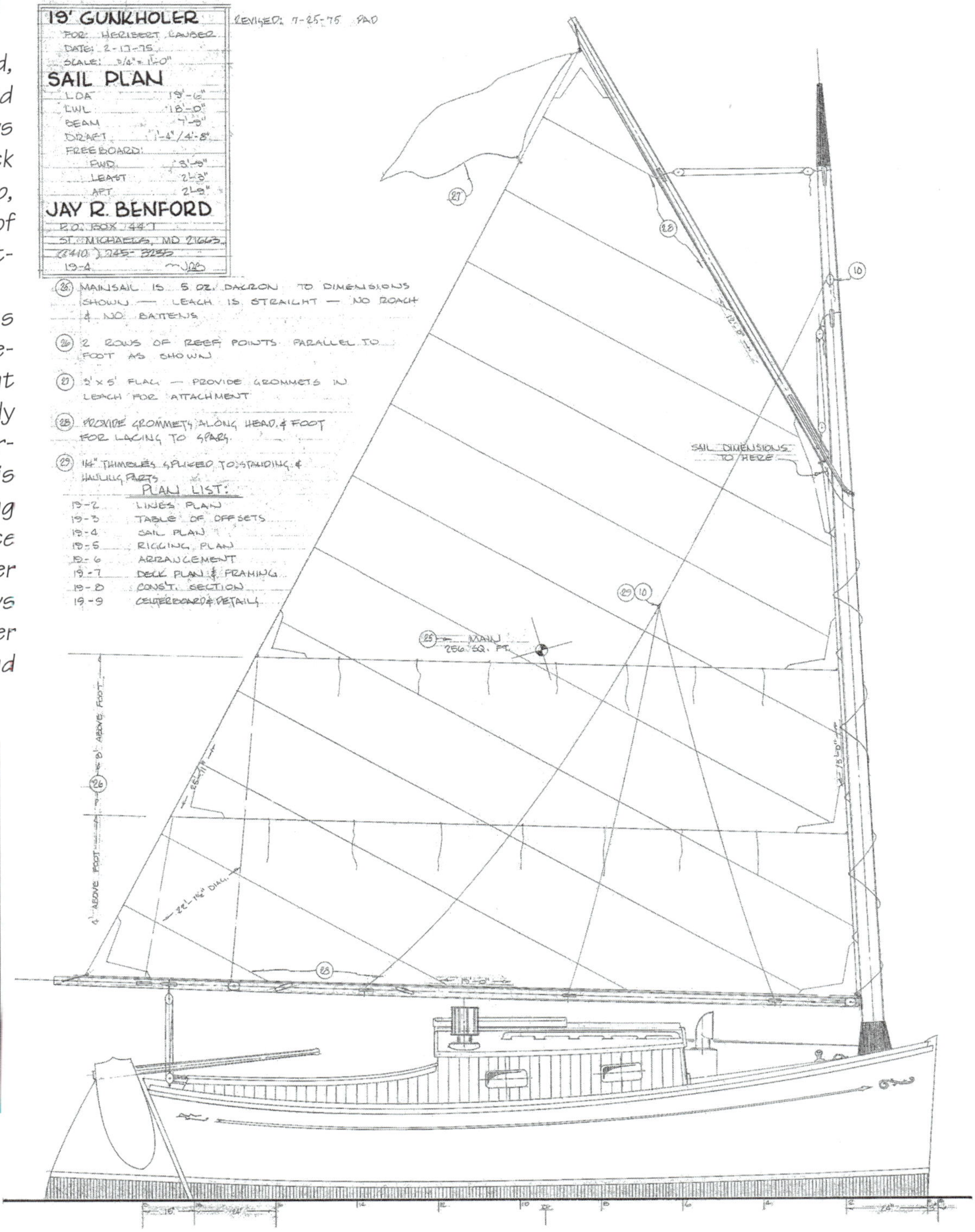

There is a single reef tucked in and she's sailing easily.

This drawing shows how to make and put the pieces of the rig together. The pieces on the boom are laid out to effectively install the sail and the reefing control lines. The sail should be stretched along the boom and gaff so that, when the halyards are properly set taut, that the sail will set with no wrinkles in it.

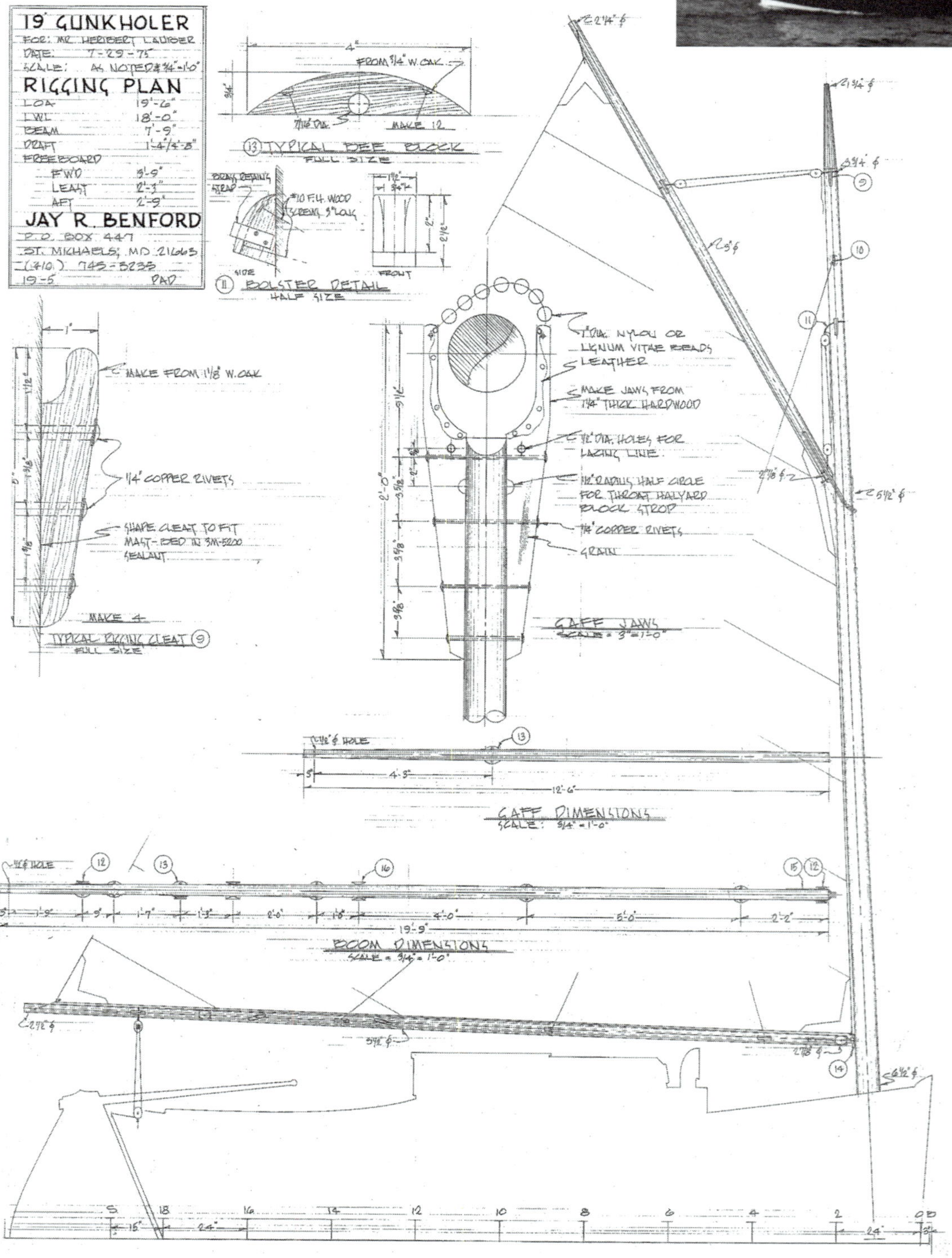

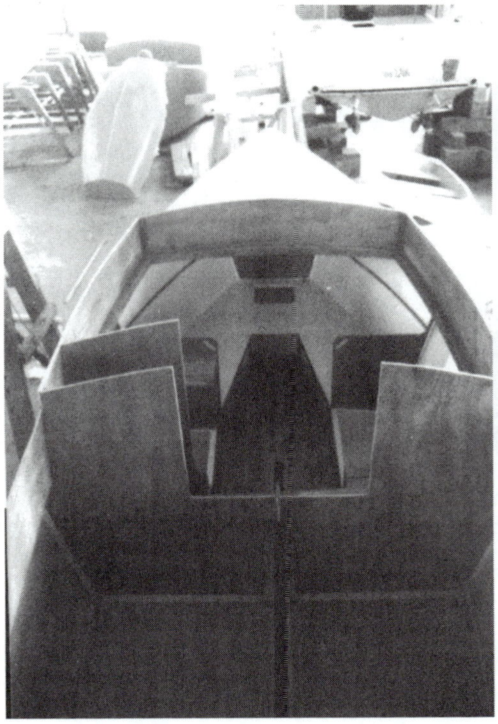

The low centerboard trunk is readily seen. And the locker, seating and galley structure. There are lift off lids to access stowage spaces under the seating and berths.

The centerboard trunk is not intrusive in the cabin and there are two comfortable seats here, one alongside the stove and the other hinged over the head. The vee of the berths can have a filler piece for couples to enjoy that space as a double berth. Note that the cockpit slopes slightly forward, to drain at the forward outboard corners through scuppers.

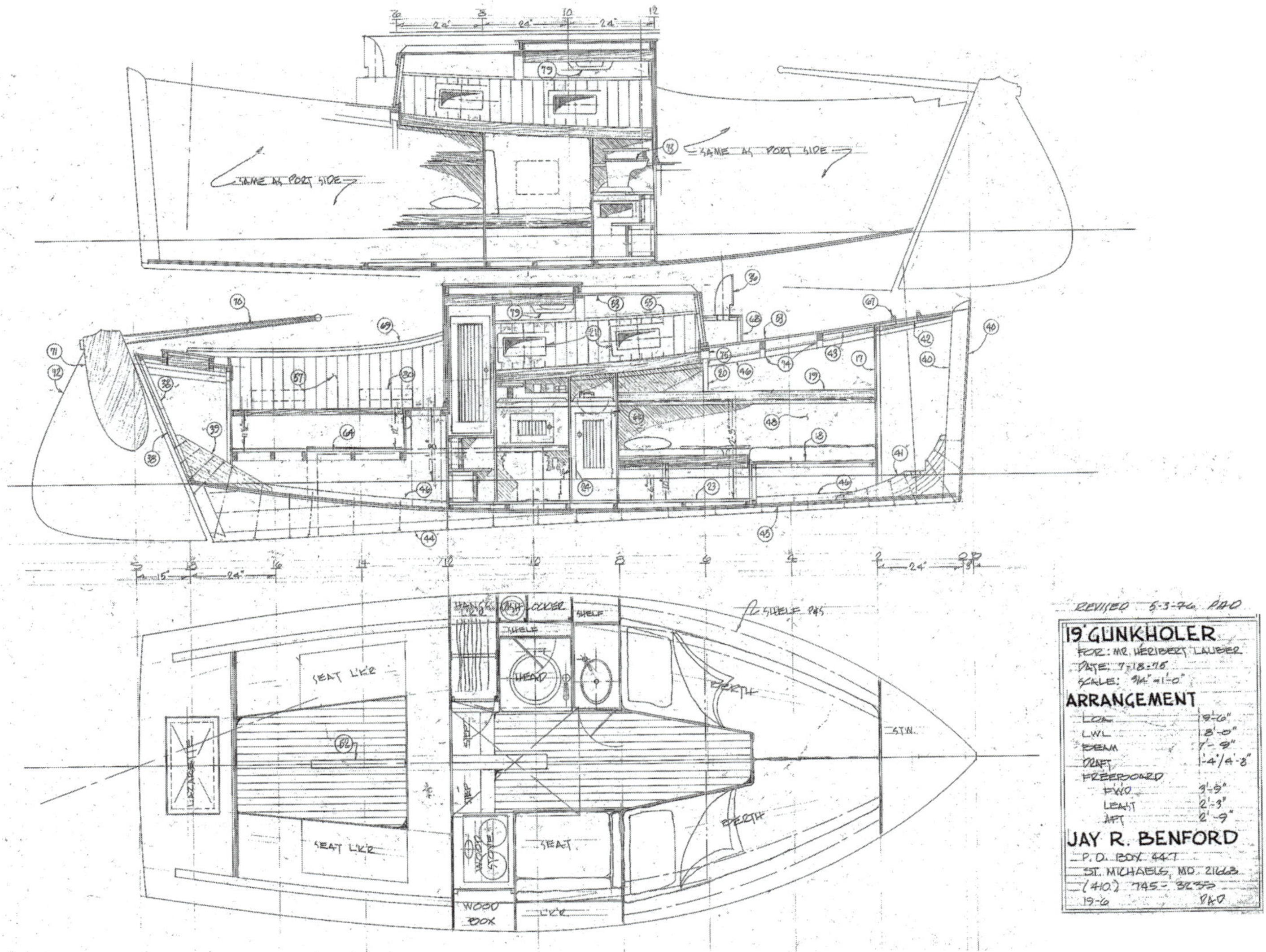

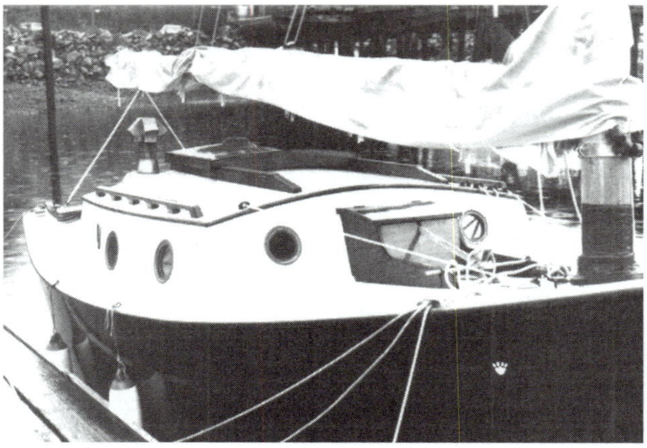

Note the running rigging shows well here, and the builder's clever box for the anchor and rode on the foredeck. The handrails would be better placed close aboard the hatch slides.

The deck plan shows the route for the running rigging leads. The handrails are set in from the house sides so that they will give you some leverage when using them.

The split view at the bottom shows the bottom and deck framing. It assumes that both sides are identical and shows half. The deck doubler for the mast and backing reinforcements for the mooring cleats are important pieces of the structure.

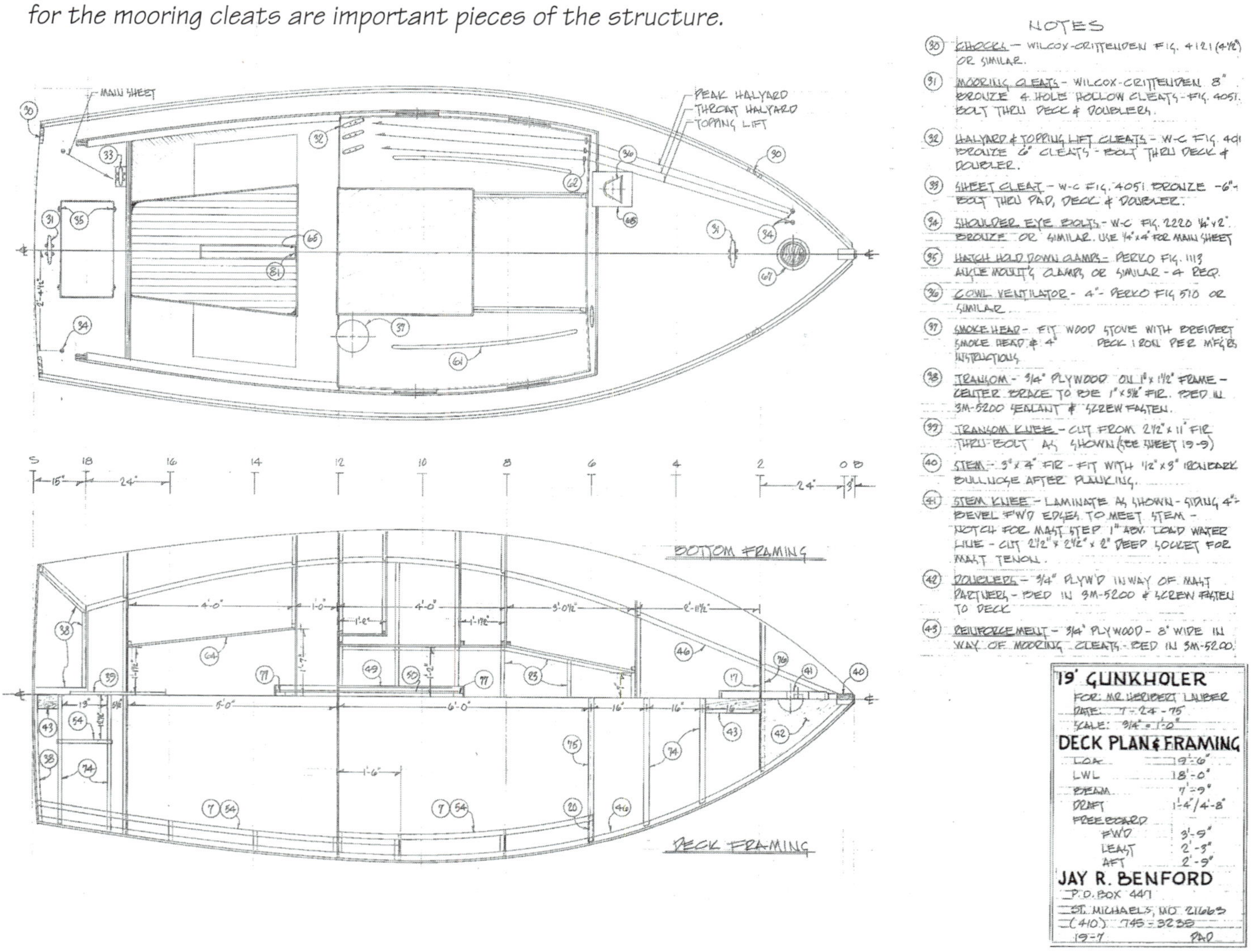

Note the handholds cut into the corners of the interior low bulkhead and stowage space subdivided under berths.

These sections show not only the structure but the outfitting of the interior. The sloped seatback on the lounging seat makes it a great place to rest and read a good book. It also is the best seat in the cabin, and positioned for the convenience of the cook in accessing the stove.

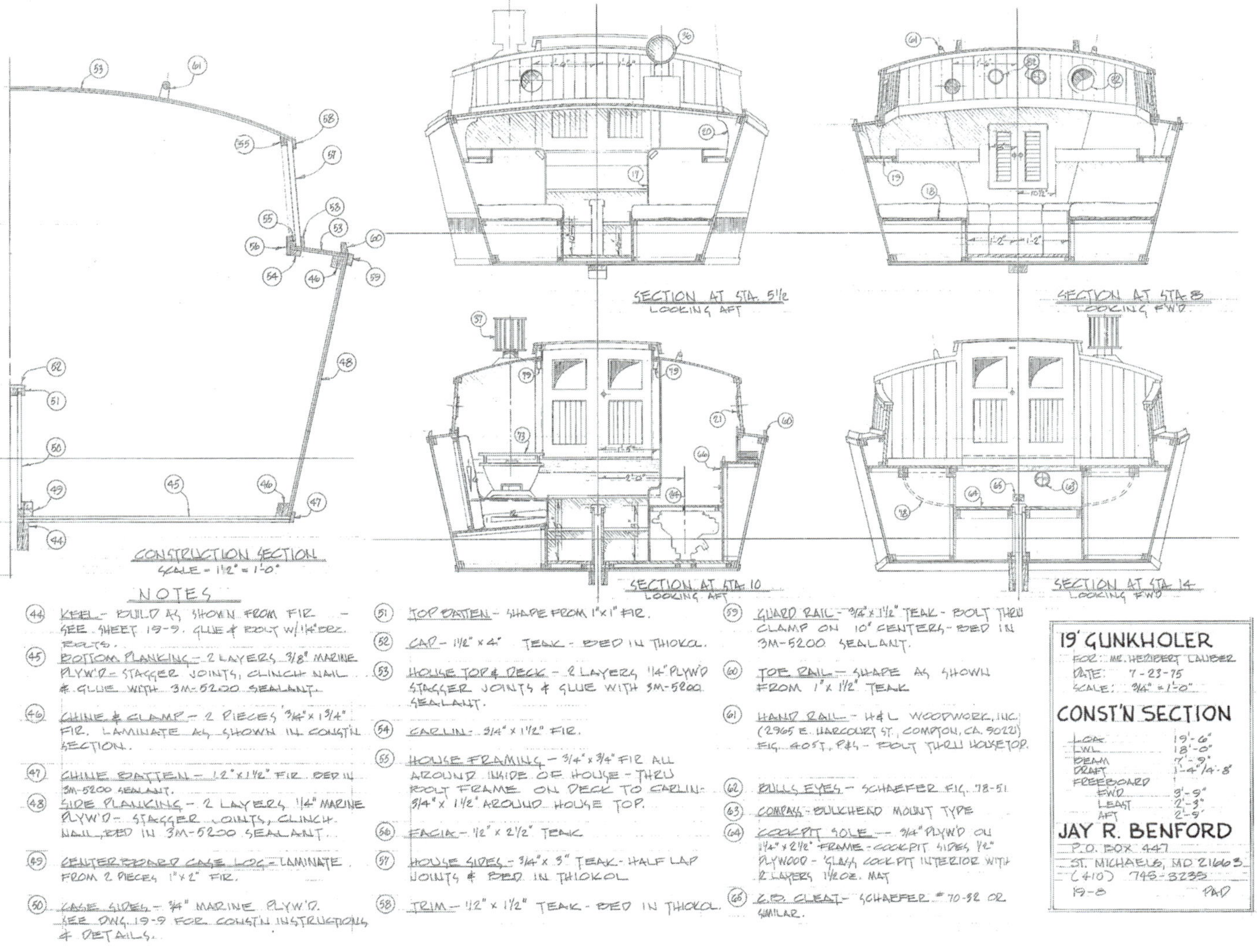

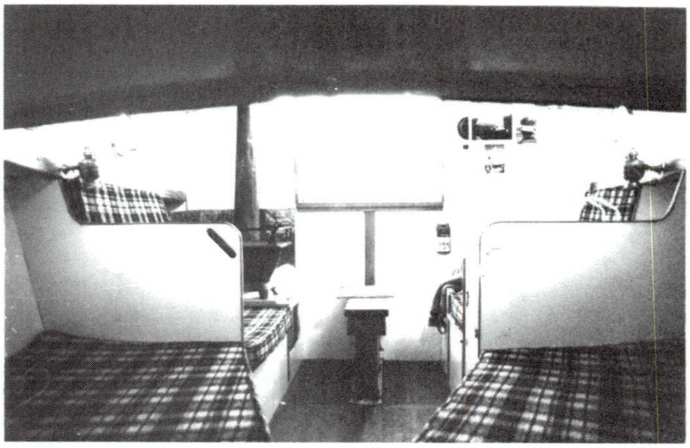

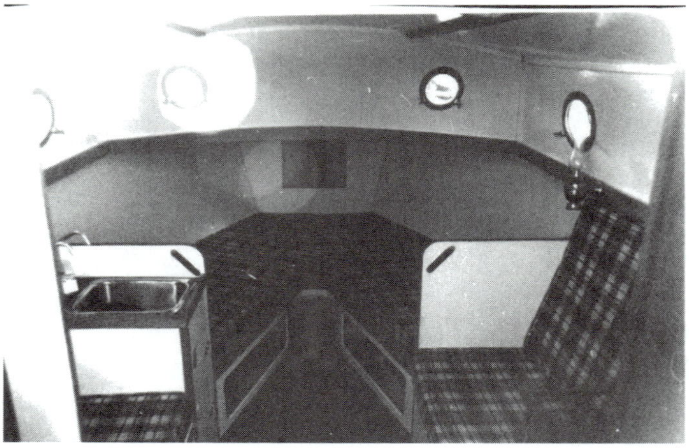

Catspaw's *completed interior.*

The ballast cast into the centerboard is important to make it sink when the pennant is released. The boats Paul Miller built had, with our concurrence, about 500 pounds of lead trim ballast secured aboard. Slightly increasing the weight cast into the board could be taken from the total of the trim ballast, but making it too heavy would make the board hard to raise.

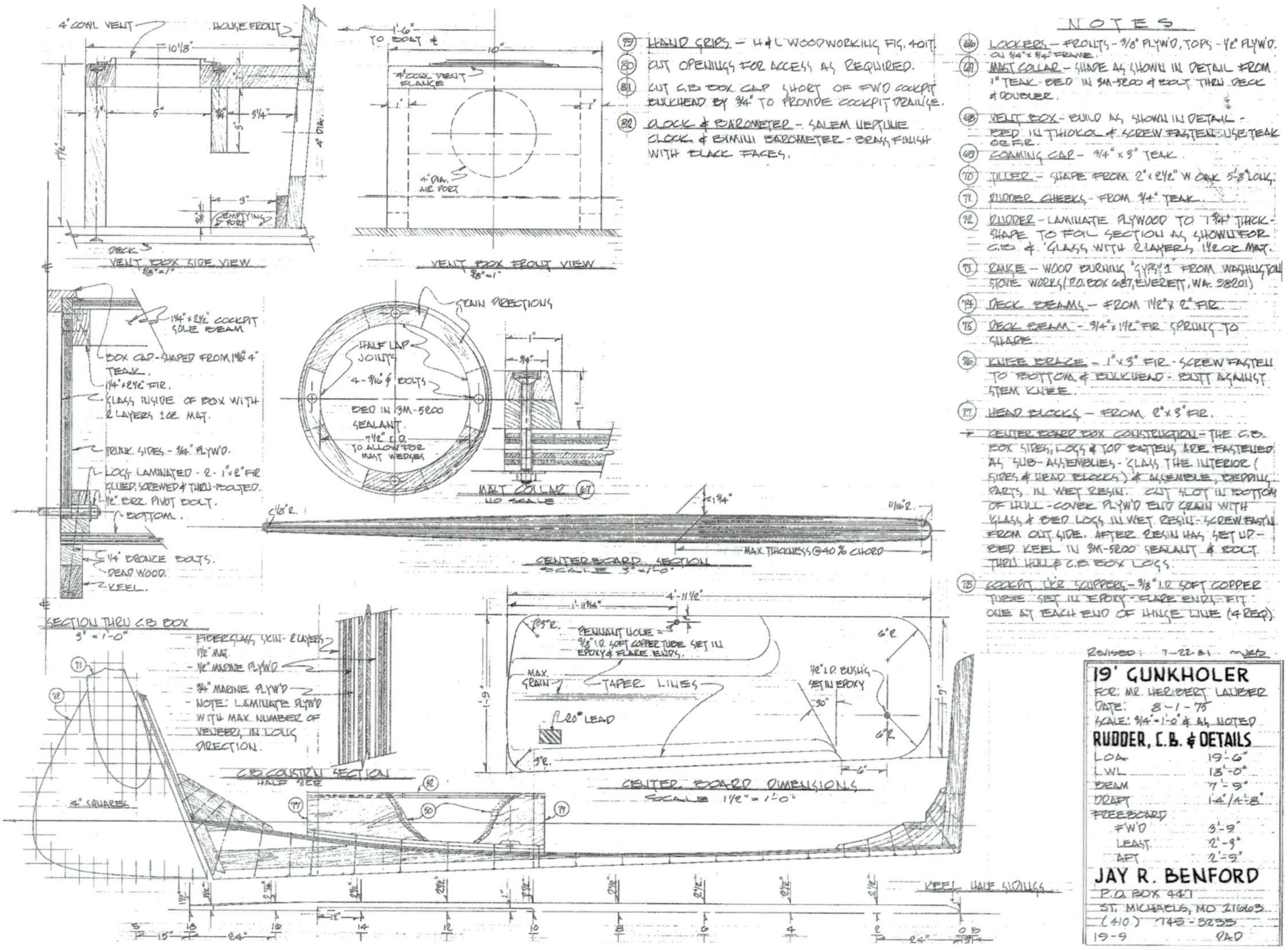

Chapter Four
20' Catboat Joy & Tug Yacht

Design Number 137
1976

Some designs seem to spring to life, inspired by just the right spark, either in the form of a suggestion from a client or an inspiration from another source. This 20' catboat is one of them.

In the spring of 1976, a very interesting letter arrived from Al Lewis. In it he said, *"I wonder if you would design a catboat for me. . . . I have been studying the books and journals for several years and have come across photographs of boats Mr. Benford has designed and they have given me great pleasure just to look at them. His* **Puffin** *has particularly delighted me. . . .*

" . . . Reflecting further, I believe that if I had a design for a neat little. . . . keel cat I might. . . . proceed with a. . . . boat. . . . I had planned to sail her in the Columbia River and perhaps in the San Juan's. . . .

"What I think I would like in a. . . . cat is a beefy, beamy keelboat big enough to sleep two in the cabin and two or three (athwartship?) in the cockpit under a boom tent; a small galley and head; an inboard diesel engine powering a nice big prop. Perhaps twin bilge keels might be considered to avoid an overly deep draft but not if it will make her too slow in stays. I read that keelcats are already bad in this regard. I am uncertain as to whether I would be trailering the boat or mooring it in Puget Sound somewhere; most likely the latter. I still like lots of sheer and tumblehome and like the looks of a round counter-stern, cruiser stern or even a fantail. Bow would be tumblehome, plumb & sharp, or canoe. A classical gaff rigged "up and down "mainsail looks beautiful, loose footed or laced, whichever. . . .

"I would think an LOA of 20' or so would be enough to incorporate the features I have mentioned, but smaller would be just fine and a little larger possibly OK too. The beam would be about half the LWL.

"The enclosed check. . . . should be applied to the cost of a preliminary design if you agree to undertake the job."

This one struck just the right note with me. A couple days later I sent back the following letter:

"Thank you for your letter of April 15th, and your expression of interest in our designs. The project you outlined took my fancy with its potential, and the enclosed print is the result. I think it ought to make a very handsome little cruiser, and we'd be delighted to have the opportunity to do up the plans! . . ."

The drawings that follow are the result of the design commission that resulted. The final boat is very close to the preliminary design I had laid out, having been so well inspired by the original request.

Quite a number of sisterships have been built and we had a mold for the hull that was taken from the original hull built for Al Lewis, from which several 'glass hulls were pulled.

Some time after we'd done the design for the original catboat version, and a mold had been made from the hull, we started thinking about what sort of pocket cruiser could be created on that hull.

The 20' Cruiser or Tug Yacht is the result of that sketching. Several have been built from the molded hulls and at least one in wood. They have proven to be capable cruisers and the reports we get back from the owners indicate that they're having a lot of fun with them.

The drawings indicate a diesel that's no longer a current model. I'd suggest the Yanmar 1GM10 with a 3.22:1 gear as the current choice for a substitution. Of course, as further years come along, the choice of current models will change. Basically, an engine in the six to ten horsepower range is what's needed and get the most reduction you can in the gearbox so that the prop size is large enough to have good thrust. The engine compartment is roomy enough to let a variety of different engines fit in it, and the beds can be modified to take into account differences in width and height of the mounts.

What about a centerboard version? We've done some sketching on a centerboard alternative version, but the drawings are not fully completed or detailed. The Marconi sloop preliminary drawing shows the direction of our thinking. We've kept the board narrower than in traditional cats, to keep the trunk unobtrusive in the cabin and not push up the engine too much into the cockpit. The board would be a thick (10-15% of chord length) NACA foil shape to provide good lift.

Particulars:	**Imperial**	**Metric**
Length overall	20'-0"	6.10 m
Length designed waterline	18'-0"	5.49 m
Beam	8'-0"	2.44 m
Draft, Catboat	3'-6"	1.07 m
Tug	2'-3"	0.69 m
Freeboard: Forward	3'-7½"	1.10 m
Least	2'-0"	0.61 m
Aft	2'-6"	0.76 m
Displacement, cruising trim, Cat	5,260 lbs.	2,385 kg
Tug	4,800 lbs.	2,177 kg
Displacement-length ratio, Cat	403	
Tug	367	
Ballast, Cat	1,600 lbs.	726 kg
Tug	750 lbs.	340 kg
Ballast ratio, Cat	30%	
Tug	16%	
Sail area	310 sq. ft.	28.8 sq. m
Sail area-displacement ratio	16.4	
Prismatic coefficient	.557	
Pounds per inch immersion	466	83 kg/cm
Entrance half-angle	26°	
Water tankage	14 Gals.	53 liters
Tug	20 Gals.	76 liters
Fuel tankage	15 Gals.	57 liters
Tug	30 Gals.	114 liters
Headroom	5'-0"	1.52 m
Tug	6'-3"	1.91 m

**CAUTION*: The figure for displacement quoted here is for the boat in cruising trim. That is, with the fuel and water tanks filled, the crew on board, as well as the crews' gear and stores in the lockers. This should not be confused with the "shipping weight" often quoted as "displacement" by some manufacturers. This should be taken into account when comparing figure and ratios between this and other designs.

Two cruiser versions, one built in Baltimore, MD, on the East Coast, left, and one at Port Townsend, WA, on the West Coast, right. Both give much pleasure and joy to their crews.

The lug rig, below left, and the sloop rig, below right, are alternative rigs that were considered for this design. The centerboard and shoal keel on the sloop show how a shallower draft version could be done.

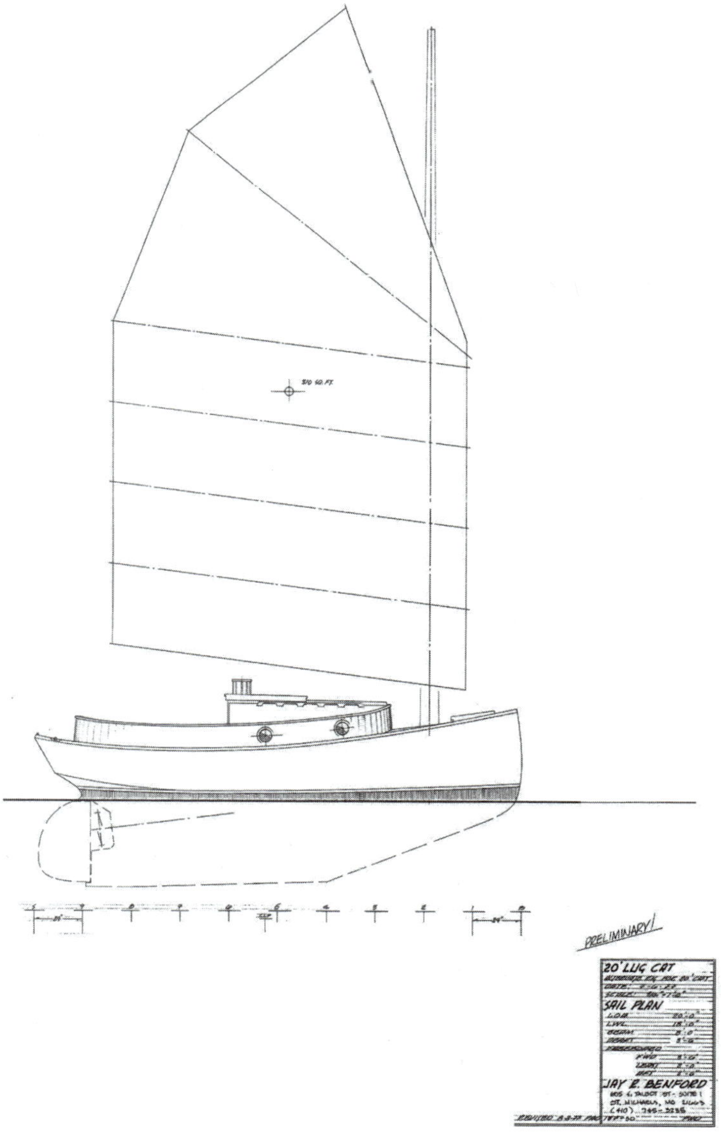

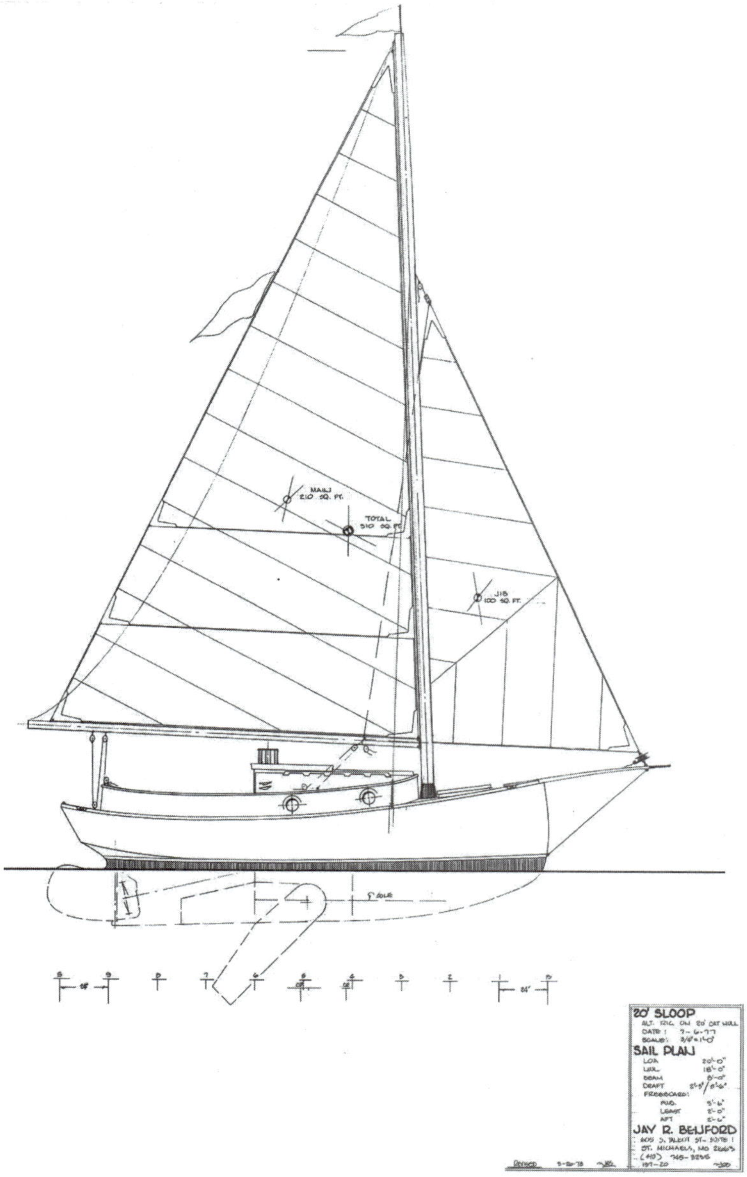

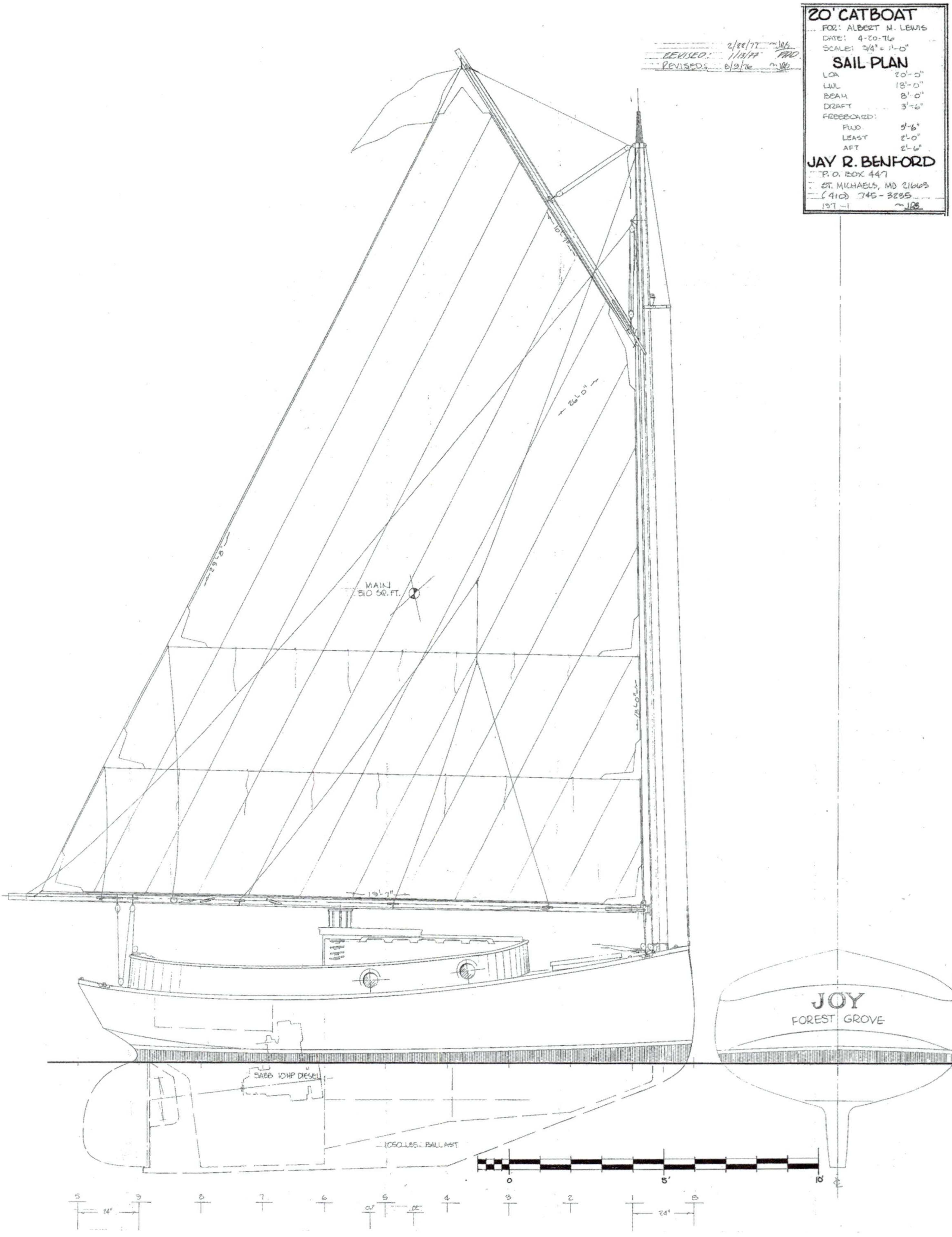
20' CATBOAT
FOR: ALBERT M. LEWIS
DATE: 4-20-76
SCALE: 3/4" = 1'-0"
SAIL PLAN
LOA 20'-0"
LWL 18'-0"
BEAM 8'-0"
DRAFT 3'-6"
FREEBOARD:
FWD 5'-6"
LEAST 2'-0"
AFT 2'-6"
JAY R. BENFORD
P. O. BOX 447
ST. MICHAELS, MD 21663
(410) 745-3235
137-1
MAIN
310 SQ. FT.
JOY
FOREST GROVE
SABB 10HP DIESEL
1050 LBS. BALLAST
0
5'
10'

STA.	HEIGHTS (℄ TO:)							
	KEEL	FAIRBODY	B1	B2	B3	BOOTTOP	KNUCKLE	SHEER
1	3-0-0	3-0-5	4-2-5	—	—	4-7-6	—	7-3-0
2	2-3-0	2-8-6	3-3-6	4-2-6	—	4-7-0	—	6-10-6
3	1-6-0	2-6-7	2-11-0	3-5-3	4-6-2	4-6-2	—	6-7-0
4	0-9-0	2-6-1	2-9-1	3-1-3	3-7-4	4-5-6	—	6-3-6
5	0-8-2	2-6-0	2-8-7	3-0-2	3-4-7	4-5-3	—	6-1-4
6	0-7-4	2-6-4	2-9-7	3-1-2	3-5-2	4-5-1+	—	6-0-3
7	0-6-6	2-10-0	3-1-2	3-4-4	3-8-7	4-5-2	4-4-4	6-0-0
8	0-6-0	3-4-6	3-7-6	3-10-6	4-4-2	4-5-4	4-6-6	6-0-4
9	0-6-0	4-4-7	4-6-2	4-3-2	—	4-6-3	4-10-1	6-2-2
S	—	5-6-6					5-3-3	6-6-0

STA.	HALF-BREADTHS (℄ TO:)										DIAGONALS		
	KEEL BOTTOM	KEEL TRACE	36" WL	42" WL	48" WL	57" WL	66" WL	KNUCKLE	SHEER	21" WL	A	B	C
1	0-0-4	0-0-4	—	0-5-6	0-10-5	1-3-0	1-5-3	—	1-7-3	—	0-8-1	1-1-7	1-5-2
2	0-0-4	0-2-0	0-6-1	1-3-1	1-9-7	2-3-5	2-6-5	—	2-8-4	—	1-1-5	1-9-5	2-4-6
3	0-0-4	0-3-3	1-2-3	2-1-2	2-7-7	3-1-4	3-3-7	—	3-4-5	0-1-2	1-5-6	2-3-2	3-1-0
4	0-0-4	0-4-1	1-8-5	2-9-7	3-3-7	3-8-4	3-9-2	—	3-8-7	0-3-1	1-7-6	2-6-5	3-7-4
5	0-3-0	0-4-4	1-11-4	3-2-3	3-8-6	3-11-6	3-11-4	—	3-10-2	0-4-1	1-8-4	2-7-6	3-11-0
6	0-4-0	0-4-2	1-8-5	3-2-1	3-9-0	4-0-0	3-11-4	—	3-10-0	0-4-2	1-7-6	2-6-7	3-11-1
7	0-3-6	0-3-6	0-7-6	2-5-2	3-4-5	3-9-7	3-9-6	3-9-0	3-8-3	0-3-6	1-3-6	2-3-1	3-7-3
8	0-2-3	0-2-3	—	0-5-2	2-3-2	3-4-0	3-4-4	3-3-6	3-4-2	0-2-3	0-8-2	1-7-4	2-11-5
9	0-0-3	0-0-3	—	—	0-0-6	1-10-6	2-5-5	2-2-4	2-7-6	0-0-3	—	0-5-6	1-10-3
S	—	—	—	—	—	—	—	—	-0-	—	—	—	

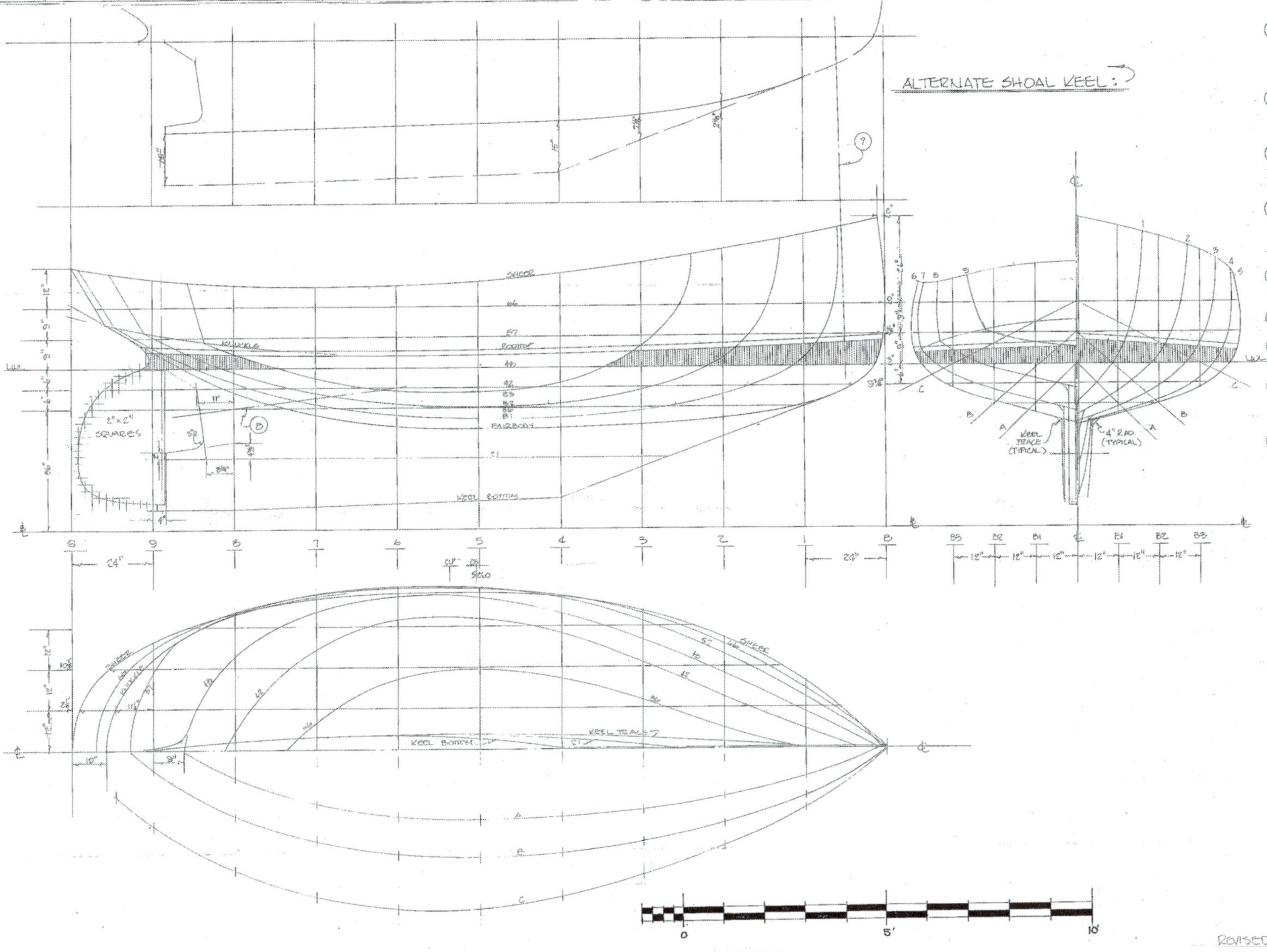

NOTES:

1. LINES & OFFSETS TO OUTSIDE OF HULL IN FEET-INCHES-EIGHTHS. DEDUCT FOR HULL THICKNESS PER CONSTRUCTION PLAN. DO NOT SCALE PRINTS!
2. ANY ALTERATION FROM THESE PLANS RELIEVES THE DESIGNERS FROM ANY FURTHER RESPONSIBILITY
3. THESE PLANS & PRINTS ARE THE PROPERTY OF THE DESIGNERS & MAY BE USED ONLY AS AUTHORIZED IN WRITING.
4. IT IS UNDERSTOOD THAT NO MORE THAN ONE BOAT WILL BE BUILT FROM THESE PLANS WITHOUT WRITTEN PERMISSION FROM THE DESIGNERS.
5. BOOTTOP STRIPE IS 1½" HIGH IN PROFILE. OFFSETS TO TOP EDGE OF STRIPES.
6. LINES MUST BE LOFTED & FAIRED FULL SIZE.
7. MAST ℄ CROSSES LWL 12" AFT OF STA. B & RAKES AFT 1" IN 36" OF HEIGHT
8. SHAFT ℄ CROSSES 42" WL AT STA 6 & IS 13¾" BELOW LWL AT CROSSING APERTURE OUTLINE
9.

REVISED: 8/5/76 ~JRB; 3/13/77 ~JRB

20' CATBOAT

FOR: ALBERT M. LEWIS
DATE: 5-6-76
SCALE: 3/4" = 1'-0"

LINES & OFFSETS

LOA	20'-0"
LWL	18'-0"
BEAM	8'-0"
DRAFT	3'-6"
FREEBOARD:	
FWD.	3'-7½"
LEAST	2'-0"
AFT	2'-6"

JAY R. BENFORD
P. O. BOX 447
ST. MICHAELS, MD 21663
(410) 745-3235
137-2 ~JRB

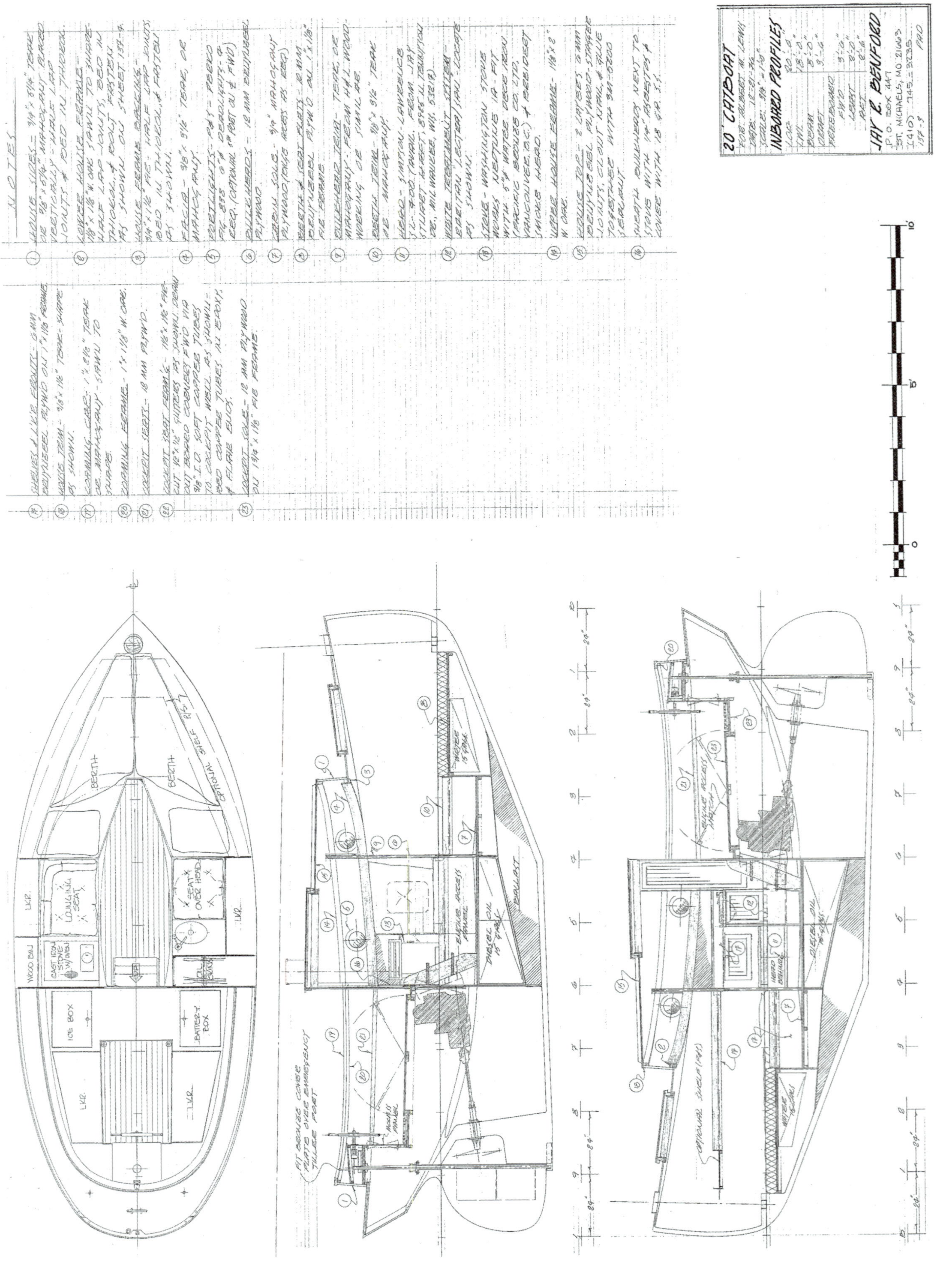
NOTES
20' CATBOAT
INBOARD PROFILES
JAY R. BENFORD
P.O. BOX 447
ST. MICHAELS, MD 21663
(410) 745-3235
BERTH
BERTH
LKR.
WOOD BIN
CAST IRON STOVE
LOUNGING SEAT
SEAT OVER HEAD
ICE BOX
BATTERY BOX
OPTIONAL SHELF
0
5
10

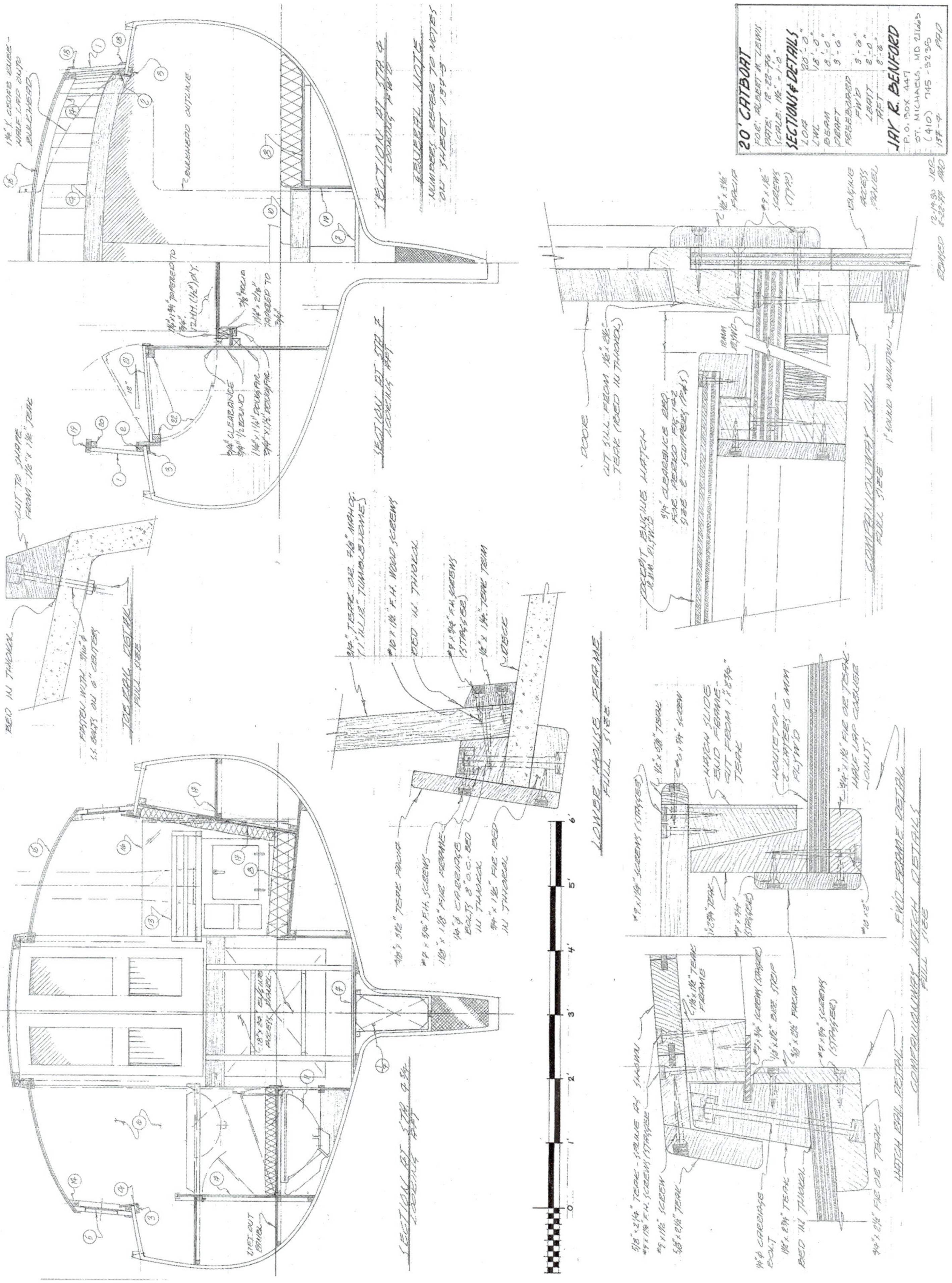

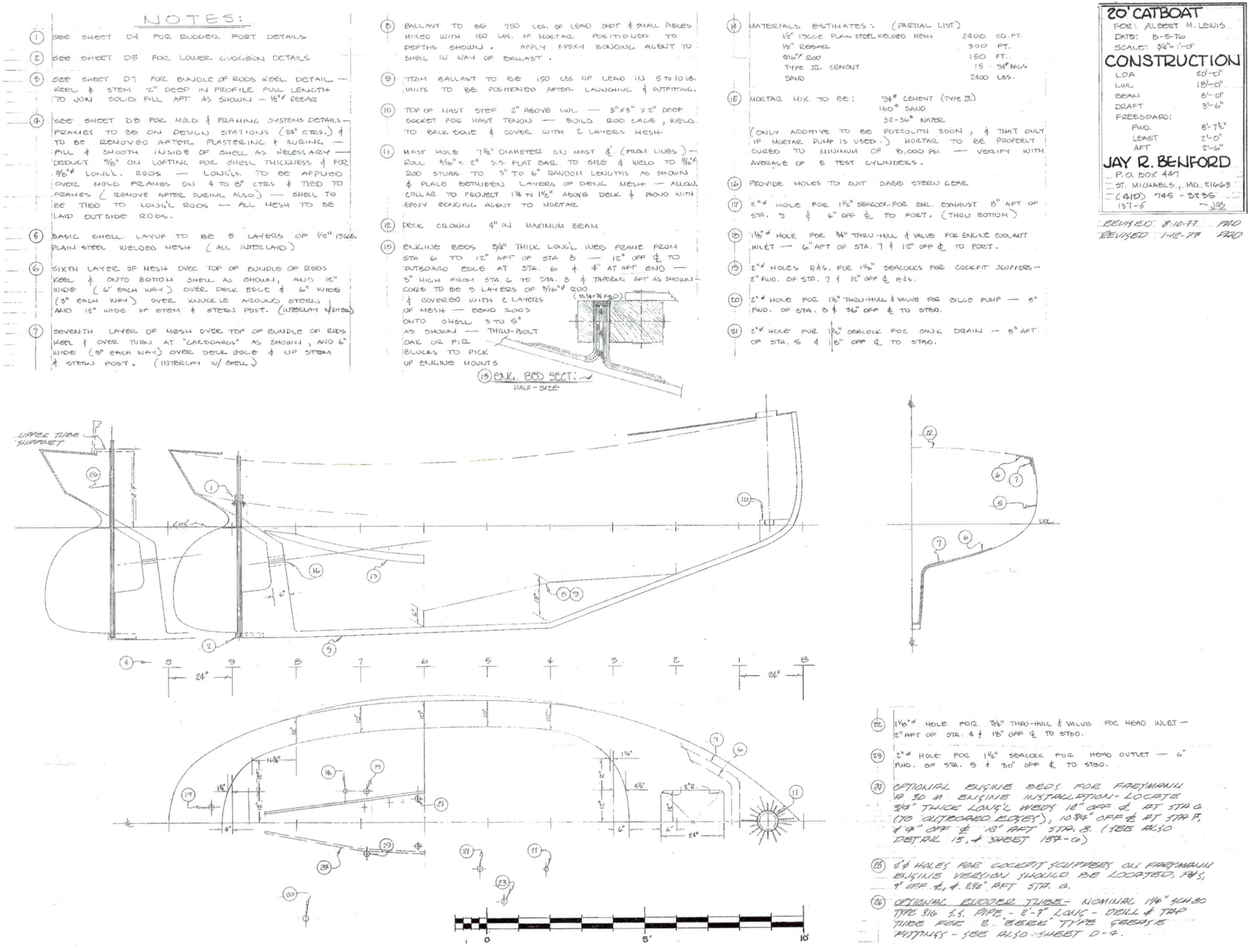
NOTES:
1 SEE SHEET D4 FOR RUDDER PORT DETAILS
2 SEE SHEET D5 FOR LOWER GUDGEON DETAILS
3 SEE SHEET D7 FOR BUNDLE OF RODS KEEL DETAIL — KEEL & STEM 2" DEEP IN PROFILE FULL LENGTH TO JOIN SOLID FILL AFT AS SHOWN — 1/2"⌀ REBAR
4 SEE SHEET D8 FOR MOLD & FRAMING SYSTEMS DETAILS — FRAMES TO BE ON DESIGN STATIONS (24" CTRS.) & TO BE REMOVED AFTER PLASTERING & CURING — FILL & SMOOTH INSIDE OF SHELL AS NECESSARY — DEDUCT 5/8" ON LOFTING FOR SHELL THICKNESS & FOR 3/8"⌀ LONG'L. RODS — LONG'LS. TO BE APPLIED OVER MOLD FRAMES ON 4 TO 8" CTRS & TIED TO FRAMES (REMOVE AFTER CURING ALSO) — SHELL TO BE TIED TO LONG'L RODS — ALL MESH TO BE LAID OUTSIDE RODS.
5 BASIC SHELL LAYUP TO BE 5 LAYERS OF 1/2" 19GA. PLAIN STEEL WELDED MESH (ALL INTERLAID)
6 SIXTH LAYER OF MESH OVER TOP OF BUNDLE OF RODS KEEL & ONTO BOTTOM SHELL AS SHOWN, AND 12" WIDE (6" EACH WAY) OVER DECK EDGE & 6" WIDE (3" EACH WAY) OVER KNUCKLE AROUND STERN AND 12" WIDE UP STEM & STERN POST. (INTERLAY W/SHELL)
7 SEVENTH LAYER OF MESH OVER TOP OF BUNDLE OF RODS KEEL & OVER TURN AT "CARDBOARDS" AS SHOWN, AND 6" WIDE (3" EACH WAY) OVER DECK EDGE & UP STEM & STERN POST. (INTERLAY W/ SHELL)
8 BALLAST TO BE 750 LBS. OF LEAD SHOT & SMALL PIGLES MIXED WITH 150 LBS. OF MORTAR POSITIONED TO DEPTHS SHOWN. APPLY EPOXY BONDING AGENT TO SHELL IN WAY OF BALLAST.
9 TRIM BALLAST TO BE 150 LBS OF LEAD IN 5 TO 10 LB. UNITS TO BE POSITIONED AFTER LAUNCHING & OUTFITTING.
10 TOP OF MAST STEP 2" ABOVE LWL — 3"x3"x2" DEEP SOCKET FOR MAST TENON — BUILD ROD CAGE, WELD TO BACK BONE & COVER WITH 2 LAYERS MESH
11 MAST HOLE 7 1/2" DIAMETER ON MAST ℄ (FROM LINES) — ROLL 3/16" x 2" S.S. FLAT BAR TO SIZE & WELD TO 3/16"⌀ ROD STUBS TO 3" TO 6" RANDOM LENGTHS AS SHOWN & PLACE BETWEEN LAYERS OF DECK MESH — ALLOW COLLAR TO PROJECT 1 1/4 TO 1 1/2" ABOVE DECK & BOND WITH EPOXY BONDING AGENT TO MORTAR
12 DECK CROWN 4" IN MAXIMUM BEAM
13 ENGINE BEDS 3/4" THICK LONG'L WEB FRAME FROM STA 6 TO 12" AFT OF STA 8 — 12" OFF ℄ TO OUTBOARD EDGE AT STA. 6 & 4" AT AFT END — 3" HIGH FROM STA. 6 TO STA. 8 & TAPERING AFT AS SHOWN — CORE TO BE 3 LAYERS OF 3/16"⌀ ROD & COVERED WITH 2 LAYERS OF MESH — BEND RODS ONTO SHELL 3 TO 5" AS SHOWN — THRU-BOLT OAK OR FIR BLOCKS TO PICK UP ENGINE MOUNTS
13 ENG. BED SECT.
HALF-SIZE
14 MATERIALS ESTIMATES: (PARTIAL LIST)
1/2" 19GA PLAIN STEEL WELDED MESH 2400 SQ.FT.
1/2" REBAR 300 FT.
3/16"⌀ ROD 150 FT.
TYPE III CEMENT 15 · 94# BAGS
SAND 2400 LBS.
15 MORTAR MIX TO BE: 94# CEMENT (TYPE III)
160# SAND
32-36# WATER
(ONLY ADDITIVE TO BE POZZOLITH 300N, & THAT ONLY IF MORTAR PUMP IS USED.) MORTAR TO BE PROPERLY CURED TO MINIMUM OF 8,000 PSI — VERIFY WITH AVERAGE OF 5 TEST CYLINDERS.
16 PROVIDE HOLES TO SUIT SABB STERN GEAR
17 2"⌀ HOLE FOR 1 1/2" SEACOCK FOR ENG. EXHAUST 8" AFT OF STA. 9 & 6" OFF ℄ TO PORT. (THRU BOTTOM)
18 1 1/8"⌀ HOLE FOR 3/4" THRU-HULL & VALVE FOR ENGINE COOLANT INLET — 6" AFT OF STA. 7 & 12" OFF ℄ TO PORT.
19 2"⌀ HOLES P&S. FOR 1 1/2" SEACOCKS FOR COCKPIT SCUPPERS — 2" FWD. OF STA. 7 & 12" OFF ℄ P&S.
20 2"⌀ HOLE FOR 1 1/2" THRU-HULL & VALVE FOR BILGE PUMP — 5" FWD. OF STA. 8 & 36" OFF ℄ TO STBD.
21 2"⌀ HOLE FOR 1 1/2" SEACOCK FOR SINK DRAIN — 8" AFT OF STA. 5 & 18" OFF ℄ TO STBD.
22 1 1/8"⌀ HOLE FOR 3/4" THRU-HULL & VALVE FOR HEAD INLET — 2" AFT OF STA. 4 & 18" OFF ℄ TO STBD.
23 2"⌀ HOLE FOR 1 1/2" SEACOCK FOR HEAD OUTLET — 6" FWD. OF STA. 5 & 30" OFF ℄ TO STBD.
24 OPTIONAL ENGINE BEDS FOR FARYMANN A 30 M ENGINE INSTALLATION - LOCATE 3/4" THICK LONG'L WEBS 12" OFF ℄ AT STA 6 (TO OUTBOARD EDGES), 10 3/4" OFF ℄ AT STA 7, & 4" OFF ℄ 12" AFT STA. 8. (SEE ALSO DETAIL 13, & SHEET 137-6)
25 2"⌀ HOLES FOR COCKPIT SCUPPERS ON FARYMANN ENGINE VERSION SHOULD BE LOCATED, P&S, 9" OFF ℄, & 21 1/2" AFT STA. 6.
26 OPTIONAL RUDDER TUBE - NOMINAL 1 1/4" SCH 80 TYPE 316 S.S. PIPE - 2'-7" LONG - DRILL & TAP TUBE FOR 2 BEEK TYPE GREASE FITTINGS - SEE ALSO SHEET D-4.
20' CATBOAT
FOR: ALBERT M. LEWIS
DATE: 8-5-76
SCALE: 3/4"=1'-0"
CONSTRUCTION
LOA 20'-0"
LWL 18'-0"
BEAM 8'-0"
DRAFT 3'-6"
FREEBOARD:
FWD. 3'-7 1/2"
LEAST 2'-0"
AFT 2'-6"
JAY R. BENFORD
P.O. BOX 447
ST. MICHAELS, MD. 21663
(410) 745-3235
137-5
REVISED 8-10-77
REVISED 1-12-77
UPPER TUBE SUPPORT
LWL
24"
0
5'
10'

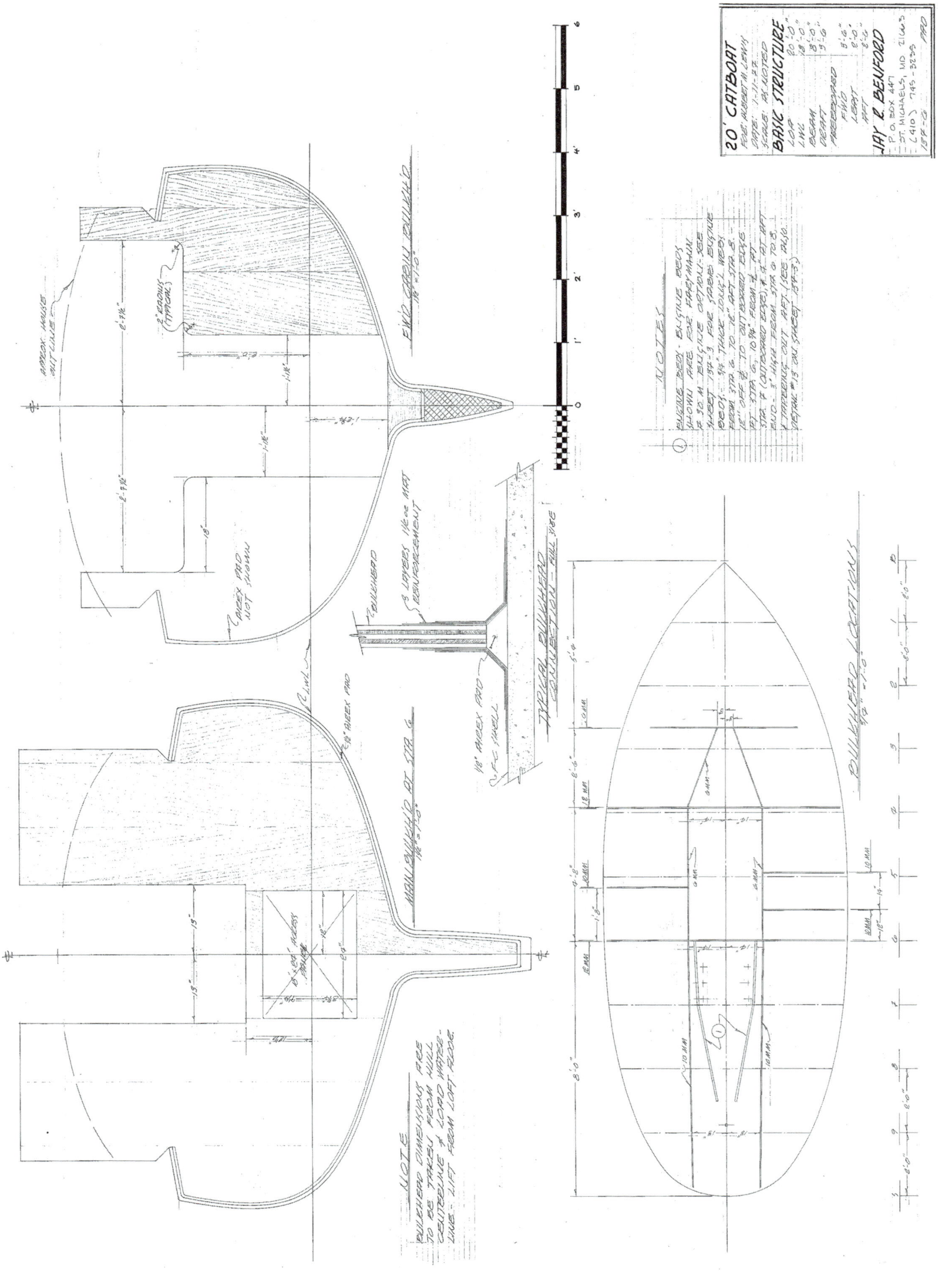
20' CATBOAT
BASIC STRUCTURE
LOA 20'-0"
LWL 18'-0"
BEAM 8'-0"
FREEBOARD
FWD
LEAST
AFT
JAY R. BENFORD
P.O. BOX 447
ST. MICHAELS, MD 21663
(410) 745-3235
NOTES
FWD CABIN BULKHD
MAIN BULKHD AT STA. 6
TYPICAL BULKHEAD CONNECTION - FULL SIZE
BULKHEAD LOCATIONS
BULKHEAD
3 LAYERS 1½ oz MAT REINFORCEMENT
1/8" AIREX PAD
F.G. SHELL
AIREX PAD NOT SHOWN
APPROX. HOUSE OUT LINE
2" RADIUS (TYPICAL)
NOTE
BULKHEAD DIMENSIONS ARE TO BE TAKEN FROM HULL CENTERLINE & LOAD WATERLINE - LIFT FROM LOFT FLOOR.

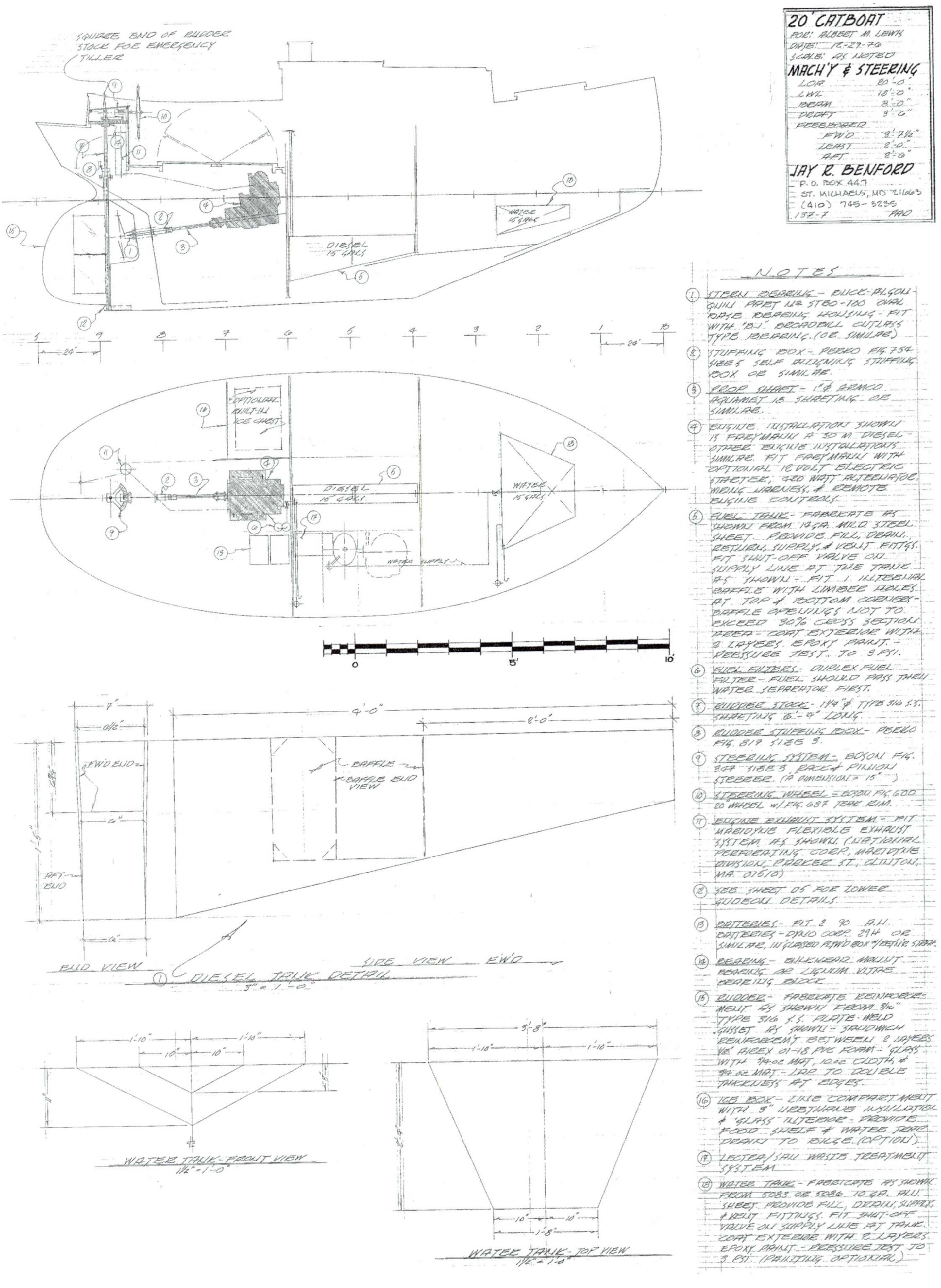
20' CATBOAT
FOR: ALBERT M. LEWIS
DATE: 12-29-76
SCALE: AS NOTED
MACH'Y & STEERING
LOA 20'-0"
LWL 18'-0"
BEAM 8'-0"
DRAFT 3'-6"
FREEBOARD
FWD 3'-7 1/16"
LEAST 2'-0"
AFT 2'-6"
JAY R. BENFORD
P.O. BOX 447
ST. MICHAELS, MD 21663
(410) 745-3235
137-7 PAD
SQUARE END OF RUDDER STOCK FOR EMERGENCY TILLER
WATER 15 GALS
DIESEL 15 GALS
OPTIONAL BUILT-IN ICE CHEST
WATER SUPPLY
0
5'
10'
4'-0"
2'-0"
7"
6 1/2"
FWD END
BAFFLE
BAFFLE END VIEW
6"
1'-5"
AFT END
END VIEW
SIDE VIEW FWD
DIESEL TANK DETAIL
3" = 1'-0"
1'-10"
10"
WATER TANK-FRONT VIEW
1 1/2" = 1'-0"
3'-8"
1'-8"
WATER TANK-TOP VIEW
1 1/2" = 1'-0"
NOTES
1 STERN BEARING - BUCK-ALGONQUIN PART No. 5T80-100 OVAL BASE BEARING HOUSING - FIT WITH "BJ" BROADBILL CUTLASS TYPE BEARING. (OR SIMILAR)
2 STUFFING BOX - PERKO FIG. 754 SIZE 5 SELF ALIGNING STUFFING BOX OR SIMILAR.
3 PROP SHAFT - 1" ɸ ARMCO AQUAMET 18 SHAFTING OR SIMILAR.
4 ENGINE INSTALLATION SHOWN IS FARYMANN A 30 M. DIESEL OTHER ENGINE INSTALLATIONS SIMILAR. FIT FARYMANN WITH OPTIONAL 12 VOLT ELECTRIC STARTER, 420 WATT ALTERNATOR, WIRING HARNESS, & REMOTE ENGINE CONTROLS.
5 FUEL TANK - FABRICATE AS SHOWN FROM 14 GA. MILD STEEL SHEET. PROVIDE FILL, DRAIN, RETURN, SUPPLY, & VENT FITTGS. FIT SHUT-OFF VALVE ON SUPPLY LINE AT THE TANK AS SHOWN - FIT 1 INTERNAL BAFFLE WITH LIMBER HOLES AT TOP & BOTTOM CORNERS - BAFFLE OPENINGS NOT TO EXCEED 30% CROSS SECTION AREA - COAT EXTERIOR WITH 2 LAYERS EPOXY PAINT. - PRESSURE TEST TO 3 PSI.
6 FUEL FILTERS - DUPLEX FUEL FILTER - FUEL SHOULD PASS THRU WATER SEPARATOR FIRST.
7 RUDDER STOCK - 1 1/4" ɸ TYPE 316 S.S. SHAFTING 6'-4" LONG.
8 RUDDER STUFFING BOX - PERKO FIG. 819 SIZE 3.
9 STEERING SYSTEM - EDSON FIG. 347 SIZE 3 RACK & PINION STEERER. ("A" DIMENSION = 15")
10 STEERING WHEEL - EDSON FIG. 600 20" WHEEL W/ FIG. 687 TEAK RIM.
11 ENGINE EXHAUST SYSTEM - FIT MARTDYNE FLEXIBLE EXHAUST SYSTEM AS SHOWN. (NATIONAL PERFORATING CORP., MARTDYNE DIVISION, PARKER ST., CLINTON, MA. 01510)
12 SEE SHEET 05 FOR LOWER GUDEON DETAILS.
13 BATTERIES - FIT 2 90 A.H. BATTERIES - DYNO CORP. 29H OR SIMILAR, IN 'GLASSED PLYWD BOX W/ RETAIN. STRAP.
14 BEARING - BULKHEAD MOUNT BEARING OR LIGNUM VITAE BEARING BLOCK.
15 RUDDER - FABRICATE REINFORCEMENT AS SHOWN FROM 3/16" TYPE 316 S.S. PLATE - WELD GUSSET AS SHOWN - SANDWICH REINFORCEM'T BETWEEN 2 LAYERS 1/2" AIREX 01-18 PVC FOAM - 'GLASS WITH 3/4 OZ MAT, 10 OZ CLOTH, & 3/4 OZ MAT - LAP TO DOUBLE THICKNESS AT EDGES.
16 ICE BOX - LINE COMPARTMENT WITH 3" URETHANE INSULATION & 'GLASS INTERIOR - PROVIDE FOOD SHELF & WATER TRAP DRAIN TO BILGE. (OPTION)
17 LECTRA/SAN WASTE TREATMENT SYSTEM
18 WATER TANK - FABRICATE AS SHOWN FROM 5083 OR 5086 10 GA. ALU. SHEET. PROVIDE FILL, DRAIN, SUPPLY, & VENT FITTINGS. FIT SHUT-OFF VALVE ON SUPPLY LINE AT TANK. COAT EXTERIOR WITH 2 LAYERS EPOXY PAINT - PRESSURE TEST TO 3 PSI. (PAINTING OPTIONAL)

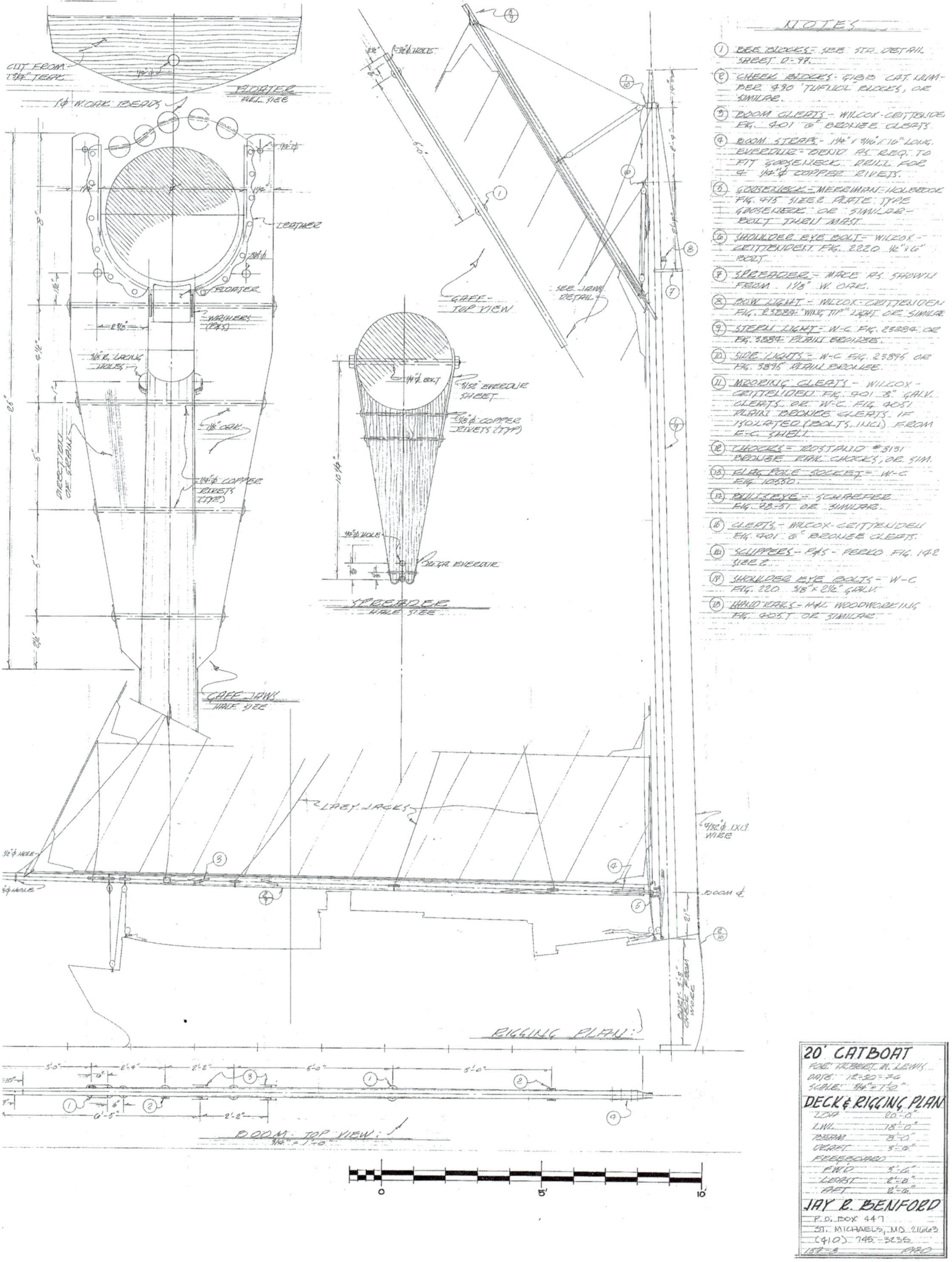
NOTES
(1) SEE STR. DETAIL SHEET D-97.
(2) CHEEK BLOCKS - GIBB CAT. NUMBER 490 "TUFNOL" BLOCKS, OR SIMILAR.
(3) BOOM CLEATS - WILCOX-CRITTENDEN FIG. 401 6" BRONZE CLEATS.
(4) BOOM STRAP - 1¼" x 3/16" x 16" LONG EVERDUR - BEND AS REQ. TO FIT GOOSENECK. DRILL FOR 4 ¼"⌀ COPPER RIVETS.
(5) GOOSENECK - MERRIMAN-HOLBROOK FIG. 475 SIZE 2 PLATE TYPE GOOSENECK OR SIMILAR - BOLT THRU MAST.
(6) SHOULDER EYE BOLT - WILCOX-CRITTENDEN FIG. 2220 ½" x 6" BOLT.
(7) SPREADER - MAKE AS SHOWN FROM 1⅛" W. OAK.
(8) BOW LIGHT - WILCOX-CRITTENDEN FIG. 23884 "WING TIP" LIGHT OR SIMILAR
(9) STERN LIGHT - W-C FIG. 23884 OR FIG. 3884 PLAIN BRONZE.
(10) SIDE LIGHTS - W-C FIG. 23895 OR FIG. 3895 PLAIN BRONZE.
(11) MOORING CLEATS - WILCOX-CRITTENDEN FIG. 401 8" GALV. CLEATS OR W-C FIG. 4051 PLAIN BRONZE CLEATS IF ISOLATED (BOLTS, INCL.) FROM F-G SHELL.
(12) CHOCKS - #3131 BRONZE CHOCKS, OR SIM.
(13) FLAG POLE SOCKET - W-C FIG. 10550.
(14) BULLSEYE - SCHAEFER FIG. 78-51 OR SIMILAR.
(15) CLEATS - WILCOX-CRITTENDEN FIG. 401 6" BRONZE CLEATS.
(16) SCUPPERS - P&S - PERKO FIG. 142 SIZE 2.
(17) SHOULDER EYE BOLTS - W-C FIG. 220 3/8" x 2½" GALV.
(18) HAND RAILS - H&L WOODWORKING FIG. 4051 OR SIMILAR.
CUT FROM 1¾" TEAK
FLOATER FULL SIZE
1" ⌀ WOOD BEADS
LEATHER
FLOATER
WASHERS (P&S)
3/16" ⌀ LACING HOLES
DIRECTION OF GRAIN
1¼" OAK
¼" ⌀ COPPER RIVETS (TYP)
GAFF JAWS HALF SIZE
GAFF - TOP VIEW
SEE JAWS DETAIL
¼" ⌀ BOLT
3/32 EVERDUR SHEET
3/16" ⌀ COPPER RIVETS (TYP)
SPREADER HALF SIZE
LAZY JACKS
9/32" ⌀ 1x19 WIRE
½" ⌀ HOLE
BOOM ℄
RIGGING PLAN
BOOM - TOP VIEW
0
5'
10'
20' CATBOAT
DATE: 12-30-76
SCALE: ¾" = 1'-0"
DECK & RIGGING PLAN
LOA 20'-0"
LWL 18'-0"
BEAM 8'-0"
DRAFT 3'-6"
FREEBOARD
FWD 3'-6"
LEAST 2'-0"
AFT 2'-6"
JAY R. BENFORD
P.O. BOX 447
ST. MICHAELS, MD 21663
(410) 745-3235

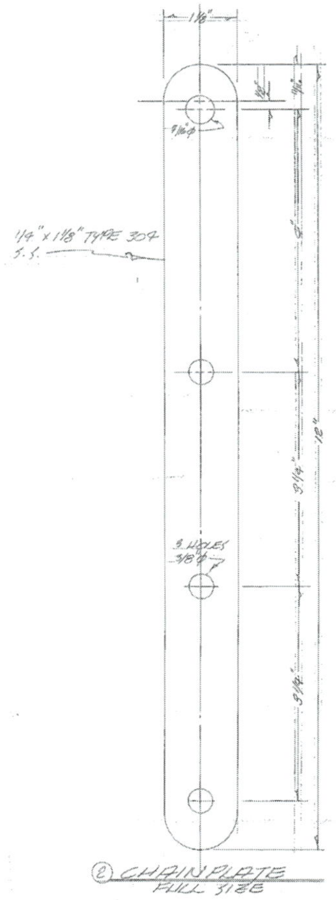

RIGGING SCHEDULE					
NAME	NUMBER REQ'D	CONFIGURATION	LINE DIA. & CONST'N	LENGTH	FITTINGS
PEAK HALYARD	1	SPLICED EYE – SINGLE – STROP – DOUBLE – DOUBLE – EYE BOLT	3/8" ϕ DAC. 3-STRAND	120'	1-SINGLE BLK 2-DOUBLE BLK'S 2-1/4" SHACKLES
THROAT HALYARD	1	TRIPLE BLOCK – STROP – SINGLE W/BECKET – EMPTY SHEAVE PEAK HALY'D DECK BLOCK	3/8" ϕ DAC. 3-STRAND	80'	1-SINGLE BLK 1-TRIPLE BLK 1-3/8" THIMBLE
TOPPING LIFT	1	EMPTY SHEAVE THROAT HALY'D TRIPLE – MATTHEW WALKER KNOT – DOUBLE – EYE BOLT	3/8" ϕ DAC. 3-STRAND	70'	1-DOUBLE
1ST REEF LUFF LINE	1	SAIL CRINGLE – EYE BOLT – EMPTY SHEAVE TOP'G LIFT DECK BLOCK	3/8" ϕ DAC. BRAID	24'	1-3/8" THIMBLE 1-SHACKLE
1ST REEF LEACH LINE	1	SAIL CRINGLE – MATTHEW WALKER KNOT – BEE BLOCK W/SHEAVE	3/8" ϕ DAC. BRAID	14'	1-BEE BLOCK W/ SHEAVE
2ND REEF LEACH LINE	1	SAIL CRINGLE – CHEEK BLOCK – MATTHEW WALKER KNOT	3/8" ϕ DAC. BRAID	19'	1-CHEEK BLK
MAIN SHEET	1	STROPS	3/8" ϕ DAC. 3-STRAND	115'	1-SINGLE W/BECKET 3-SINGLE BLK'S 1-SINGLE W/CAM CLEAT 1-3/8" THIMBLE
LAZY-JACKS		SEE SAIL PLAN FOR ARRANGEMENT.	1/4" ϕ NYLON BRAID	180'	6-1/4" THIMBLES
HEAD STAY	1	LENGTH LISTED INCLUDES T'BUCKLE 1/2 OPEN	7/32" ϕ 1X19 S.S.	APPX 26'	1-3/8" TURNBUCKLE
DOCK LINES	4		3/8" ϕ NYLON 3-STRAND	40'	
LIGHT GROUND TACKLE	1	6' CHAIN – 12-H DANFORTH	1/2" NYLON 3-STRAND	100'	1-1/2" THIMBLE 2- SHACKLES

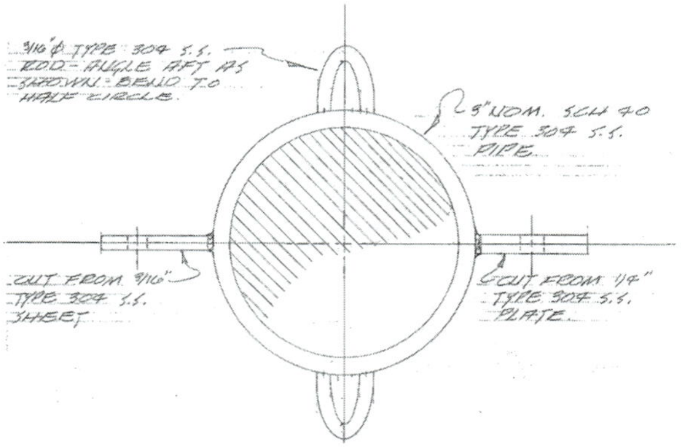

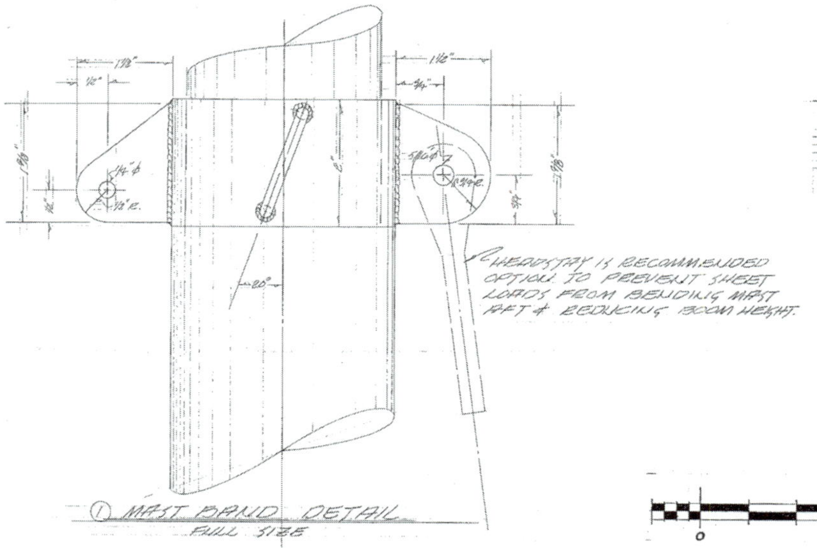

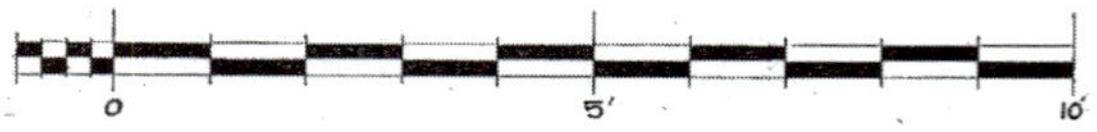

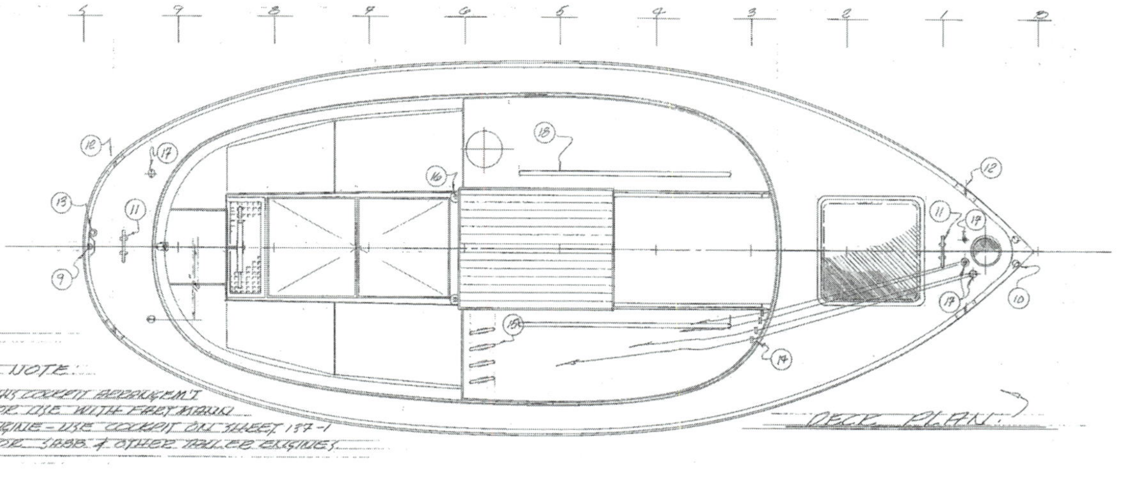

NOTE:
THIS COCKPIT ARRANGEMENT FOR USE WITH FARYMANN ENGINE - USE COCKPIT ON SHEET 157-1 FOR SABB & OTHER LARGER ENGINES.

20' CATBOAT
FOR: ALBERT M. LEWIS
DATE: 1-6-97
SCALE: AS NOTED

RIG DETAILS

LOA	20'-0"
LWL	18'-0"
BEAM	8'-0"
DRAFT	3'-6"
FREEBOARD	
FWD	3'-6"
LEAST	2'-0"
AFT	2'-0"

JAY R. BENFORD
P. O. BOX 447
ST. MICHAELS, MD 21663
(410) 745-3235
157-10 PAD

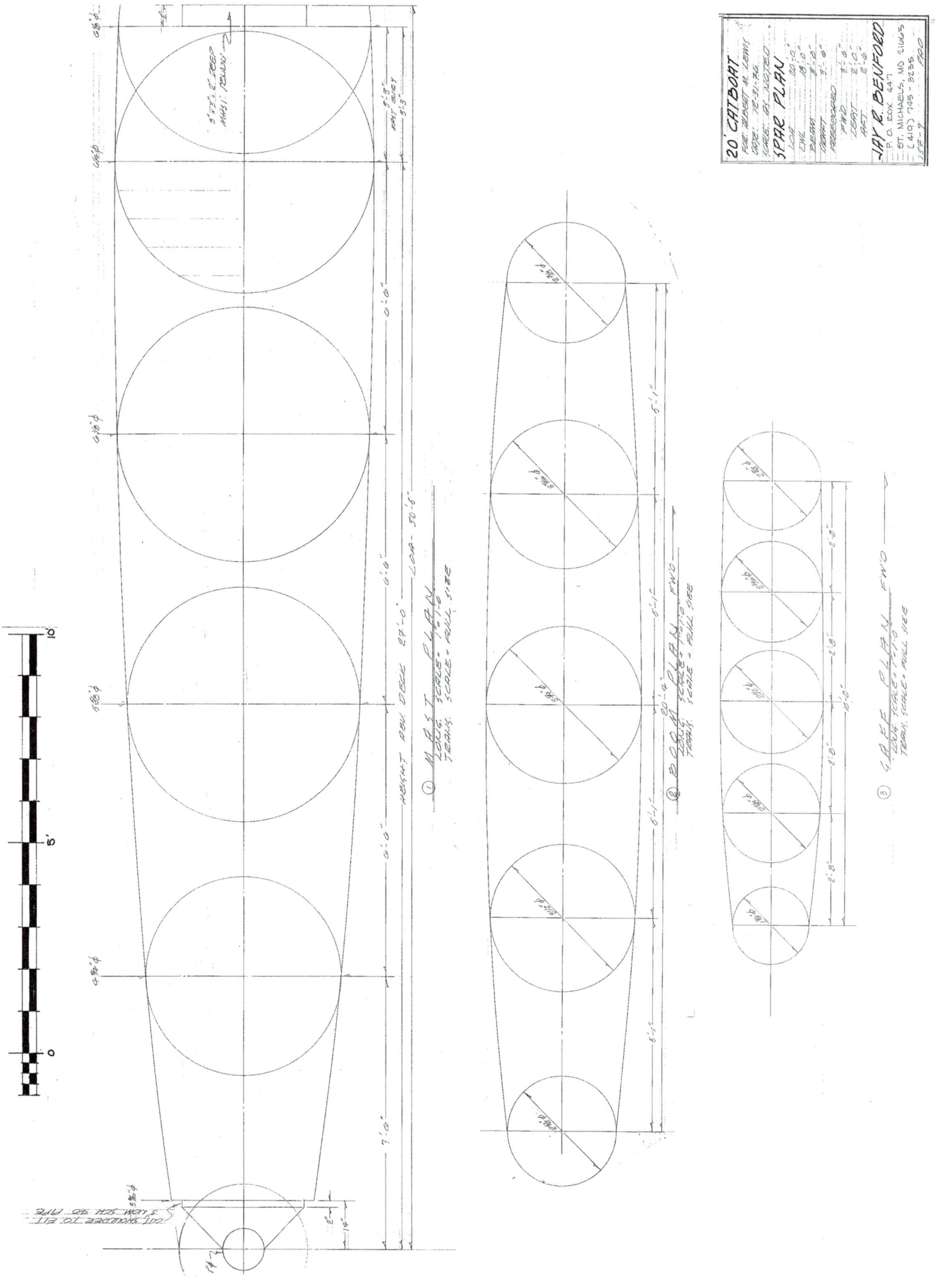
20' CATBOAT
SPAR PLAN
JAY R. BENFORD
P. O. BOX 447
ST. MICHAELS, MD 21663
(410) 745-3235
MAST PLAN
BOOM PLAN
GAFF PLAN
FWD
10'
5'
0

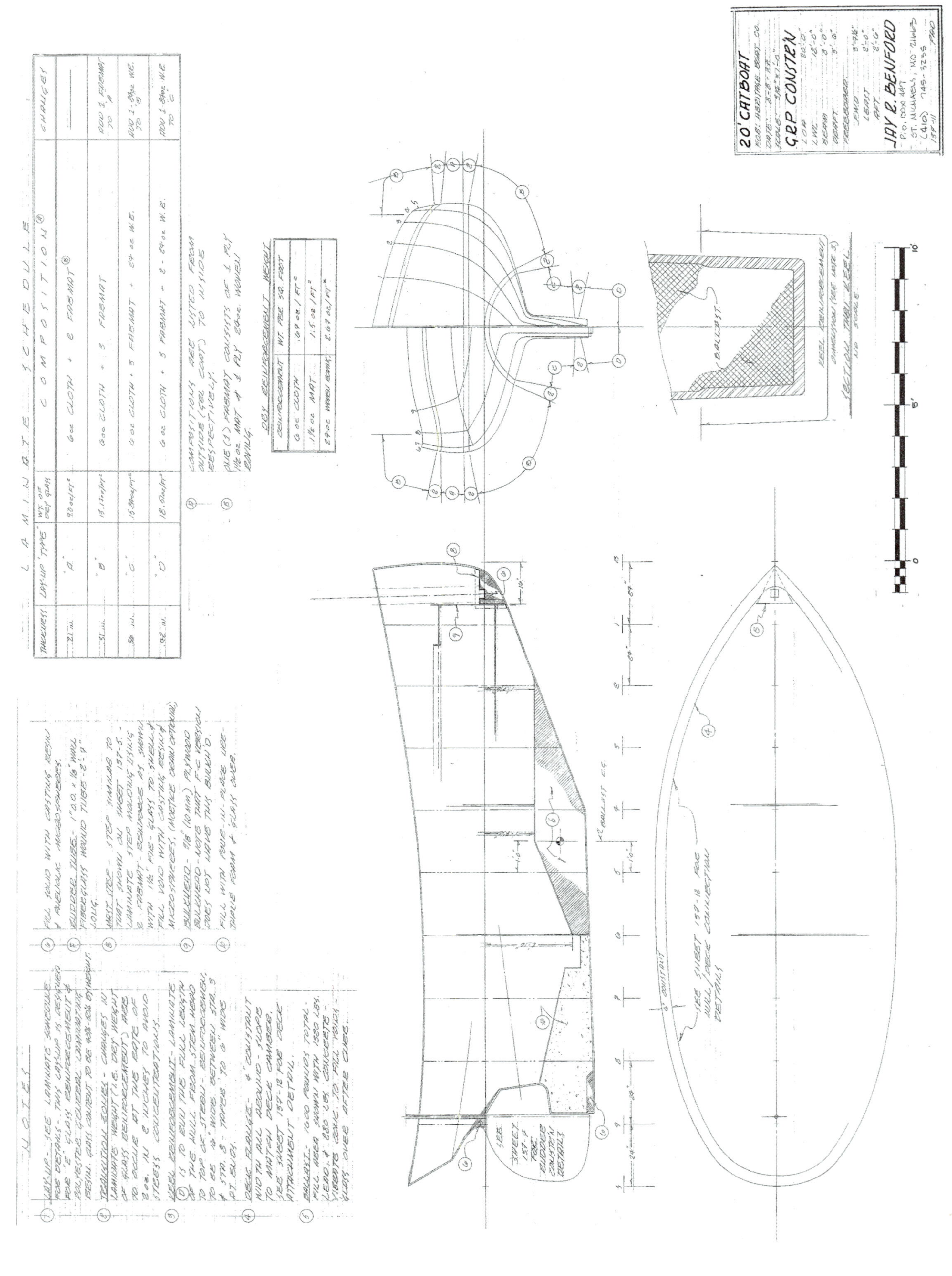
NOTES
LAMINATE SCHEDULE
DRY REINFORCEMENT WEIGHT
BALLAST
KEEL REINFORCEMENT DIMENSION (SEE NOTE 3)
SECTION THRU KEEL
NO SCALE
SEE SHEET 137-12 FOR HULL/DECK CONNECTION DETAILS
20' CATBOAT
FOR: HERITAGE BOAT CO.
DATE: 8-8-77
GRP CONSTR'N
LOA 20'-0"
LWL 18'-0"
BEAM 8'-0"
DRAFT 3'-6"
JAY R. BENFORD
P.O. BOX 447
ST. MICHAELS, MD 21663
(410) 745-3235
137-11
0
5'
10'

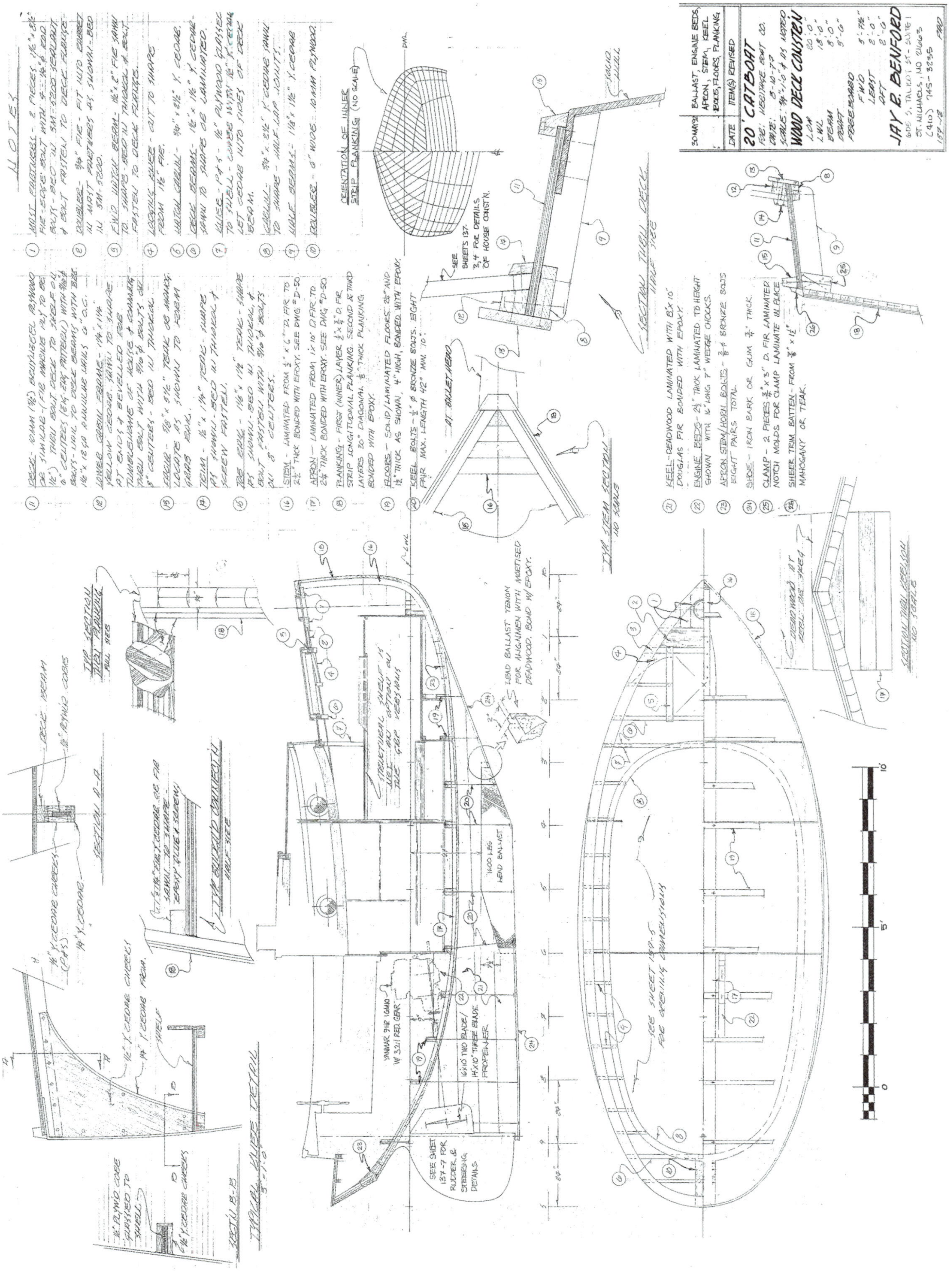
NOTES
(1) MAST PARTNERS - 2 PIECES 1½" x 8½" FIR-EDGE BOLT WITH 4 - ¼"ø ROD BOLTS - BED IN 3M-5200 SEALANT & BOLT FASTEN TO DECK FLANGE.
(2) DOUBLER - ¾" FIR - FIT INTO RABBET IN MAST PARTNERS AS SHOWN - BED IN 3M-5200.
(3) FWD HATCH BEAM - 1½" x 2" FIR SAWN TO SHAPE - BED IN THIOKOL & BOLT FASTEN TO DECK FLANGE.
(4) LODGING KNEE - CUT TO SHAPE FROM 1½" FIR.
(5) HATCH CARLIN - ¾" x 2½" Y. CEDAR.
(6) DECK BEAMS - 1½" x 1½" Y. CEDAR - SAWN TO SHAPE OR LAMINATED.
(7) KNEE - P & S - ½" PLYWOOD GLASSED TO SHELL - COVERED WITH ½" Y. CEDAR. LET CEDAR INTO SIDES OF DECK BEAM.
(8) CARLIN - ¾" x 2½" Y. CEDAR SAWN TO SHAPE - HALF-LAP JOINTS.
(9) HALF BEAMS - 1⅛" x 1½" Y. CEDAR
(10) DOUBLER - 6" WIDE - 10MM PLYWOOD.
(11) DECK - 10MM (⅜") BRUYNZEEL PLYWOOD OR SIMILAR - (FIR MARINE PLY TO BE ½") THRU BOLT DECK TO SHELF ON 6" CENTERS (ZIG-ZAG PATTERN) WITH 5/16"ø BOLTS - NAIL TO DECK BEAMS WITH BRZ. 1½" x 12 GA ANNULAR NAILS 6" O.C.
(12) LOWER CABIN FRAME - 1¼" x 1¼" YELLOW CEDAR. SAWN TO SHAPE AT ENDS & BEVELLED FOR TUMBLEHOME OF HOUSE & COAMING - THRU BOLT WITH 5/16"ø BOLTS ON 8" CENTERS - BED IN THIOKOL.
(13) TRIM - ⅝" x 3½" TEAK OR MAHOG. LOCATE AS SHOWN TO FORM GRAB RAIL.
(14) TRIM - ½" x 1¼" TEAK - SHAPE AS SHOWN - BED IN THIOKOL & SCREW FASTEN.
(15) TOE RAIL - 1⅛" x 1½" TEAK - SHAPE AS SHOWN - BED IN THIOKOL & BOLT FASTEN WITH 5/16"ø BOLTS ON 8" CENTERS.
(16) STEM - LAMINATED FROM ½" x 6" D. FIR TO 2½" THICK BONDED WITH EPOXY. SEE DWG #D-50
(17) APRON - LAMINATED FROM 1x10 D.FIR TO 2¼" THICK BONDED WITH EPOXY. SEE DWG #D-50
(18) PLANKING - FIRST (INNER) LAYER ½" x ¾" D. FIR STRIP LONGITUDINAL PLANKING. SECOND & THIRD LAYERS 30° DIAGONAL ⅛" THICK PLANKING BONDED WITH EPOXY.
(19) FLOORS - SOLID/LAMINATED FLOORS 2¼" AND 1½" THICK AS SHOWN, 4" HIGH, BONDED WITH EPOXY.
(20) KEEL BOLTS - ½"ø BRONZE BOLTS. EIGHT PAIR MAX. LENGTH 42" MIN. 10".
(21) KEEL - DEADWOOD LAMINATED WITH 8x10 DOUGLAS FIR BONDED WITH EPOXY.
(22) ENGINE BEDS - 2½" THICK LAMINATED TO HEIGHT SHOWN WITH 16" LONG 7" WEDGE CHOCKS.
(23) APRON STEM/HORN BOLTS - ⅜"ø BRONZE BOLTS EIGHT PAIRS TOTAL.
(24) SHOE - IRON BARK OR GUM ¾" THICK.
(25) CLAMP - 2 PIECES ¾" x 3" D. FIR LAMINATED. NOTCH MOLDS FOR CLAMP LAMINATE IN PLACE.
(26) SHEER TRIM BATTEN - FROM ⅞" x 1½" MAHOGANY OR TEAK.
ORIENTATION OF INNER STRIP PLANKING (NO SCALE)
DWL
SEE SHEETS 137-3,4 FOR DETAILS OF HOUSE CONST'N.
SOLID HULL
SECTION THRU DECK
HALF SIZE
TYP. STEM SECTION
NO SCALE
AT BILLET HEAD
LWL
YANMAR 9HP 1GM10 W/ 3.2:1 RED. GEAR
STRUCTURAL SHELF IS NOT AN OPTION ON THE GRP VERSIONS
SEE SHEET 137-7 FOR RUDDER & STEERING DETAILS
16x10 TWO BLADE/ 14x10 THREE BLADE PROPELLER
1600 LBS LEAD BALLAST
LEAD BALLAST TENON FOR ALIGNMEN WITH MORTISED DEADWOOD. BOND W/ EPOXY.
SEE SHEET 137-5 FOR OPENING DIMENSIONS
DEADWOOD AT KEEL OR SKEG
SECTION THRU KEELSON
NO SCALE
TYPICAL KNEE DETAIL
3" = 1'-0"
SECT'N B-B
½" PLYWD CORE GLASSED TO SHELL
⅛" Y. CEDAR CHEEKS
½" Y. CEDAR CHEEKS
¼" Y. CEDAR FRAM.
SHELF
¼" Y. CEDAR CHEEKS (P&S)
¼" Y. CEDAR
DECK BEAM
½" PLYWD CORE
SECTION A-A
TYP. BULKH'D CONNECT'N
HALF SIZE
1" x 1¼" FIR Y. CEDAR OR FIR SAWN TO SHAPE - EPOXY GLUE & SCREW
TYP. SECTION THRU PLANKING
FULL SIZE
0 5' 10'
30MAY92 BALLAST, ENGINE BEDS, APRON, STEM, KEEL BOLTS, FLOORS, PLANKING
DATE ITEM(S) REVISED
20' CATBOAT
FOR: HERITAGE BOAT CO.
DATE: 8-10-77
SCALE: ¾" = 1'-0" & AS NOTED
WOOD DECK CONSTR'N
LOA 20'-0"
LWL 18'-0"
BEAM 8'-0"
DRAFT 3'-6"
FREEBOARD
FWD 3'-7½"
LEAST 2'-0"
AFT 2'-6"
JAY R. BENFORD
606 S. TALBOT ST - SUITE 1
ST. MICHAELS, MD 21663
(410) 745-3235
137-12 PRO

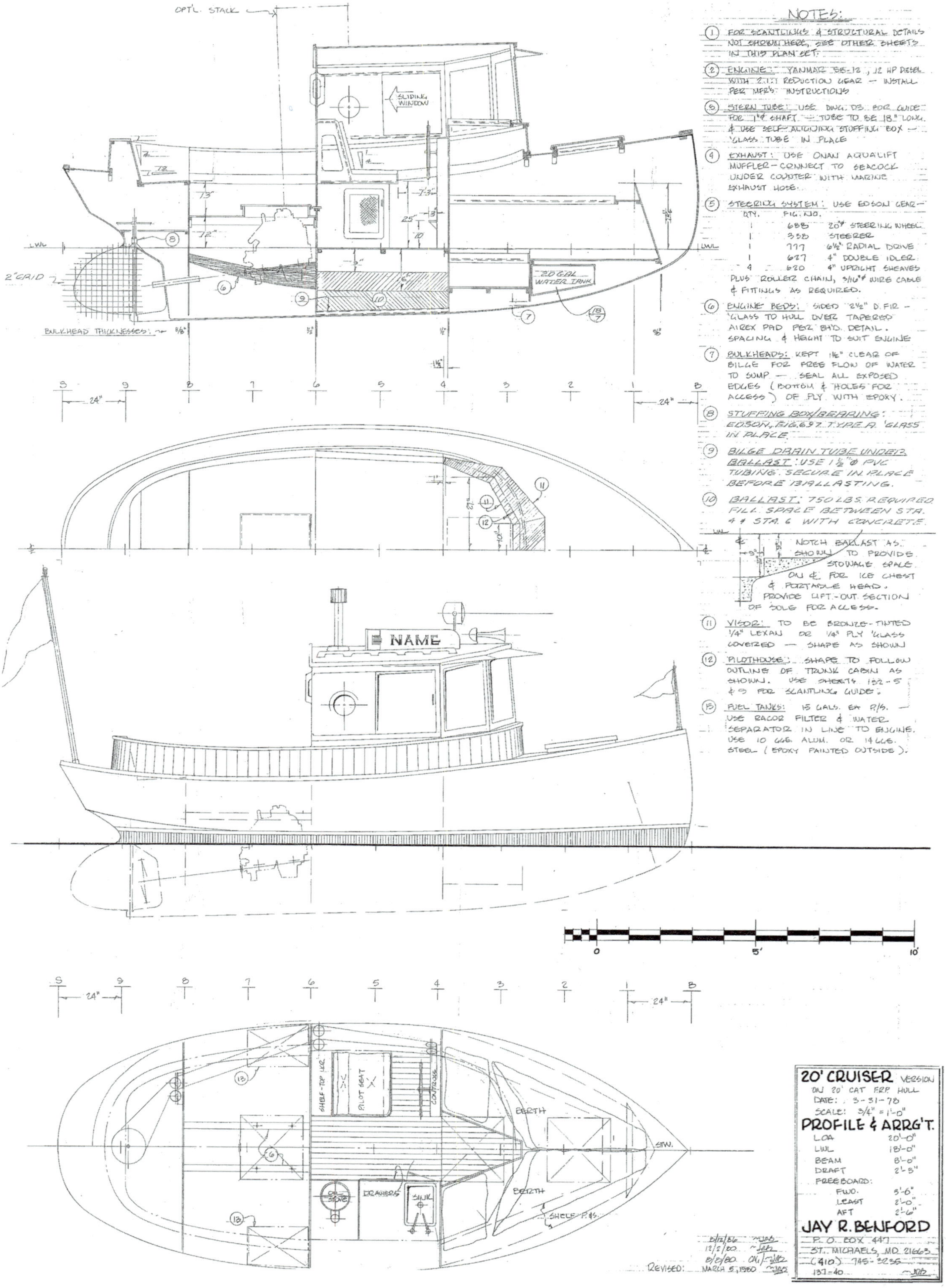
NOTES:
(1) FOR SCANTLINGS & STRUCTURAL DETAILS NOT SHOWN HERE, SEE OTHER SHEETS IN THIS PLAN SET.
(2) ENGINE: YANMAR SB-12, 12 HP DIESEL WITH 2.1:1 REDUCTION GEAR — INSTALL PER MFR'S INSTRUCTIONS
(3) STERN TUBE: USE DWG. D3 FOR GUIDE FOR 1"Ø SHAFT — TUBE TO BE 18" LONG. & USE SELF-ALIGNING STUFFING BOX — 'GLASS TUBE IN PLACE
(4) EXHAUST: USE ONAN AQUALIFT MUFFLER — CONNECT TO SEACOCK UNDER COUNTER WITH MARINE EXHAUST HOSE.
(5) STEERING SYSTEM: USE EDSON GEAR —
QTY. FIG. NO.
1 688 20"Ø STEERING WHEEL
1 358 STEERER
1 777 6½" RADIAL DRIVE
1 627 4" DOUBLE IDLER
4 620 4" UPRIGHT SHEAVES
PLUS ROLLER CHAIN, 3/16"Ø WIRE CABLE & FITTINGS AS REQUIRED.
(6) ENGINE BEDS: SIDED 2½" D. FIR — 'GLASS TO HULL OVER TAPERED AIREX PAD PER BH'D. DETAIL. SPACING & HEIGHT TO SUIT ENGINE
(7) BULKHEADS: KEPT 1½" CLEAR OF BILGE FOR FREE FLOW OF WATER TO SUMP — SEAL ALL EXPOSED EDGES (BOTTOM & HOLES FOR ACCESS) OF PLY WITH EPOXY.
(8) STUFFING BOX/BEARING: EDSON, FIG. 697 TYPE A. 'GLASS IN PLACE.
(9) BILGE DRAIN TUBE UNDER BALLAST: USE 1½"Ø PVC TUBING. SECURE IN PLACE BEFORE BALLASTING.
(10) BALLAST: 750 LBS. REQUIRED. FILL SPACE BETWEEN STA. 4 & STA. 6 WITH CONCRETE.
NOTCH BALLAST AS SHOWN TO PROVIDE STOWAGE SPACE ON ℄ FOR ICE CHEST & PORTABLE HEAD. PROVIDE LIFT-OUT SECTION OF SOLE FOR ACCESS.
(11) VISOR: TO BE BRONZE-TINTED 1/4" LEXAN OR 1/4" PLY 'GLASS COVERED — SHAPE AS SHOWN
(12) PILOTHOUSE: SHAPE TO FOLLOW OUTLINE OF TRUNK CABIN AS SHOWN. USE SHEETS 132-5 & 9 FOR SCANTLING GUIDE.
(13) FUEL TANKS: 15 GALS. EA P/S. — USE RACOR FILTER & WATER SEPARATOR IN LINE TO ENGINE. USE 10 GAGE ALUM. OR 14 GAGE STEEL (EPOXY PAINTED OUTSIDE).
OPT'L. STACK
SLIDING WINDOW
2"GRID
LWL
20 GAL WATER TANK
BULKHEAD THICKNESSES
NAME
SHELF-TOP LKR
PILOT SEAT
CONTROLS
BERTH
STW.
DRAWERS
SINK
SHELF P/S
20' CRUISER VERSION ON 20' CAT F.R.P. HULL
DATE: 3-31-78
SCALE: 3/4" = 1'-0"
PROFILE & ARRG'T.
LOA 20'-0"
LWL 18'-0"
BEAM 8'-0"
DRAFT 2'-3"
FREEBOARD:
FWD. 3'-6"
LEAST 2'-0"
AFT 2'-6"
JAY R. BENFORD
P. O. BOX 447
ST. MICHAELS, MD 21663
(410) 745-3235
137-40
REVISED: MARCH 5, 1980
8/8/80
12/2/80
5/12/86

Chapter 5
22' Gunkholer

Design Number 15
1965

When I apprenticed with John Atkin, in 1962 to 1963, one of their designs that I thought was a very practical and affordable solution to a modest cruising boat was a slightly larger sailing skiff like this. After I left, this memory provided the inspiration for this new design.

With the hull form like a large sailing skiff, these simple cruising vessels are intended for plywood construction. Carvel planking is also an option. The 22' Gunkholer offers five rig options with sail areas ranging from 199 to 330 square feet. She sleeps two or three. The keel versions could be fitted with a small inboard diesel. An outboard of about five to ten horsepower will also provide adequate and inexpensive power.

Particulars:	22'	22'	22'V	22'V
Units	Imperial	Metric	Imperial	Metric
Length overall	22'-0"	6.71 m	22'-0"	6.71 m
Length designed waterline	20'-0"	6.10 m	20'-0"	6.10 m
Beam	8'-0"	2.44 m	8'-0"	2.44 m
Draft (cb up/down)	1'-4"/3'-10"	0.41/1.17 m	3'-0"	0.91 m
Freeboard: Forward	3'-10"	1.17 m	4'-0"	1.22 m
Least	2'-0"	0.61 m	2'-4"	0.71 m
Aft	2'-6"	0.76 m	3'-0"	0.91 m
Displacement*, cruising trim.	4,750 lbs.	2,155 kg.	4,500 lbs.	2,041 kg.
Displacement-length ratio	265		253	
Ballast, pounds	900	408 kg.	1,260	572 kg.
Ballast ratio	19%		28%	
Sail area	199 to 330 ft.^{2}	18.49 to 30.66 m^2	310 ft.^{2}	28.80 m^2
Sail area-displacement ratio	11.27 to 18.69		18.20	
Prismatic coefficient			.547	
Water tankage, gallons	20	76 liters	20	76 liters
Fuel tankage, gallons	10	38 liters	10	38 liters
Headroom	4'-3" to 5'-1"		5'-1" to 5'-3"	

*CAUTION: The figure for displacement quoted here is for the boat in cruising trim. That is, with the fuel and water tanks filled, the crew on board, as well as the crews' gear and stores in the lockers. This should not be confused with the "shipping weight" often quoted as "displacement" by some manufacturers. This should be taken into account when comparing figure and ratios between this and other designs.

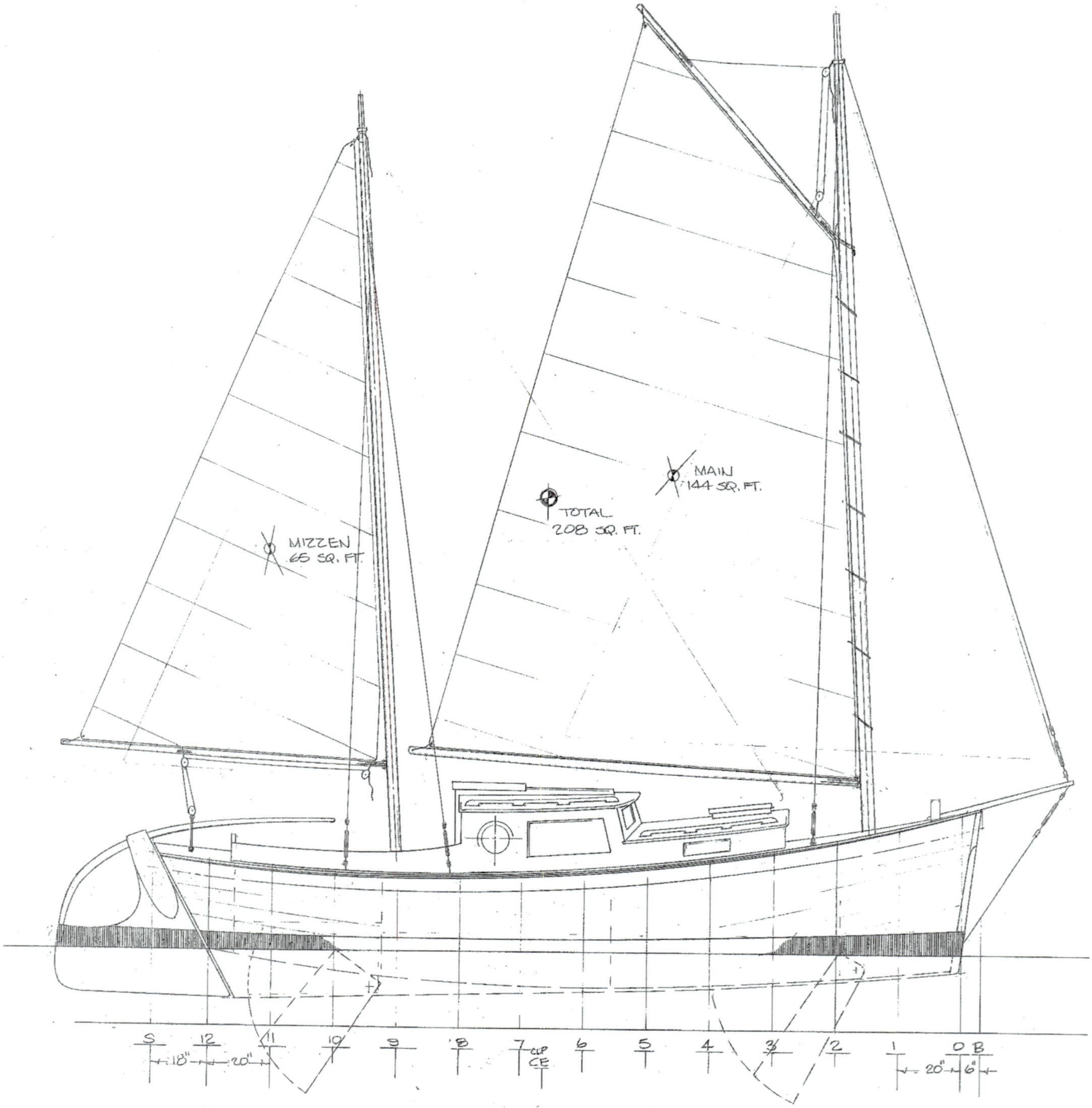
MIZZEN
65 SQ. FT.
TOTAL
208 SQ. FT.
MAIN
144 SQ. FT.
S
12
11
10
9
8
7
CLP
CE
6
5
4
3
2
1
0
B
18"
20"
20"
6"

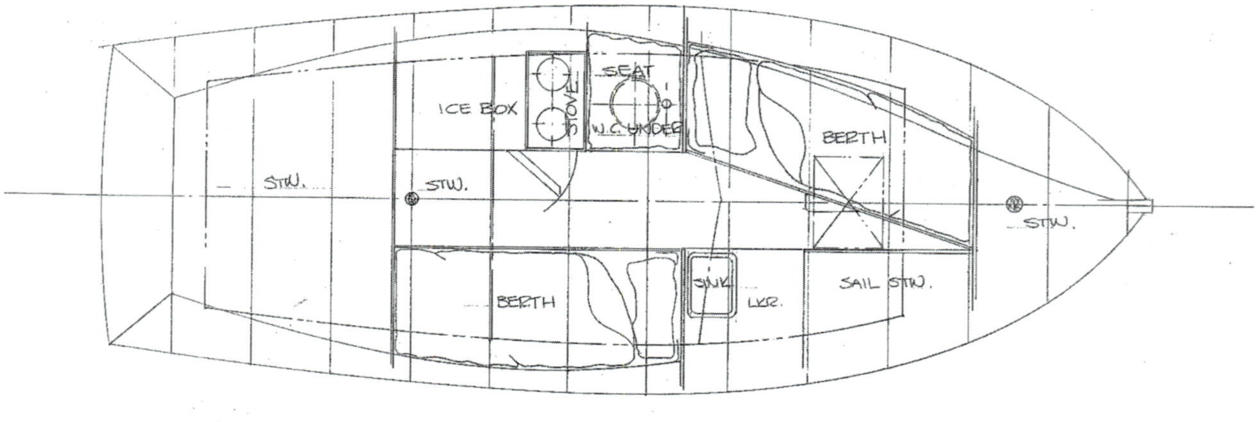
ICE BOX
STOVE
SEAT
W.C. UNDER
BERTH
STW.
STW.
STW.
BERTH
SINK
LKR.
SAIL STW.

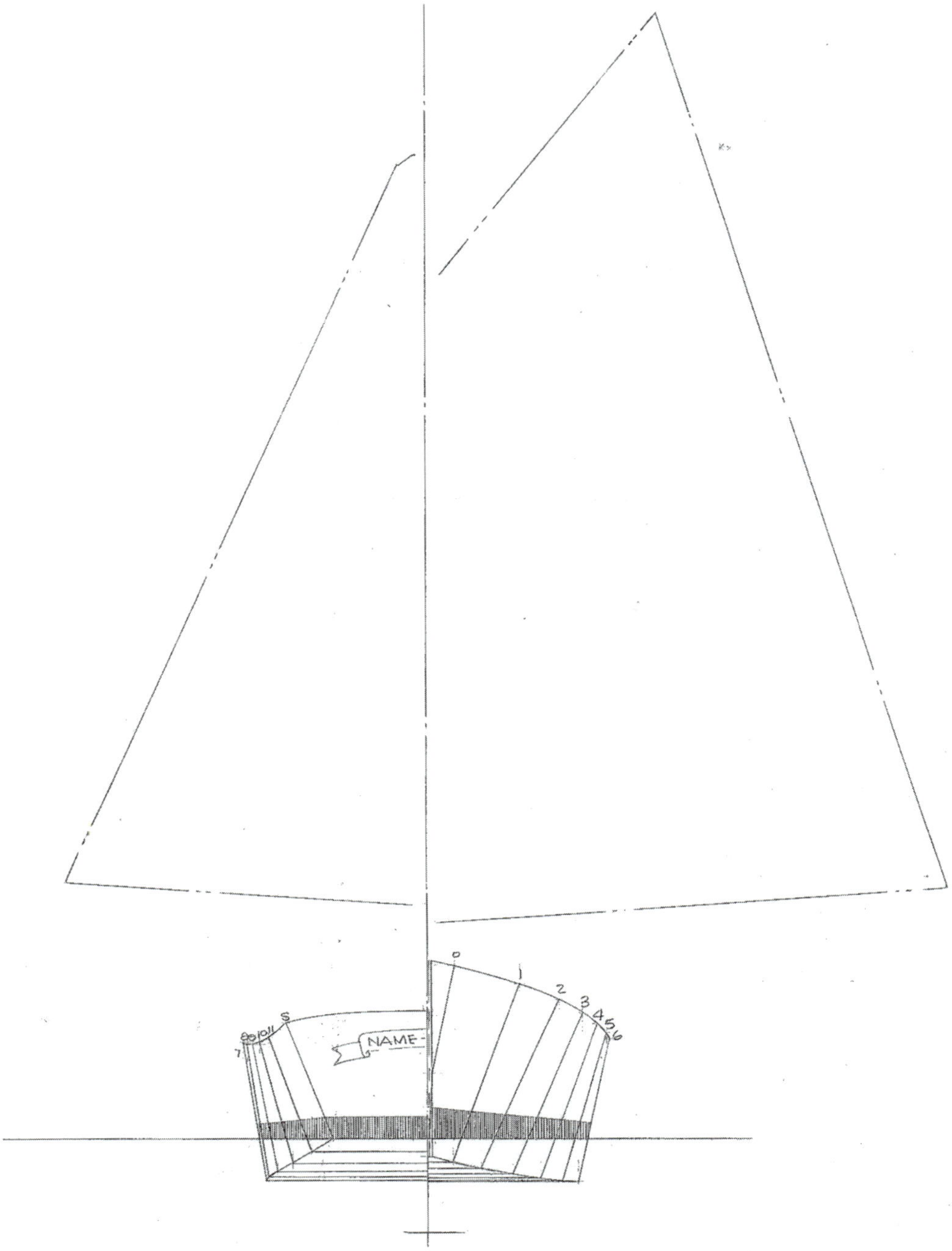

UBO *built in the Pacific Northwest from the V-Bottom version. See page 96 for drawings.*

22' GUNKHOLER
PRELIMINARY STUDY
DATE: SEPT. 15, 1965
SCALE: ½" = 1'-0"

PROFILE & ARRG'T.

LOA	22'-0"
LWL	20'-0"
BEAM	8'-0"
DRAFT	15"/3'-10"
FREEBOARD:	
FWD.	3'-10"
LEAST	2'-0"
AFT	2'-6"

JAY R. BENFORD
605 S. TALBOT ST, - SUITE 1
ST. MICHAELS, MD 21663
(410) 745-3235

1-15 JRB

STATION	B	0	1	2	3	4	5	6	7	8	9	10	11	12	S
HEIGHTS															
BASE TO SKEG	-	1'-6½"	1'-4½"	1'-3"	1'-1¾"	1'-0¾"	1'-0"	–		STRAIGHT		–	0'-8½"		
" " BOTTOM		1'-7⅜"	1'-5⅜"	1'-3⅞"	1'-2⅝"	1'-1⅝"	1'-0⅞"	1'-0¾"	1'-1"	1'-1¾"	1'-3¼"	1'-5¼"	1'-8¼"	2'-0"	
LWL TO BOOTTOP		0'-9¾"		0'-8¼"		0'-7⅜"		0'-6⅝"		0'-6⅛"		0'-6⅛"		0'-6¾"	0'-7¼"
" " WALE		3'-2¾"	2'-10¼"	2'-6⅛"	2'-2¾"	2'-0⅛"	1'-9¾"	1'-8⅛"	1'-7⅛"	1'-6¾"	1'-6⅞"	1'-7½"	1'-9"	1'-11"	2'-1½"
" " SHEER	3'-10"	3'-8¾"	3'-4⅛"	3'-0⅛"	2'-8⅝"	2'-5¾"	2'-3⅜"	2'-1¾"	2'-0¾"	2'-0¼"	2'-0¼"	2'-0⅞"	2'-2¼"	2'-4⅛"	2'-6⅜"
(OPTIONAL KEEL)		1'-6½"	1'-4½"	1'-3"	1'-1¾"	ST.	-0'-6"		- STRAIGHT -		-0'-6"	ST.	0'-0⅝"		
HALF-BREADTHS															
CHINE		0'-1¾"	0'-6¾"	1'-2"	1'-10⅜"	2'-5¾"	2'-11⅝"	3'-3¾"	3'-6"	3'-5⅜"	3'-3"	2'-11"	2'-5⅞"	2'-0"	
EDGE OF DECK (SHEER)		0'-6⅛"	2'-0½"	2'-10⅜"	3'-5"	3'-9⅛"	3'-11¼"	4'-0"	3'-11⅞"	3'-11"	3'-9½"	3'-7¾"	3'-5½"	3'-3"	3'-0⅜"

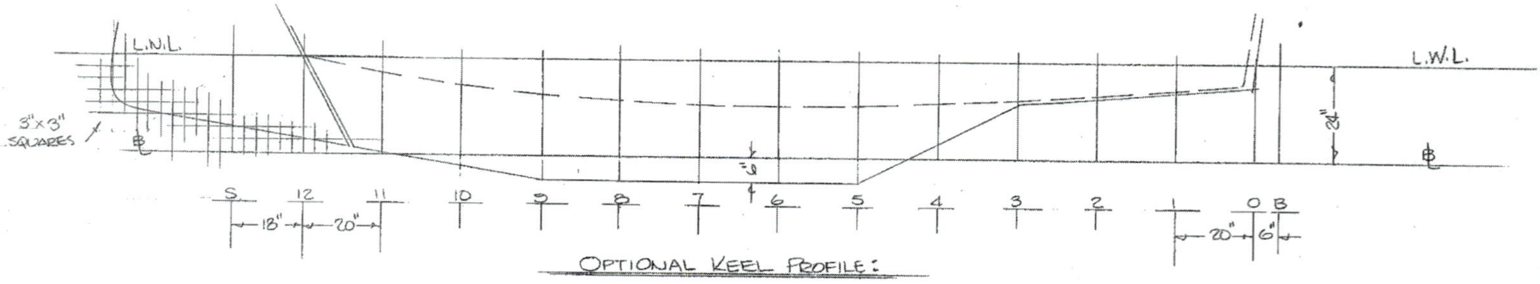

OPTIONAL KEEL PROFILE:

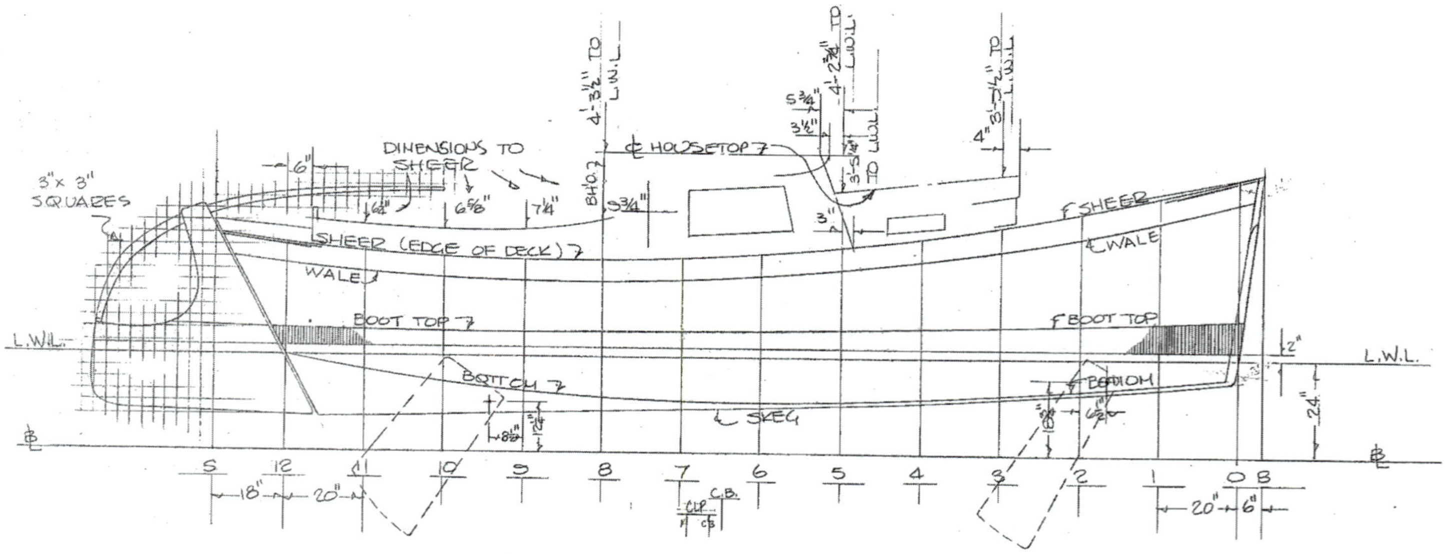

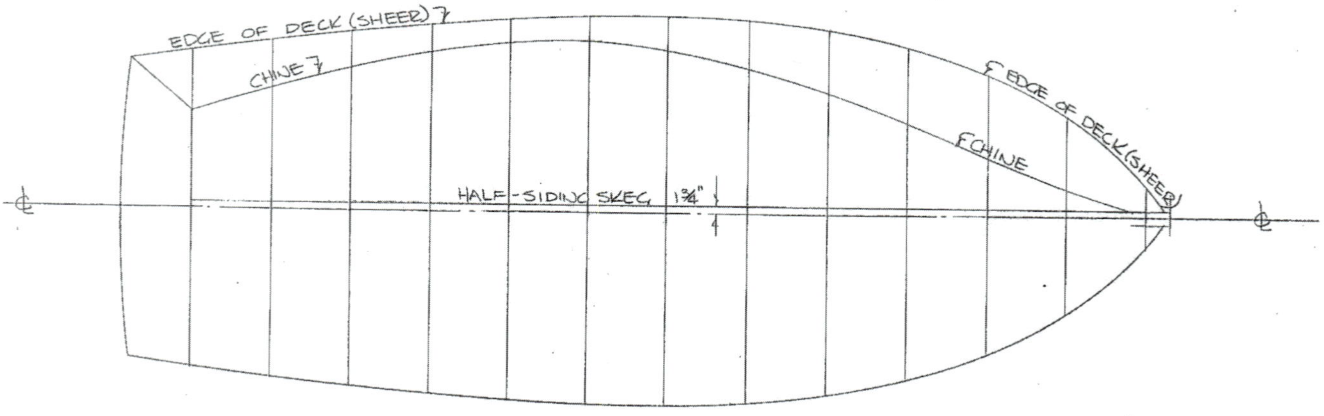

Bow view of **UBO.** *She has had bulwarks added, necessitating the channels to hold the chainplates from pressing inward on the bulwarks.*

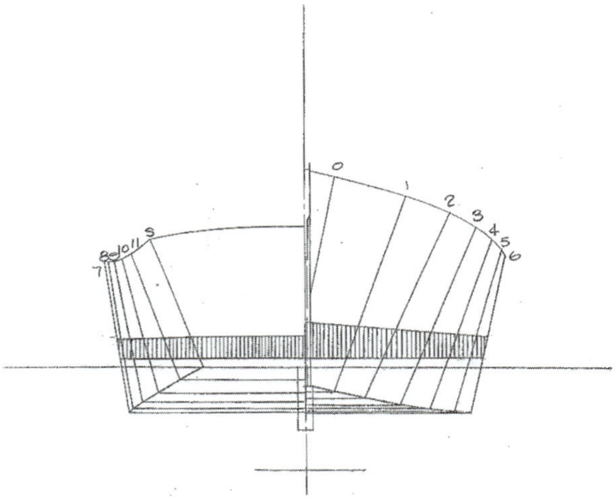

NOTES:

1. LINES MUST BE LOFTED FULL SIZE - DO NOT SCALE PRINT & USE TABLE OF OFFSETS & DIMENSIONS ON DRAWING
2. ANY ALTERATION TO THIS DRAWING RELIEVES THE DESIGNER FROM RESPONSIBILITY
3. IT IS UNDERSTOOD THAT NO MORE THAN ONE BOAT WILL BE BUILT FROM THIS SET OF PLANS WITHOUT THE CONSENT OF THE DESIGNER
4. LINES & OFFSETS TO OUTSIDE OF PLANKING IN FEET & INCHES

DISPL ~ 4750 LBS.

22' GUNKHOLER
KEEL & C.B. VERSIONS
DATE: NOV. 2, 1965
SCALE: ½" = 1'-0"

LINES & OFFSETS

LOA	22'-0"
LWL	20'-0"
BEAM	8'-0"
DRAFT	15" OR 30"
FREEBOARD:	
FWD.	3'-10"
LEAST	2'-0"
AFT	2'-6"

JAY R. BENFORD
605 S. TALBOT ST, SUITE 1
ST. MICHAELS, MD 21663
(410) 745-3235

2-15 ~JRB

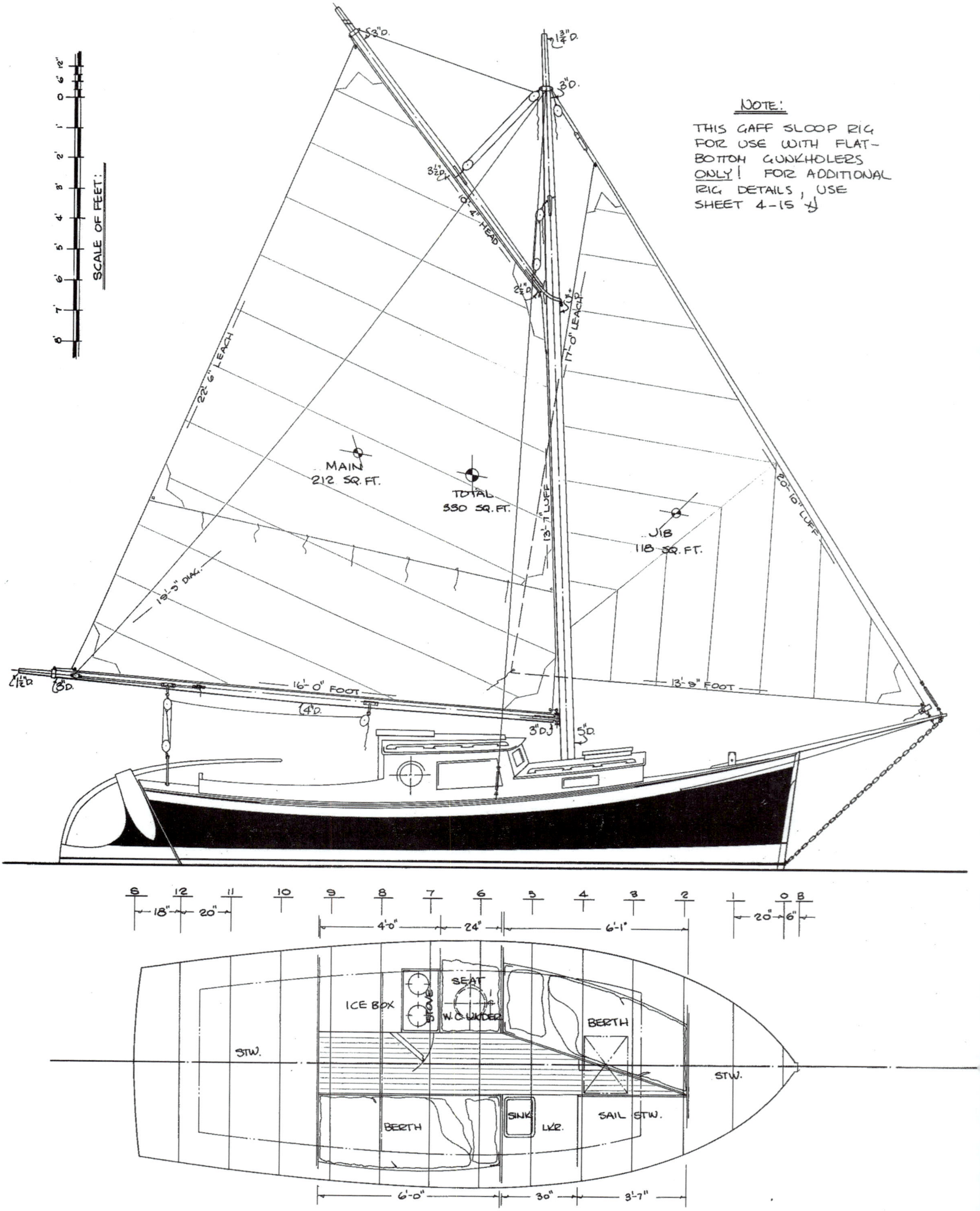

SCALE OF FEET:
NOTE:
THIS GAFF SLOOP RIG FOR USE WITH FLAT-BOTTOM GUNKHOLERS ONLY! FOR ADDITIONAL RIG DETAILS, USE SHEET 4-15
MAIN 212 SQ. FT.
TOTAL 330 SQ. FT.
JIB 118 SQ. FT.
22'-6" LEACH
19'-5" DIAG.
16'-0" FOOT
13'-9" FOOT
13'-7" LUFF
17'-0" LEACH
20'-10" LUFF
10'-4" HEAD
ICE BOX
STOVE
SEAT
W.C. UNDER
BERTH
STW.
SINK
LKR.
SAIL STW.
6'-0"
30"
3'-7"
4'-0"
24"
6'-1"
18"
20"

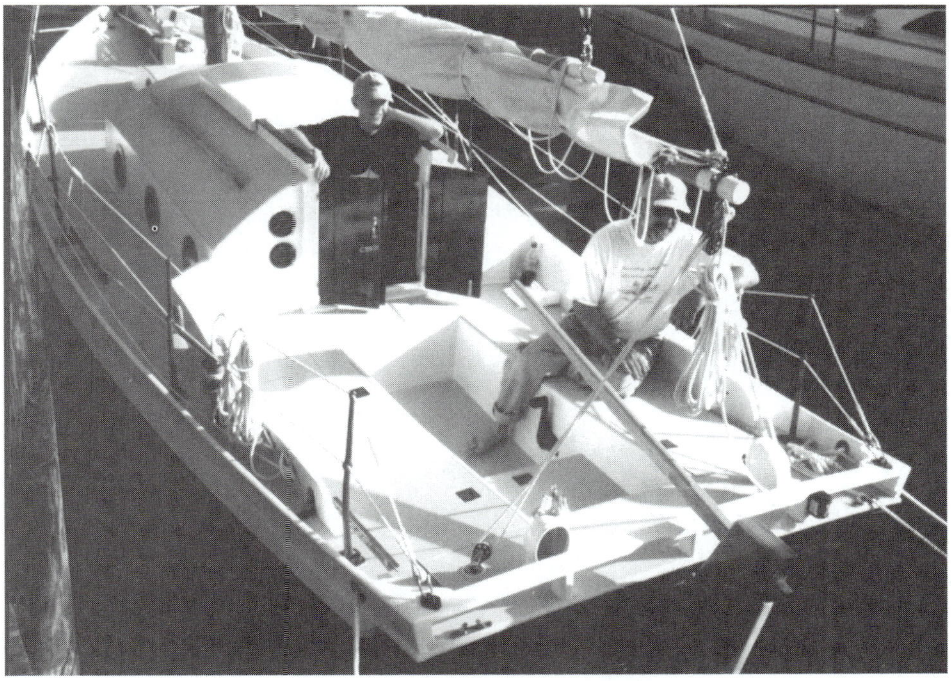

Miss Becky *with captain and crew aboard in Michigan.*

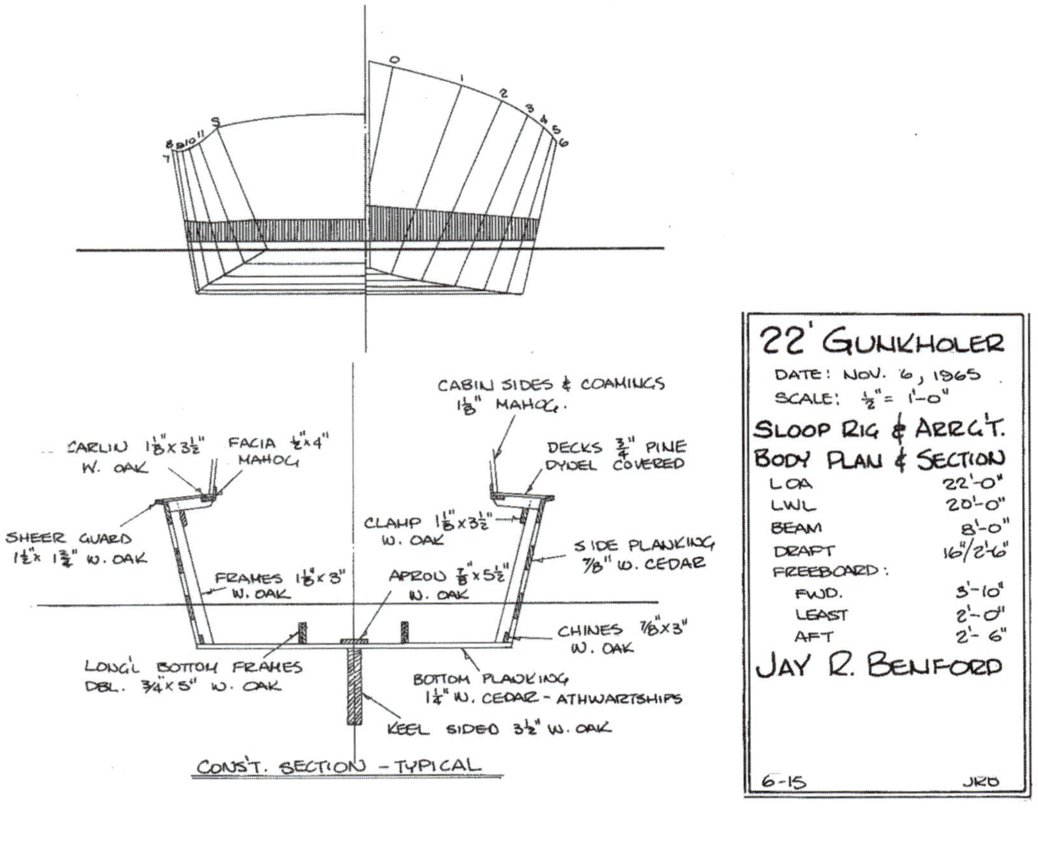

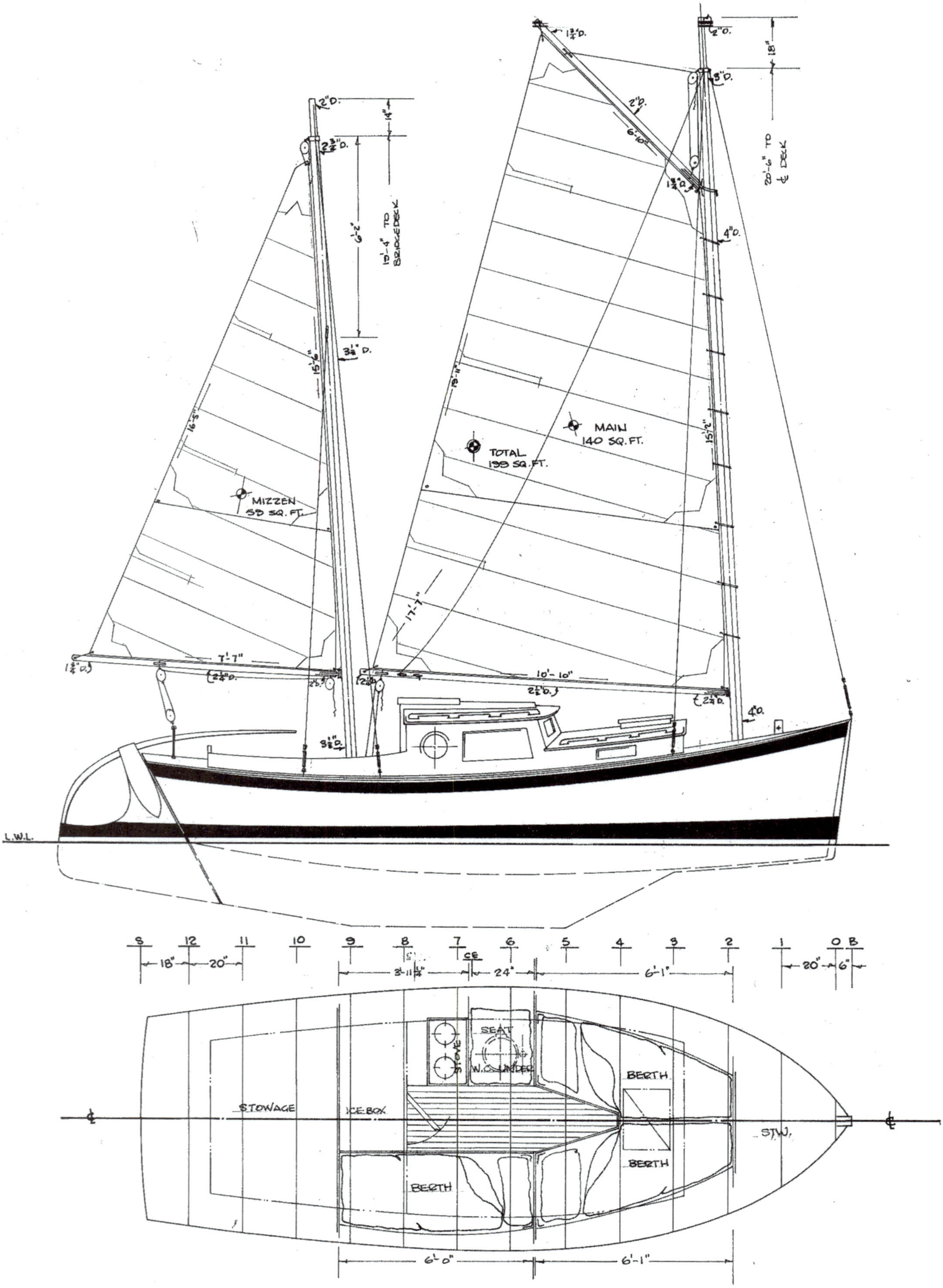

2"D.
14"
2 3/4"D.
6'-2"
13'-4" TO BRIDGEDECK
3 1/2" D.
15'1/2"
16'5"
MIZZEN
58 SQ. FT.
7'-7"
1 3/4"D.
2 1/4"D.
3 1/2"D.
1 3/4"D.
2"D.
6'-10"
2"D.
3"D.
18"
20'-6" TO ℄ DECK
1 3/4"D.
4"D.
19'-11"
MAIN
140 SQ. FT.
TOTAL
198 SQ. FT.
15'1/2"
17'-7"
10'-10"
2 1/4"D.
2 1/4"D.
4"D.
L.W.L.
S
12
11
10
9
8
7
6
5
4
3
2
1
0
B
18"
20"
3'-11 1/4"
CE
24"
6'-1"
20"
6"
SEAT
STOVE
W.C. UNDER
BERTH
STOWAGE
ICE BOX
STW.
BERTH
BERTH
℄
℄
6'-0"
6'-1"

NOTES:

1. MAST RAKES ARE: (IN HEIGHT ABOVE DECK)

CATBOAT	13½"
PIROGUE MAIN	14"
" MIZZEN	15"

2. STICK LENGTHS ARE:

CATBOAT BOOM	17'-6"
" GAFF	12'-2"
PIROGUE MAIN BOOM	11'-4"
" " GAFF	7'-4"
" MIZZEN BOOM	8'-2"

3. ALL SHROUDS ARE 5/32" ϕ S.S. 1 X 19 WIRE WITH SWAGED ENDS – TURNBUCKLES 5/16" BRZ.
4. SHEETS & HALYARDS 3/8" ϕ DACRON – ABOUT 240' REQUIRED FOR CATBOAT – 260' FOR PIROGUE
5. SPARS ARE SOLID SITKA SPRUCE – CIRCULAR IN CROSS SECTION
6. SAILS ARE 5 OZ. DACRON – REEF POINTS AS SHOWN – NO ROACH

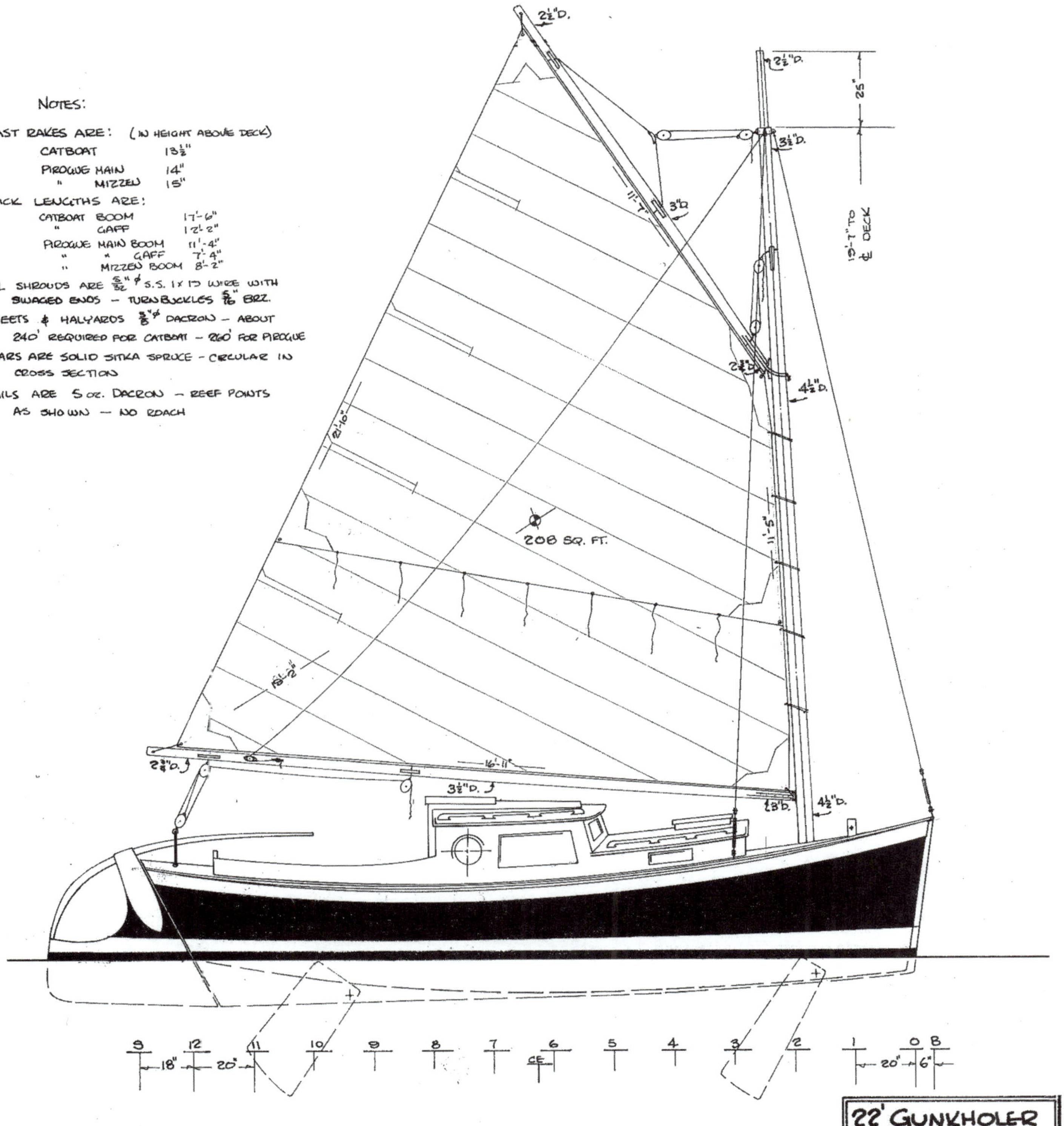

NOTES:

1. STOVE IS WILCOX-CRITTENDEN FIG. B256 OR EQUAL
2. W.C. IS WILCOX CRITTENDEN FIG. 1450 "THIRD-MATE" OR EQUAL
3. BERTH CUSHIONS 4" FOAM – "NAUGHYDE" OR FABRIC COVERS TO SUIT
4. ICE BOX MAY BE BUILT IN WITH 2" STYROFOAM INSULATION OR A PORTABLE CAMP CHEST MAY BE USED

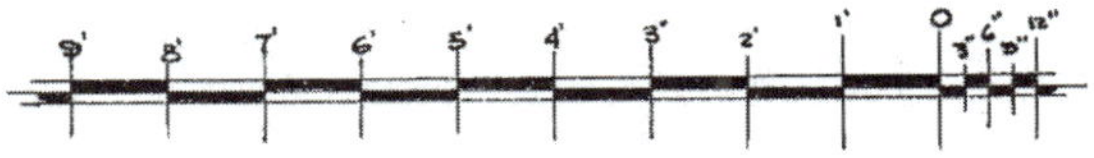

SCALE IN FEET:

22' GUNKHOLER

PIROGUE & CAT RIGS

DATE: JAN. 21, 1966

SCALE: ½" = 1'-0"

SAIL PLANS & ARRG'T.

LOA	22'-0"
LWL	20'-0"
BEAM	8'-0"
DRAFT	16"/2'-6"
FREEBOARD:	
FWD.	3'-10"
LEAST	2'-0"
AFT	2'-6"

JAY R. BENFORD

NAVAL ARCHITECT

9-15 JRB

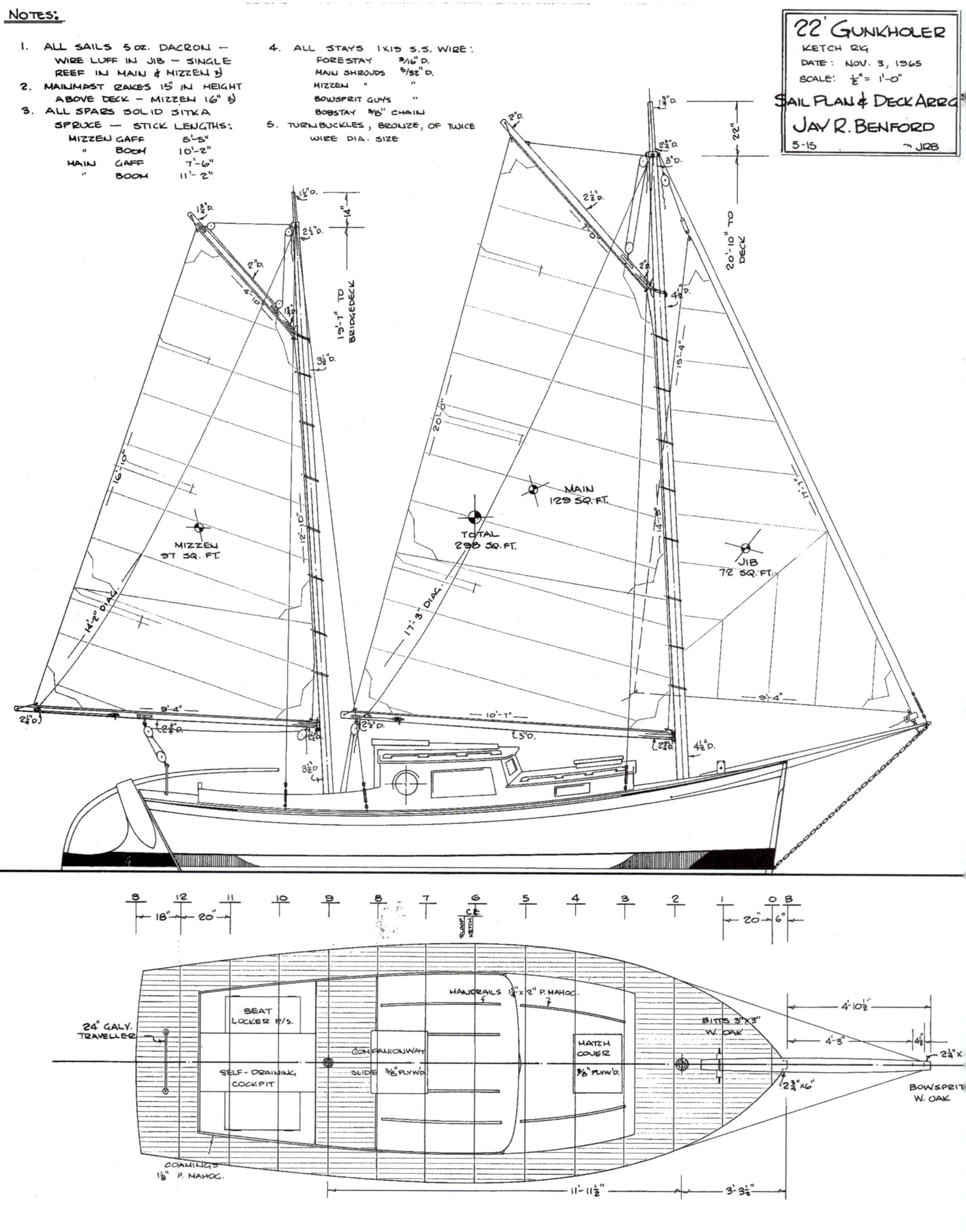
NOTES:
1. ALL SAILS 5 OZ. DACRON — WIRE LUFF IN JIB — SINGLE REEF IN MAIN & MIZZEN
2. MAINMAST RAKES 15" IN HEIGHT ABOVE DECK — MIZZEN 16"
3. ALL SPARS SOLID SITKA SPRUCE — STICK LENGTHS:
MIZZEN GAFF 5'-5"
" BOOM 10'-2"
MAIN GAFF 7'-6"
" BOOM 11'-2"
4. ALL STAYS 1X19 S.S. WIRE:
FORESTAY 3/16" D.
MAIN SHROUDS 5/32" D.
MIZZEN " "
BOWSPRIT GUYS "
BOBSTAY 3/8" CHAIN
5. TURNBUCKLES, BRONZE, OF TWICE WIRE DIA. SIZE
22' GUNKHOLER
KETCH RIG
DATE: NOV. 3, 1965
SCALE: 1/2" = 1'-0"
SAIL PLAN & DECK ARR'G
JAY R. BENFORD
5-15
JRB
MIZZEN
97 SQ. FT.
MAIN
129 SQ. FT.
TOTAL
298 SQ. FT.
JIB
72 SQ. FT.
19'-7" TO BRIDGEDECK
20'-10" TO DECK
14'-2" DIAG.
17'-3" DIAG.
HANDRAILS 1 1/2" x 2" P. MAHOG.
SEAT LOCKER P/S.
24" GALV. TRAVELLER
SELF-DRAINING COCKPIT
COMPANIONWAY SLIDE 3/8" PLYW'D.
HATCH COVER 3/8" PLYW'D.
BITTS 3" x 3" W. OAK
BOWSPRIT W. OAK
COAMINGS 1 1/8" P. MAHOG.
11'-11 1/2"
3'-3 1/2"
4'-10 1/2"
4'-3"

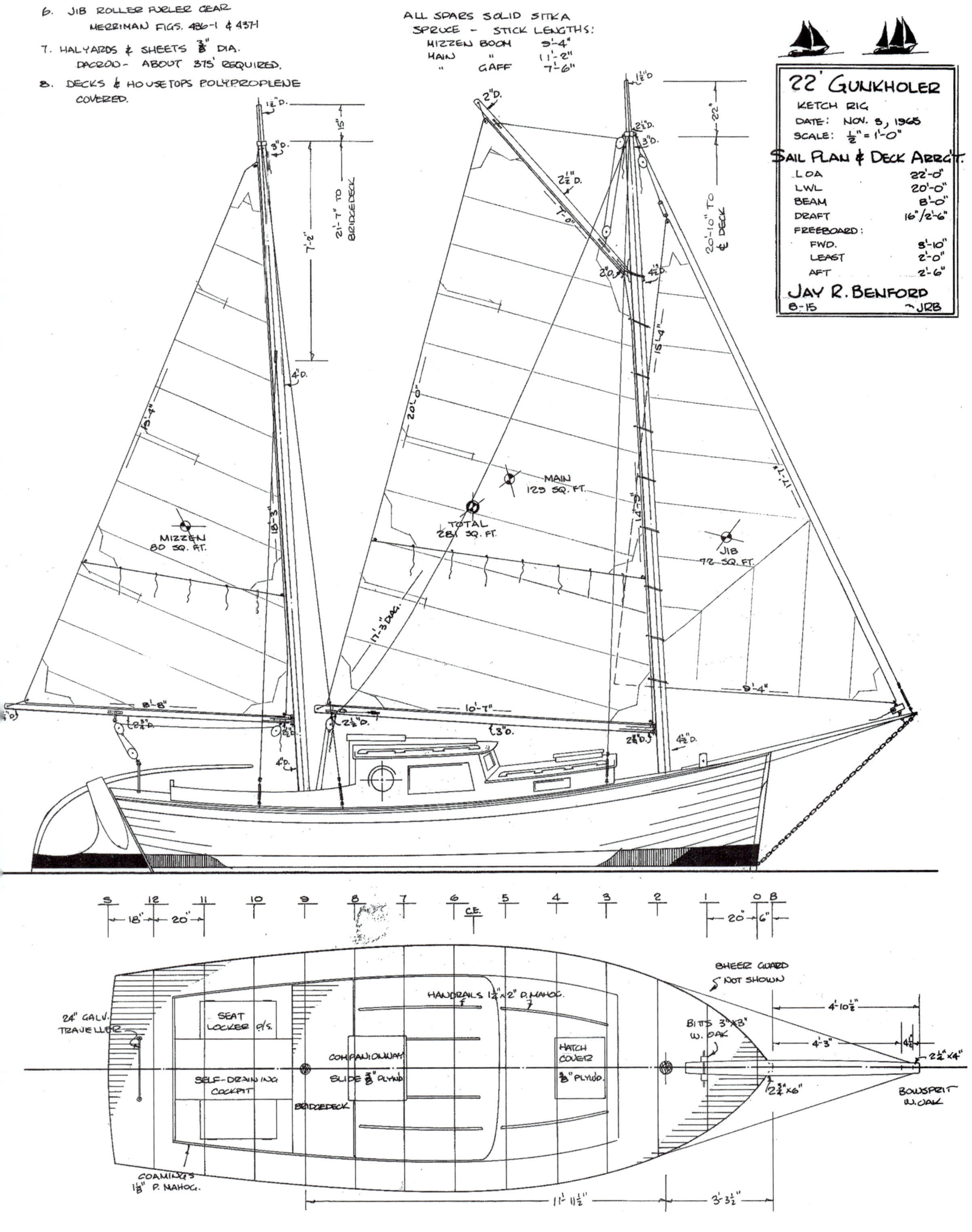
6. JIB ROLLER FURLER GEAR
MERRIMAN FIGS. 486-1 & 4371
7. HALYARDS & SHEETS 3/8" DIA.
DACRON - ABOUT 375' REQUIRED.
8. DECKS & HOUSETOPS POLYPROPLENE
COVERED.
ALL SPARS SOLID SITKA
SPRUCE - STICK LENGTHS:
MIZZEN BOOM 9'-4"
MAIN " 11'-2"
" GAFF 7'-6"
22' GUNKHOLER
KETCH RIG
DATE: NOV. 8, 1965
SCALE: 1/2" = 1'-0"
SAIL PLAN & DECK ARR'G'T.
LOA 22'-0"
LWL 20'-0"
BEAM 8'-0"
DRAFT 16"/2'-6"
FREEBOARD:
FWD. 3'-10"
LEAST 2'-0"
AFT 2'-6"
JAY R. BENFORD
8-15 ~JRB
MIZZEN
80 SQ. FT.
MAIN
125 SQ. FT.
TOTAL
281 SQ. FT.
JIB
72 SQ. FT.
21'-7" TO BRIDGEDECK
20'-10" TO ℄ DECK
17'-3" DIAG.
8'-8"
10'-7"
9'-4"
SHEER GUARD
NOT SHOWN
HANDRAILS 1 1/4" x 2" P. MAHOG.
24" GALV. TRAVELLER
SEAT LOCKER P/S.
SELF-DRAINING COCKPIT
COMPANIONWAY
SLIDE 3/8" PLYWD
BRIDGEDECK
HATCH COVER
3/8" PLYWD.
BITTS 3" X 3" W. OAK
BOWSPRIT W. OAK
COAMINGS 1 1/8" P. MAHOG.
11'-11 1/2"
3'-3 1/2"
4'-10 1/2"
4'-3"
C.E.

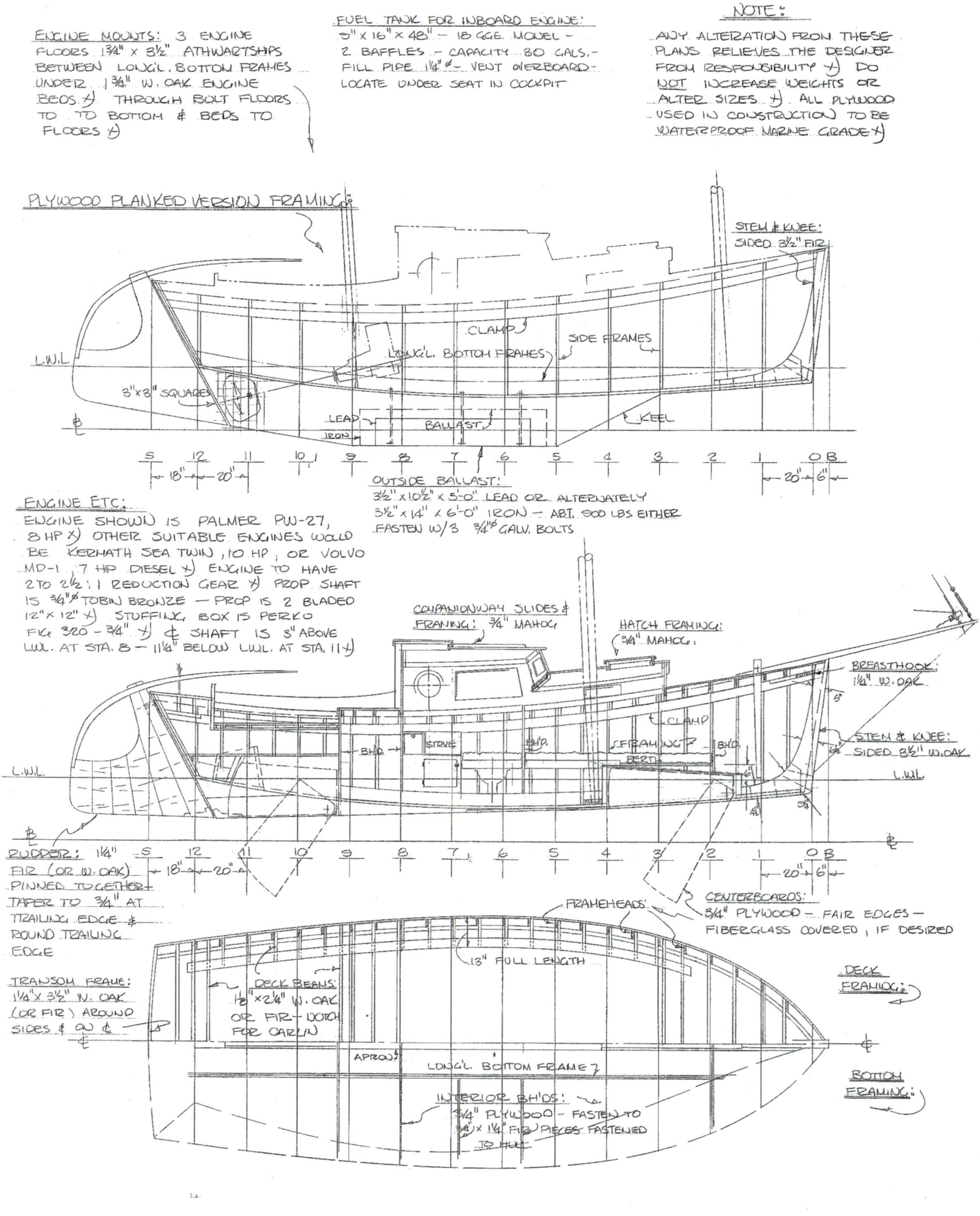

ENGINE MOUNTS: 3 ENGINE FLOORS 1¾" x 3½" ATHWARTSHIPS BETWEEN LONG'L. BOTTOM FRAMES UNDER 1¾" W. OAK ENGINE BEDS ✗) THROUGH BOLT FLOORS TO TO BOTTOM & BEDS TO FLOORS ✗)
FUEL TANK FOR INBOARD ENGINE: 5" x 16" x 48" — 18 GGE. MONEL — 2 BAFFLES — CAPACITY 30 GALS. — FILL PIPE 1¼"Ø — VENT OVERBOARD — LOCATE UNDER SEAT IN COCKPIT
NOTE: ANY ALTERATION FROM THESE PLANS RELIEVES THE DESIGNER FROM RESPONSIBILITY ✗) DO NOT INCREASE WEIGHTS OR ALTER SIZES ✗) ALL PLYWOOD USED IN CONSTRUCTION TO BE WATERPROOF MARINE GRADE ✗)
PLYWOOD PLANKED VERSION FRAMING:
STEM & KNEE: SIDED 3½" FIR
CLAMP
SIDE FRAMES
LONG'L. BOTTOM FRAMES
L.W.L.
3"x3" SQUARES
LEAD
IRON
BALLAST
KEEL
S 12 11 10 9 8 7 6 5 4 3 2 1 0 B
18" 20" 20" 6"
OUTSIDE BALLAST: 3½" x 10½" x 5'-0" LEAD OR ALTERNATELY 3½" x 14" x 6'-0" IRON — ABT. 900 LBS EITHER FASTEN W/3 ¾"Ø GALV. BOLTS
ENGINE ETC: ENGINE SHOWN IS PALMER PW-27, 8 HP ✗) OTHER SUITABLE ENGINES WOULD BE KERMATH SEA TWIN, 10 HP, OR VOLVO MD-1, 7 HP DIESEL ✗) ENGINE TO HAVE 2 TO 2½ : 1 REDUCTION GEAR ✗) PROP SHAFT IS ¾"Ø TOBIN BRONZE — PROP IS 2 BLADED 12" x 12" ✗) STUFFING BOX IS PERKO FIG. 320 - ¾" ✗) ℄ SHAFT IS 3" ABOVE LWL. AT STA. 8 — 11¼" BELOW LWL. AT STA. 11 ✗)
COMPANIONWAY SLIDES & FRAMING: ¾" MAHOG.
HATCH FRAMING: ¾" MAHOG.
BREASTHOOK: 1¼" W. OAK
CLAMP
5°
STEM & KNEE: SIDED 3½" W. OAK
BH'D.
STOVE
BH'D.
FRAMING
BH'D.
BERTH
6"
4"
L.W.L.
RUDDER: 1¼" FIR (OR W. OAK) PINNED TOGETHER TAPER TO ¾" AT TRAILING EDGE & ROUND TRAILING EDGE
S 12 11 10 9 8 7 6 5 4 3 2 1 0 B
18" 20" 20" 6"
CENTERBOARDS: ¾" PLYWOOD — FAIR EDGES — FIBERGLASS COVERED, IF DESIRED
FRAMEHEADS
13" FULL LENGTH
TRANSOM FRAME: 1¼" x 3½" W. OAK (OR FIR) AROUND SIDES & ON ℄
DECK BEAMS: 1½" x 2¼" W. OAK OR FIR — NOTCH FOR CARLIN
DECK FRAMING:
APRON
LONG'L. BOTTOM FRAME
BOTTOM FRAMING:
INTERIOR BH'DS: ¾" PLYWOOD — FASTEN TO ¾" x 1¼" FIR PIECES FASTENED TO HULL

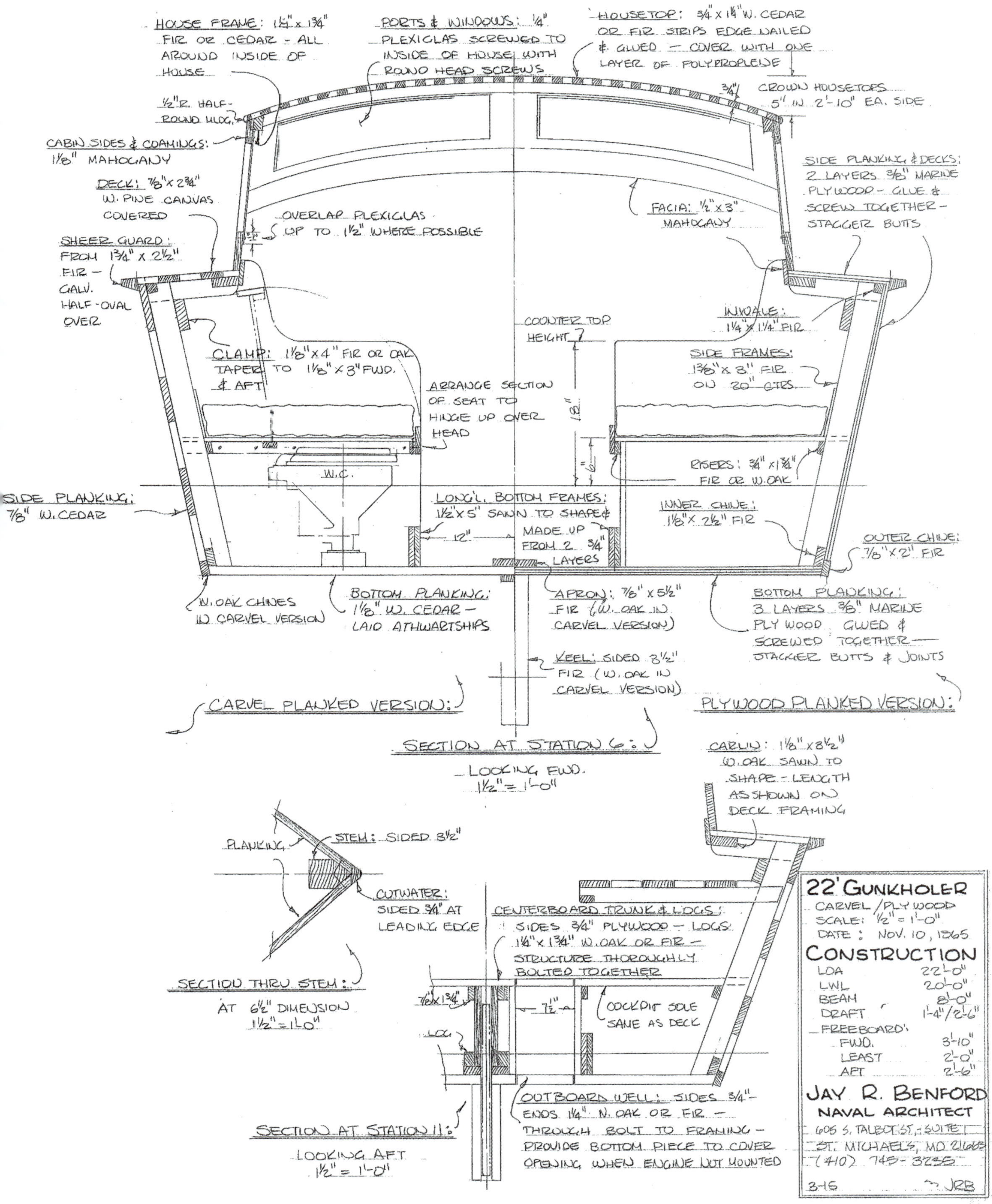
HOUSE FRAME: 1¼" x 1¾" FIR OR CEDAR - ALL AROUND INSIDE OF HOUSE
PORTS & WINDOWS: ¼" PLEXIGLAS SCREWED TO INSIDE OF HOUSE WITH ROUND HEAD SCREWS
HOUSETOP: ¾" x 1¼" W. CEDAR OR FIR STRIPS EDGE NAILED & GLUED - COVER WITH ONE LAYER OF POLYPROPLENE
½"R. HALF-ROUND MLDG.
¾"
CROWN HOUSETOPS 5" IN 2'-10" EA. SIDE
CABIN SIDES & COAMINGS: 1⅛" MAHOGANY
DECK: ⅞" x 2¾" W. PINE CANVAS COVERED
SIDE PLANKING & DECKS: 2 LAYERS ⅜" MARINE PLYWOOD - GLUE & SCREW TOGETHER - STAGGER BUTTS
FACIA: ½" x 3" MAHOGANY
OVERLAP PLEXIGLAS UP TO 1½" WHERE POSSIBLE
SHEER GUARD: FROM 1¾" x 2½" FIR - GALV. HALF-OVAL OVER
INWALE: 1¼" x 1¼" FIR
COUNTER TOP HEIGHT
CLAMP: 1⅛" x 4" FIR OR OAK TAPER TO 1⅛" x 3" FWD. & AFT
SIDE FRAMES: 1⅜" x 3" FIR ON 20" CTRS.
ARRANGE SECTION OF SEAT TO HINGE UP OVER HEAD
18"
RISERS: ¾" x 1¾" FIR OR W. OAK
W.C.
SIDE PLANKING: ⅞" W. CEDAR
LONG'L. BOTTOM FRAMES: 1½" x 5" SAWN TO SHAPE & MADE UP FROM 2 ¾" LAYERS
12"
INNER CHINE: 1⅛" x 2½" FIR
OUTER CHINE: ⅞" x 2" FIR
W. OAK CHINES IN CARVEL VERSION
BOTTOM PLANKING: 1⅛" W. CEDAR - LAID ATHWARTSHIPS
APRON: ⅞" x 5½" FIR (W. OAK IN CARVEL VERSION)
BOTTOM PLANKING: 3 LAYERS ⅜" MARINE PLY WOOD GLUED & SCREWED TOGETHER - STAGGER BUTTS & JOINTS
KEEL: SIDED 3½" FIR (W. OAK IN CARVEL VERSION)
CARVEL PLANKED VERSION:
PLYWOOD PLANKED VERSION:
SECTION AT STATION 6:
LOOKING FWD.
1½" = 1'-0"
CARLIN: 1⅛" x 8½" W. OAK SAWN TO SHAPE - LENGTH AS SHOWN ON DECK FRAMING
PLANKING
STEM: SIDED 3½"
CUTWATER: SIDED ¾" AT LEADING EDGE
SECTION THRU STEM:
AT 6½" DIMENSION
1½" = 1'-0"
CENTERBOARD TRUNK & LOGS: SIDES ¾" PLYWOOD - LOGS 1¼" x 1¾" W. OAK OR FIR - STRUCTURE THOROUGHLY BOLTED TOGETHER
⅞" x 1¾"
7½"
COCKPIT SOLE SAME AS DECK
LOG
OUTBOARD WELL: SIDES ¾" - ENDS 1¼" W. OAK OR FIR - THROUGH BOLT TO FRAMING - PROVIDE BOTTOM PIECE TO COVER OPENING WHEN ENGINE NOT MOUNTED
SECTION AT STATION 11:
LOOKING AFT
1½" = 1'-0"
22' GUNKHOLER
CARVEL/PLYWOOD
SCALE: ½" = 1'-0"
DATE: NOV. 10, 1965
CONSTRUCTION
LOA 22'-0"
LWL 20'-0"
BEAM 8'-0"
DRAFT 1'-4"/2'-6"
FREEBOARD:
FWD. 3'-10"
LEAST 2'-0"
AFT 2'-6"
JAY R. BENFORD
NAVAL ARCHITECT
605 S. TALBOT ST. - SUITE 1
ST. MICHAELS, MD 21663
(410) 745-3235
3-15
JRB

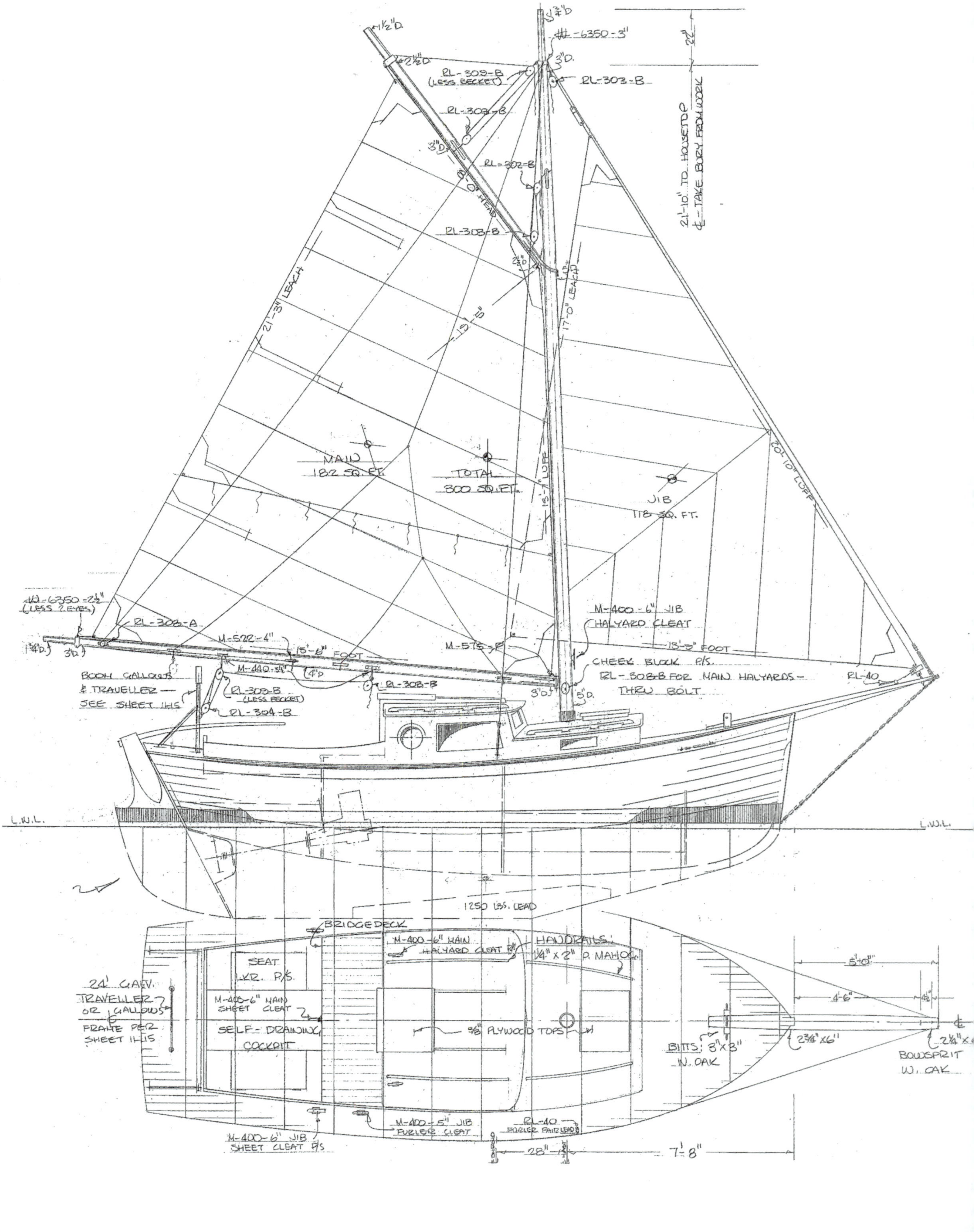
MAIN
182 SQ. FT.
TOTAL
300 SQ. FT.
JIB
118 SQ. FT.
21'-3" LEACH
17'-0" LEACH
20'-10" LUFF
15'-5" FOOT
13'-3" FOOT
8'-0" HEAD
21'-10" TO HOUSETOP
℄ - TAKE BURY FROM WORK
RL-303-B
RL-302-B
RL-303-B (LESS BECKET)
RL-308-A
RL-304-B
RL-40
M-522-4"
M-440-3¼"
M-575-F
M-400-6" JIB HALYARD CLEAT
CHEEK BLOCK P/S.
RL-308-B FOR MAIN HALYARDS - THRU BOLT
BOOM GALLOWS & TRAVELLER - SEE SHEET 11.15
L.W.L.
1250 LBS. LEAD
BRIDGEDECK
M-400-6" MAIN HALYARD CLEAT P/S
HANDRAILS: 1¼" x 2" P. MAHOG.
SEAT LKR. P/S.
24' GALV. TRAVELLER OR GALLOWS FRAME PER SHEET 11.15
M-400-6" MAIN SHEET CLEAT
SELF-DRAINING COCKPIT
5/8" PLYWOOD TOPS
BITTS: 3"x3" W. OAK
2¾"x6"
BOWSPRIT W. OAK
M-400-5" JIB FURLER CLEAT
RL-40 FURLER FAIRLEAD
M-400-6" JIB SHEET CLEAT P/S
28"
7'-8"
5'-10"
4'-6"

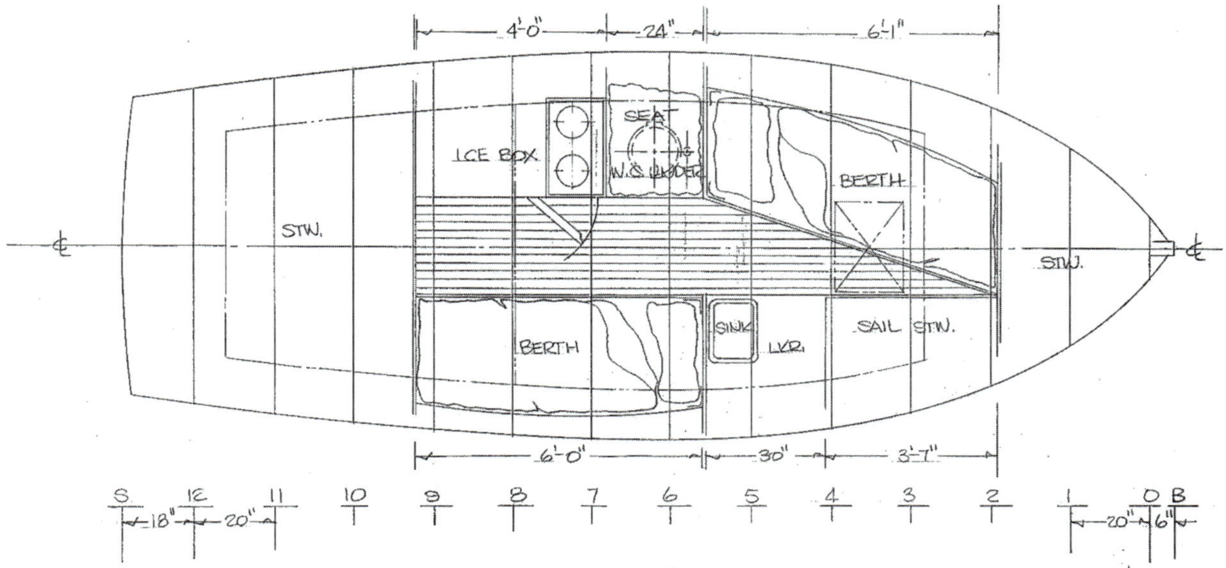

NOTES:

1. SPARS SOLID SITKA SPRUCE - CIRCULAR IN SECTION.
2. MAST RAKES 12" IN HEIGHT ABOVE DECK.
3. STICK LENGTHS ARE:
   GAFF 8' 8"
   BOOM 17'-6"
4. SAILS 3.8 OZ. DACRON - WIRE LUFF IN JIB — NO ROACH IN MAINSAIL — SINGLE ROW OF REEF POINTS IN MAIN.
5. FORESTAY, SHROUDS & BOWSPRIT STAYS 3/16"⌀ S.S. 1X19 WIRE - SWAGED ENDS — 3/8"⌀ FORGED BRONZE TURNBUCKLES.
6. JIB ROLLER FURLER GEAR MERRIMAN FIGS. 436-1 & 437-1.
7. HALYARDS & SHEETS 3/8"⌀ DACRON - ABOUT 325' REQUIRED.
8. 5/8" SAILTRACK ON MAST, BOOM, & GAFF
9. GAFF JAWS, BOOM GALLOWS & TRAVELLER & SPAR ON SHEET 11-15
10. TOPPING LIFT & LAZY JACKS 1/4" NYLON
11. HARDWARE FIGURE NUMBERS FROM MERRIMAN (M), RACE-LITE (RL) & WILCOX-CRITTENDEN (#)
12. ALL FITTINGS ON DECK THRU-BOLTED TO DOUBLERS

CONSTRUCTION:

KEEL & STEM: W. OAK SIDED 3½"

PLANKING & DECK: 2 LAYERS OF 1/4" MARINE PLYWOOD GLUED & SCREWED TOGETHER — FASTENED TO FRAMING W/ ANCHORFAST NAILS

FRAMES: 1¼" x 2¼" W. OAK ON 20" CENTERS

BALLAST: 1250 LBS. LEAD ON KEEL — 300 TO 500 LBS. INSIDE BETWEEN FRAMES 3 & 7 — POSITION & AMOUNT TO BE DETERMINED AFTER CRUISING GEAR IS ABOARD

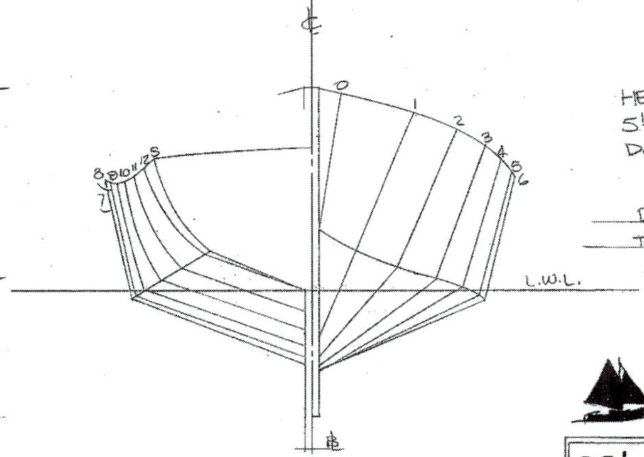

HEADROOM 5'-1" UNDER DOGHOUSE

DISPLACEMENT TO LWL — 4500 LBS

ENGINE: PALMER PW-27 8HP — 2:1 REVERSE & REDUCTION GEAR

PROP: 2 BLD 12"x12" LH - BORE FOR 3/4"⌀ SHAFT

PROP SHAFT: 3/4"⌀ TOBIN BRZ. — SHAFT ⊄ CROSSES LWL AT FR. 8 & 15 14" BELOW LWL AT FR. 12

FUEL TANK: TEMPO - BMT-12-L - 7½"x18"x24" - 12 GALS — UNDER STBD. COCKPIT SEAT — FILL THRU SIDE DECK — VENT OVERBOARD

22' GUNKHOLER

FOR: DR. C.N. STEWARD
DATE: JAN. 6, 1967
SCALE: ½" = 1'-0"

SAIL PLAN & DECK ARR'G'T.

LOA	22'-0"
LWL	20'-0"
BEAM	8'-0"
DRAFT	1'6" / 2'-6"
FREEBOARD:	
FWD	3'-10"
LEAST	2'-0"
AFT	2'-6"

JAY R. BENFORD
605 S. TALBOT ST, SUITE 1
ST. MICHAELS, MD 21663
(410) 745-3235

4-15 ~JRB

REVISED: 8/74 ~JRB

STATION	0	1	2	3	4	5	6	7	8	9	10	11	12
BL TO RABBET	4'-1¼"	2'-0¾"	1'-8¾"	1'-7"	1'-5¾"	1'-5½"	1'-5¾"	1'-6¾"	1'-8¼"	1'-10¾"	2'-2¼"	2'-6¾"	3'-0"
BL TO CHINE	4'-1¼"	3'-9"	3'-5½"	3'-2½"	3'-0¼"	2'-10¾"	2'-9¾"	2'-9½"	2'-10¼"	2'-11½"	3'-1¾"	3'-4¼"	3'-7¾"
℄ TO CHINE	0'-1¾"	0'-10¾"	1'-8¾"	2'-5¼"	2'-11½"	3'-3¾"	3'-5½"	3'-5¾"	3'-5¼"	3'-2¾"	2'-10¾"	2'-5¾"	2'-0"

NOTE: LOFT SHEER HEIGHTS & HALF BREADTHS PER SHEET 2-15

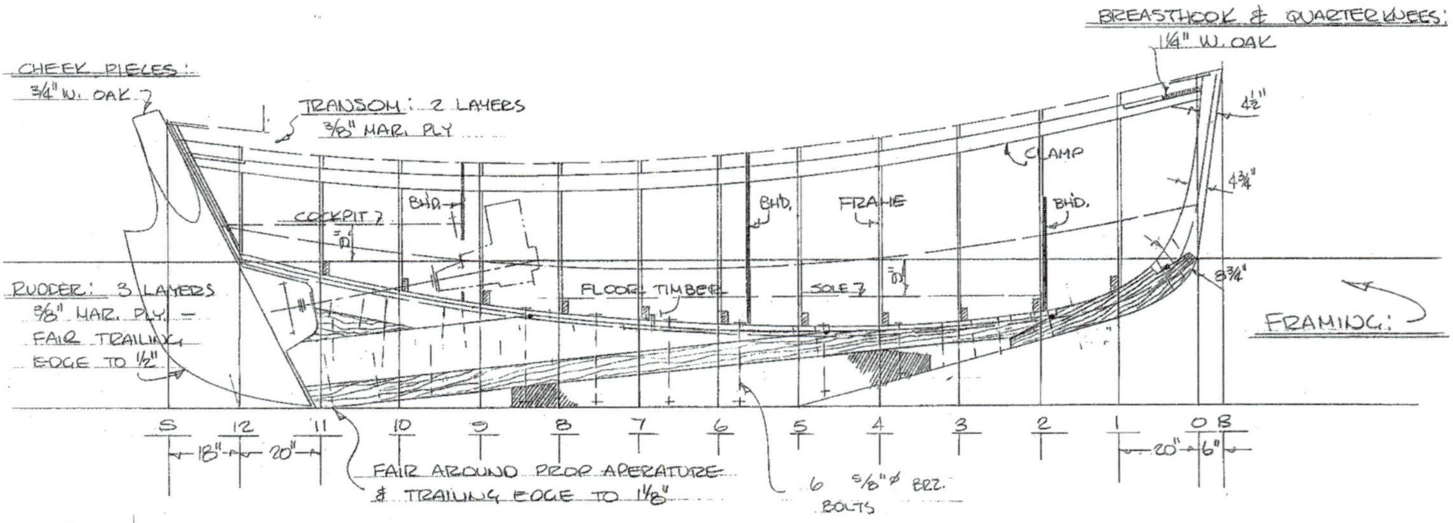

STA.	CURVATURE
8	¼"
9	⅝"
10	1¼"
11	1¾"
12	2⅛"
S	2¼"

OUTWARD CURVATURE OF STERN FRAMES MEASURED THUS:

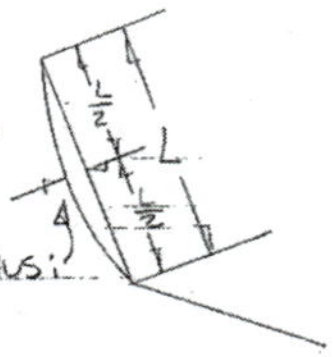

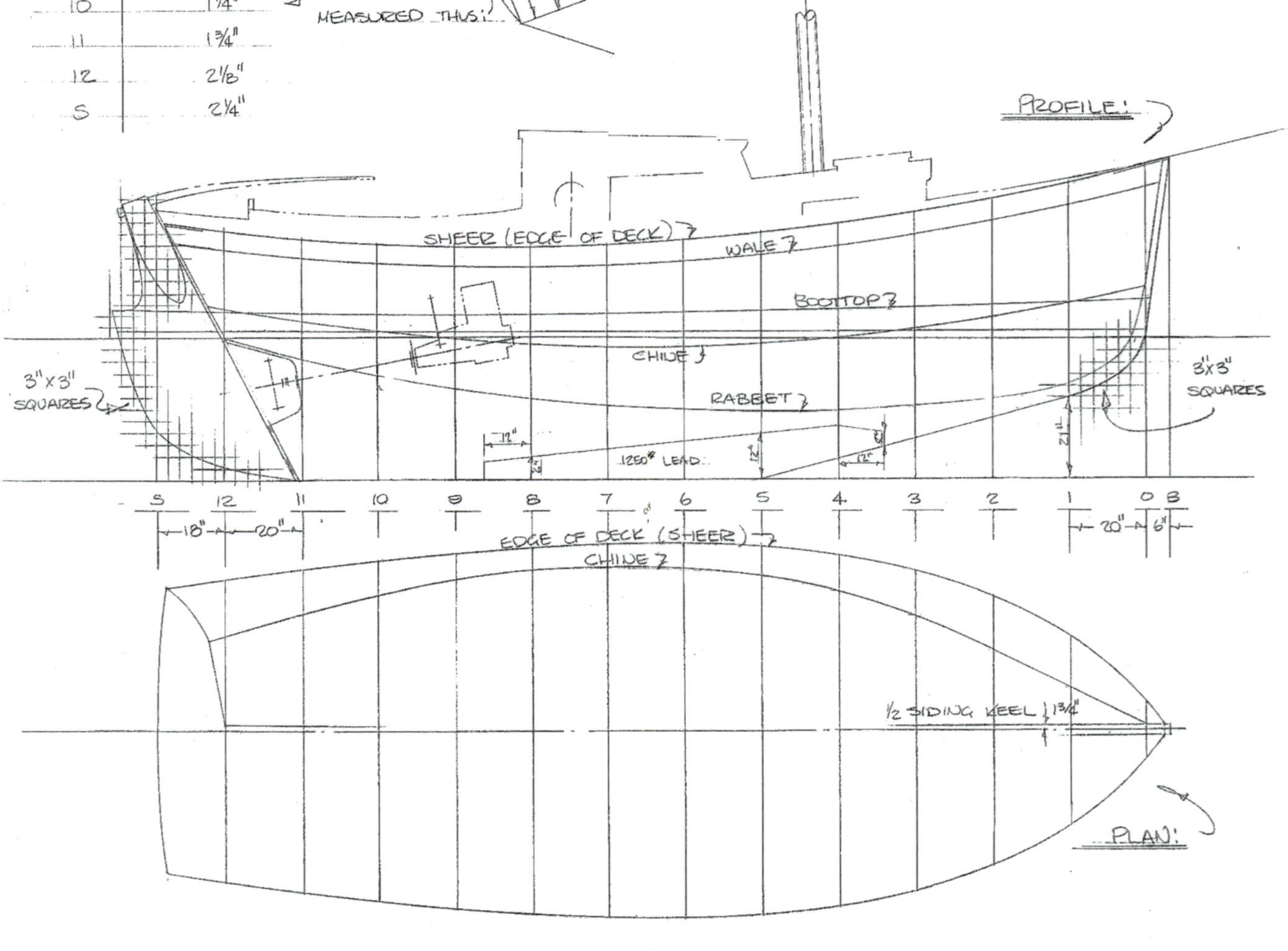

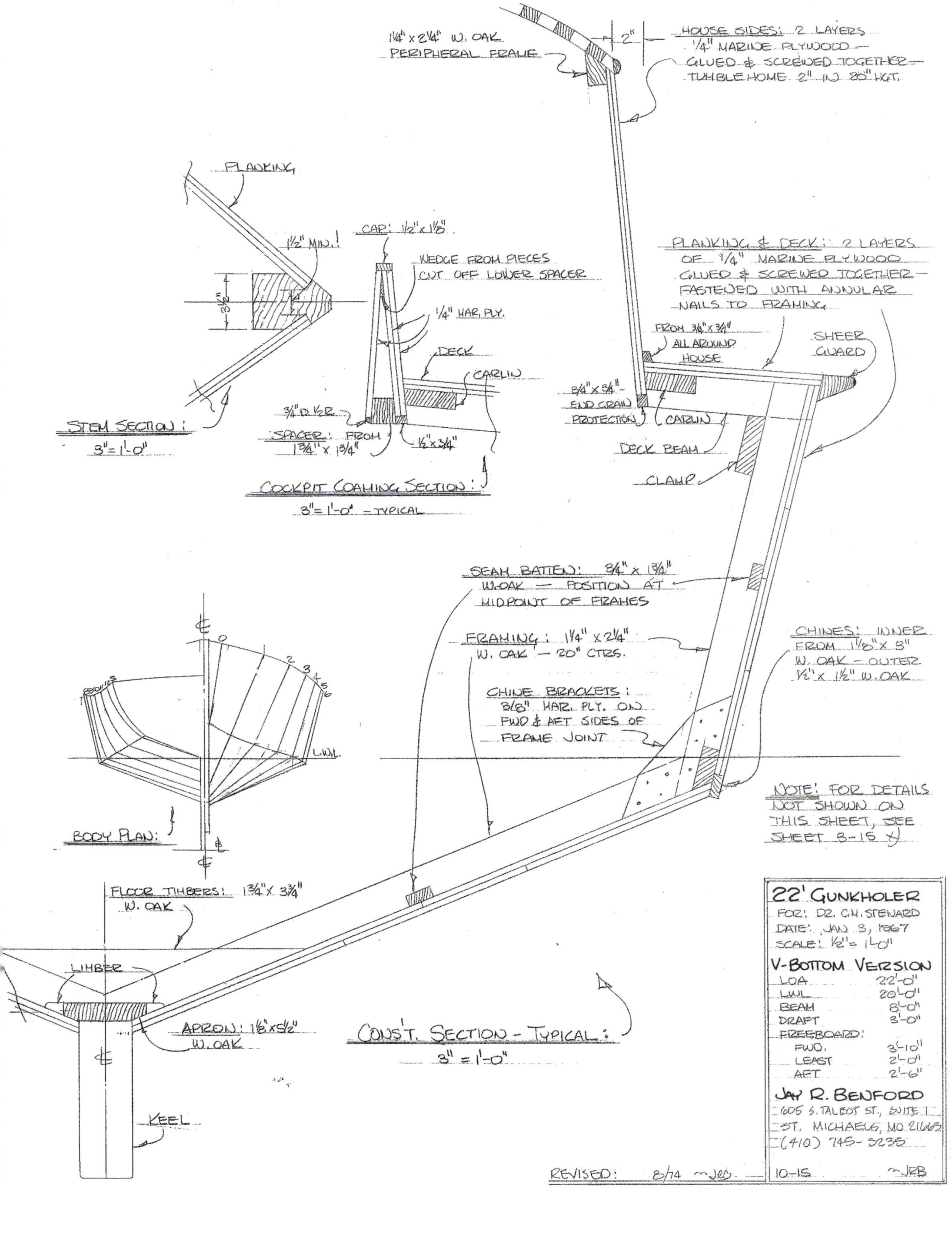
1¼" x 2¼" W. OAK
PERIPHERAL FRAME
2"
HOUSE SIDES: 2 LAYERS
¼" MARINE PLYWOOD —
GLUED & SCREWED TOGETHER —
TUMBLEHOME 2" IN 20" HGT.
PLANKING
1½" MIN.!
3½"
CAP: ½" x 1⅛"
WEDGE FROM PIECES
CUT OFF LOWER SPACER
¼" MAR. PLY.
DECK
CARLIN
¾" D. ½R.
SPACER: FROM
1¾" x 1¾"
½" x ¾"
STEM SECTION:
3" = 1'-0"
COCKPIT COAMING SECTION:
3" = 1'-0" — TYPICAL
PLANKING & DECK: 2 LAYERS
OF ¼" MARINE PLYWOOD
GLUED & SCREWED TOGETHER —
FASTENED WITH ANNULAR
NAILS TO FRAMING
FROM ¾" x ¾"
ALL AROUND
HOUSE
SHEER
GUARD
¾" x ¾" —
END GRAIN
PROTECTION
CARLIN
DECK BEAM
CLAMP
SEAM BATTEN: ¾" x 1¾"
W. OAK — POSITION AT
MIDPOINT OF FRAMES
FRAMING: 1¼" x 2¼"
W. OAK — 20" CTRS.
CHINE BRACKETS:
⅜" MAR. PLY. ON
FWD & AFT SIDES OF
FRAME JOINT
CHINES: INNER
FROM 1⅛" x 3"
W. OAK — OUTER
½" x 1½" W. OAK
NOTE: FOR DETAILS
NOT SHOWN ON
THIS SHEET, SEE
SHEET 3-15 X
L.W.L.
BODY PLAN:
FLOOR TIMBERS: 1¾" x 3¾"
W. OAK
LIMBER
APRON: 1⅛" x 5½"
W. OAK
CONS'T. SECTION - TYPICAL:
3" = 1'-0"
KEEL
REVISED: 8/74 ~JRB
22' GUNKHOLER
FOR: DR. C.M. STEWARD
DATE: JAN 3, 1967
SCALE: ½" = 1'-0"
V-BOTTOM VERSION
LOA 22'-0"
LWL 20'-0"
BEAM 8'-0"
DRAFT 3'-0"
FREEBOARD:
FWD. 3'-10"
LEAST 2'-0"
AFT 2'-6"
JAY R. BENFORD
605 S. TALBOT ST., SUITE 1
ST. MICHAELS, MD 21663
(410) 745-3235
10-15 ~JRB

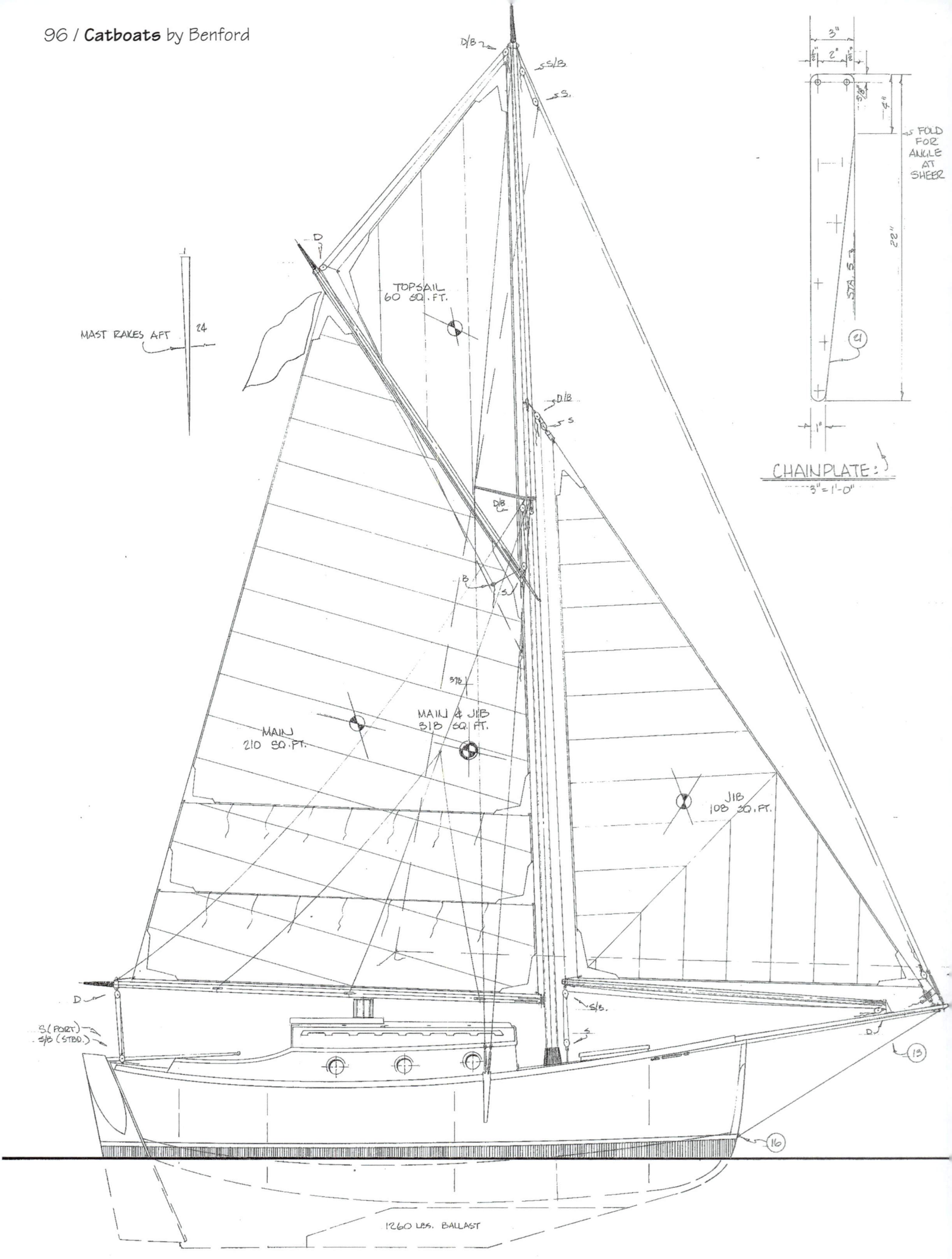
TOPSAIL
60 SQ. FT.
MAST RAKES AFT
24
MAIN
210 SQ. FT.
MAIN & JIB
318 SQ. FT.
JIB
108 SQ. FT.
D/B
S/B
S
D
B
S (PORT)
S/B (STBD.)
13
16
1260 LBS. BALLAST
3"
2"
FOLD FOR ANGLE AT SHEER
22"
STA. 5
21
1"
CHAINPLATE:
3" = 1'-0"

SAILS:

SAIL	AREA	WEIGHT	FOOT	LUFF	LEACH	HEAD	DIAGONAL	NOTES
MAIN	210	5 oz.	14'-0"	14'-0"	24'-3"	12'-0"	19'-0"	(1)
JIB	108	4 oz.	11'-11"	22'-6"	18'-3"	—	—	(2)
TOPSAIL	60	3 oz.	14'-5"	20'-0"	9'-2"	-	—	(3)
GENOA	260	3 oz.	18'-1"	32'-0"	29'-1"	-	-	

NOTES:

(1) NO ROACH IN LEACH & NO BATTENS — 2 ROWS OF REEF POINTS, 3'-0" & 6'-0" ABOVE & PARALLEL TO FOOT.

(2) WIRE LUFF FOR ROLLER FURLING.

(3) ALL 3 SIDES ROPED — SAIL IS SET FLYING.

(4) ALL SAILS DACRON — DARKER COLORED CLOTH PREFERED FOR LESS GLARE.

(5) HALYARDS 5/16"⌀ DACRON:

PEAK	110'
THROAT	65'
JIB	80'
GENOA	95

ALSO 5/16"⌀ DACRON:

GENOA DOWNHAUL	60'
" SHEETS	80'
JIB SHEET	50'
MAIN "	60'
TOTAL 5/16"⌀	600'

(6) IN 1/4"⌀ DACRON:

MAIN TOPPING LIFT	50
JIB " "	50'
TOPSAIL HALYARD	70
" OUTHAUL	60'
" DOWNHAUL	15'
JIB OUTHAUL	45'
MAIN LAZYJACKS	85'
TOTAL 1/4"⌀	375'

(7) JIB FURLING LINE TO BE 35' OF 3/16"⌀ DACRON BRAID.

(8) ABOVE LENGTHS ASSUME ALL RUNNING RIGGING LEADS TO CLEATS AT AFT END OF HOUSE, EXCEPT LAZY JACKS & TOPSAIL LINES (CLEAT ON MAST.)

(9) BOOM: 14'-6" GOOSENECK TO BAND, 12" TIP, 2 3/4"⌀ AT CENTER, 2 1/8"⌀ AT MAST END, 2"⌀ AT BAND.

(10) GAFF: 12'-6" JAWS TO BAND, 12" TIP, 2 1/2"⌀ AT CENTER, 2"⌀ AT ENDS.

(11) JIB CLUB: 11'-2" STICK LENGTH, 2 1/2"⌀ AT CENTER, 2"⌀ AT ENDS.

(12) ALL SPARS TO BE SITKA SPRUCE OR D. FIR, CIRCULAR IN SECTION.

Miss Becky's *swept back spreaders.*

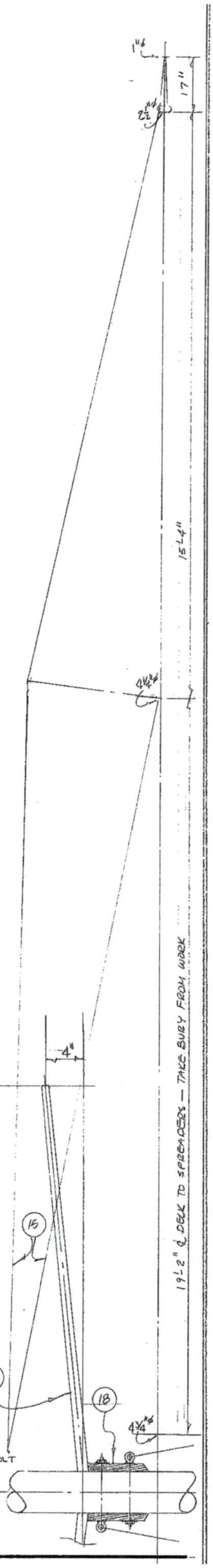

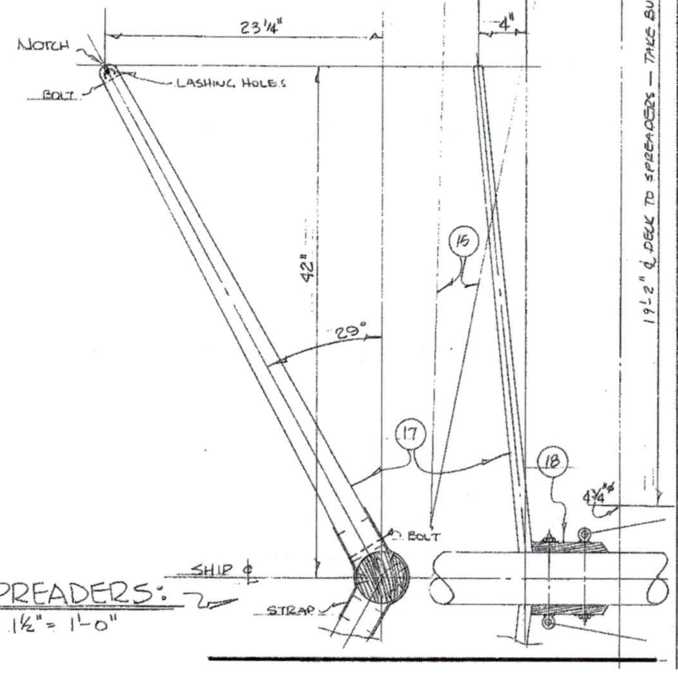

SPREADERS: 1½" = 1'-0"

22' GUNKHOLER
FOR: PETER A. ZIMMER
DATE: 11-24-76
SCALE: 1/2" = 1'-0"

SAIL PLAN ~ CUTTER

LOA	22'-0"
LWL	20'-0"
BEAM	8'-0"
DRAFT	3'-0"
FREEBOARD:	
FWD.	4'-0"
LEAST	2'-4"
AFT	3'-0"

JAY R. BENFORD
605 S. TALBOT ST, SUITE 1
ST. MICHAELS, MD 21663
(410) 745-3235
12-15 ~JRB

REVISED: 2-16-77 ~JRB

(13) SEE SHEET D 100 FOR PEDESTAL GOOSENECK FOR JIB CLUB.

(14) SEE SHEET D 97 FOR DETAILS OF BEE BLOCKS FOR LAZY JACKS ETC.

(15) UPPER & LOWER SHROUDS & HEADSTAY 3/16"ϕ WIRE — BOBSTAY 5/16"ϕ

(16) 1/2" x 6" WILCOX-CRITTENDEN FIG. 2220 BRZ. SHOULDER EYE BOLT

(17) SPREADERS FROM 1 1/8" x 3" D. FIR; TAPERED TO 3/4" x 1 1/2" AT OUTBOARD ENDS — NOTCH OUTBOARD END FOR SHROUD & PROVIDE SMALL HOLES FOR LASHING LINE AS SHOWN — 3/16"ϕ BOLT THRU-SPREADER AT EACH END TO PREVENT SPLITTING — 1/8" x 1/2" STRAP (S.S. OR BRZ.) AROUND FORE & AFT SIDE OF MAST & SCREWED TO MAST & SPREADERS.

(18) 3/4" x 1 1/2" x 6 1/2" BLOCK P. & S. ON MAST WITH 3/8" SHOULDER EYEBOLTS THRU MAST FOR LOWER SHROUDS — BEVEL TOP TO SUPPORT SPREADERS

(19) MAST BOOT COLLAR FROM 1 1/8" D. FIR TO SECTION SHOWN — ALL AROUND MAST HOLE — ALLOW FOR 1/2" WEDGES ALL AROUND.

(20) BOWSPRIT FROM 1 1/2" x 7" D. FIR — CONSTANT SECTION FROM HATCH FRAME TO STEM — 1 3/8" x 6" HALF-WAY FROM STEM TO TIP — 1 1/8" x 4" AT TIP — THRU-BOLT TO BREASTHOOK & DECK FRAME

(21) CHAINPLATES FROM 3/16" PLATE TO SHAPE SHOWN (S.S. OR EPOXIED STEEL) — 2 HOLES FOR 3/8"ϕ TOGGLE PINS LOCATED AS SHOWN — BOLTS THRU CLAMP & LONG'L STRINGER & DOUBLER PAD (1/2" D. FIR OR W. OAK) LOCATED APPROX. AS SHOWN — 1/4"ϕ BOLTS.

(22) BOWSPRIT PLATES (1/8" x 2" x 7") TOP & BOTTOM OF BOWSPRIT — HOLE IN CENTER FOR 5/8" & IN ENDS FOR 3/8" SHOULDER EYE BOLTS

(23) BLOCK CODE
S = SINGLE
S/B SINGLE W/ BECKET
D DOUBLE
D/B DOUBLE W/ BECKET
B BULLSEYE

Builder-owner-captain and crew rightly proud of their work in putting **Miss Becky** *together. Their primary cruising ground is Lake Michigan.*

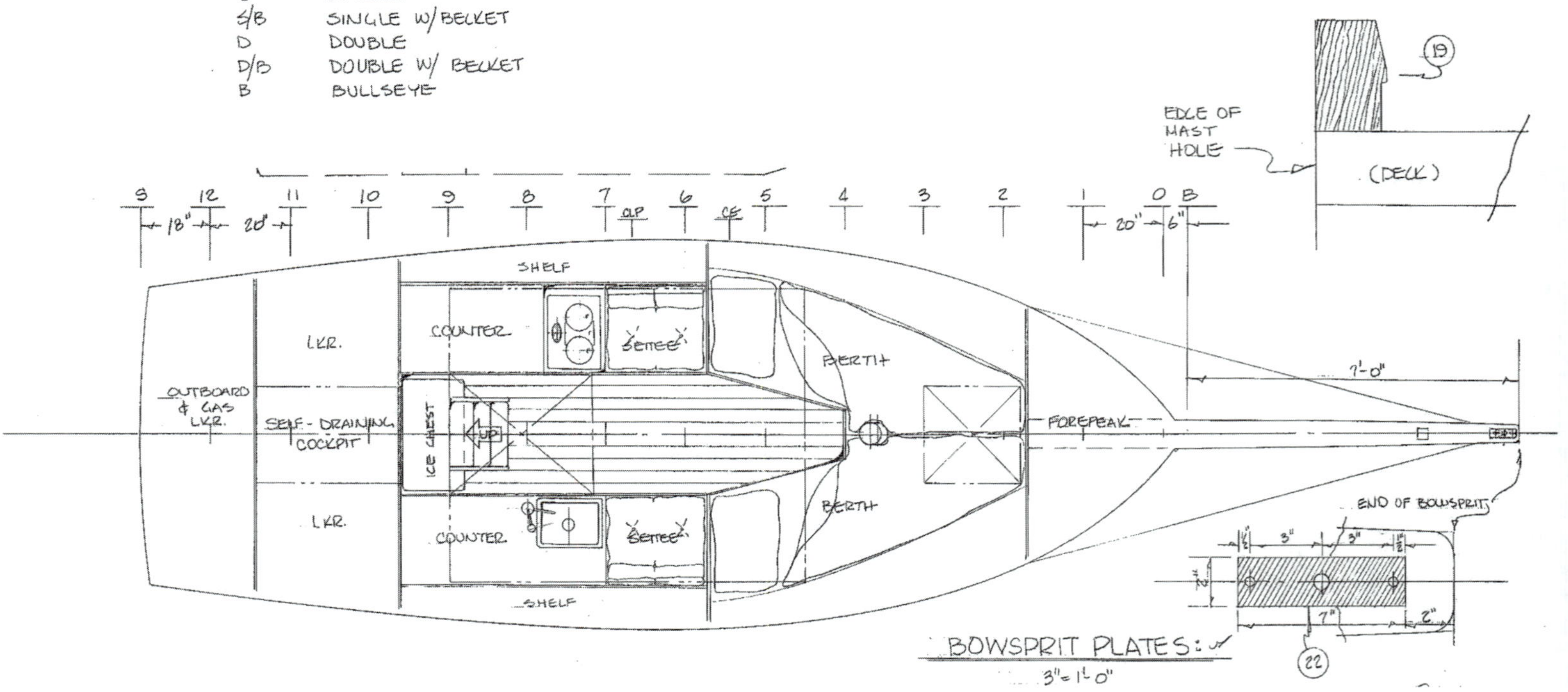

(34) RUDDER: FROM PLY LAMINATED TO 1½" THICKNESS & FAIRED DOWN TO ¼" AT TRAILING EDGE — CHEEKS FROM ¾" D. FIR OR W. OAK TO SHAPE SHOWN.

(35) TILLER: FROM 1½" x 1½" W. OAK — TAPER TO 1⅛" ⌀ AT FWD END — ¼" ⌀ BOLT THRU RUDDER CHEEKS FOR PIVOT

(36) BREASTHOOK: FROM 1⅛" D. FIR OR W. OAK

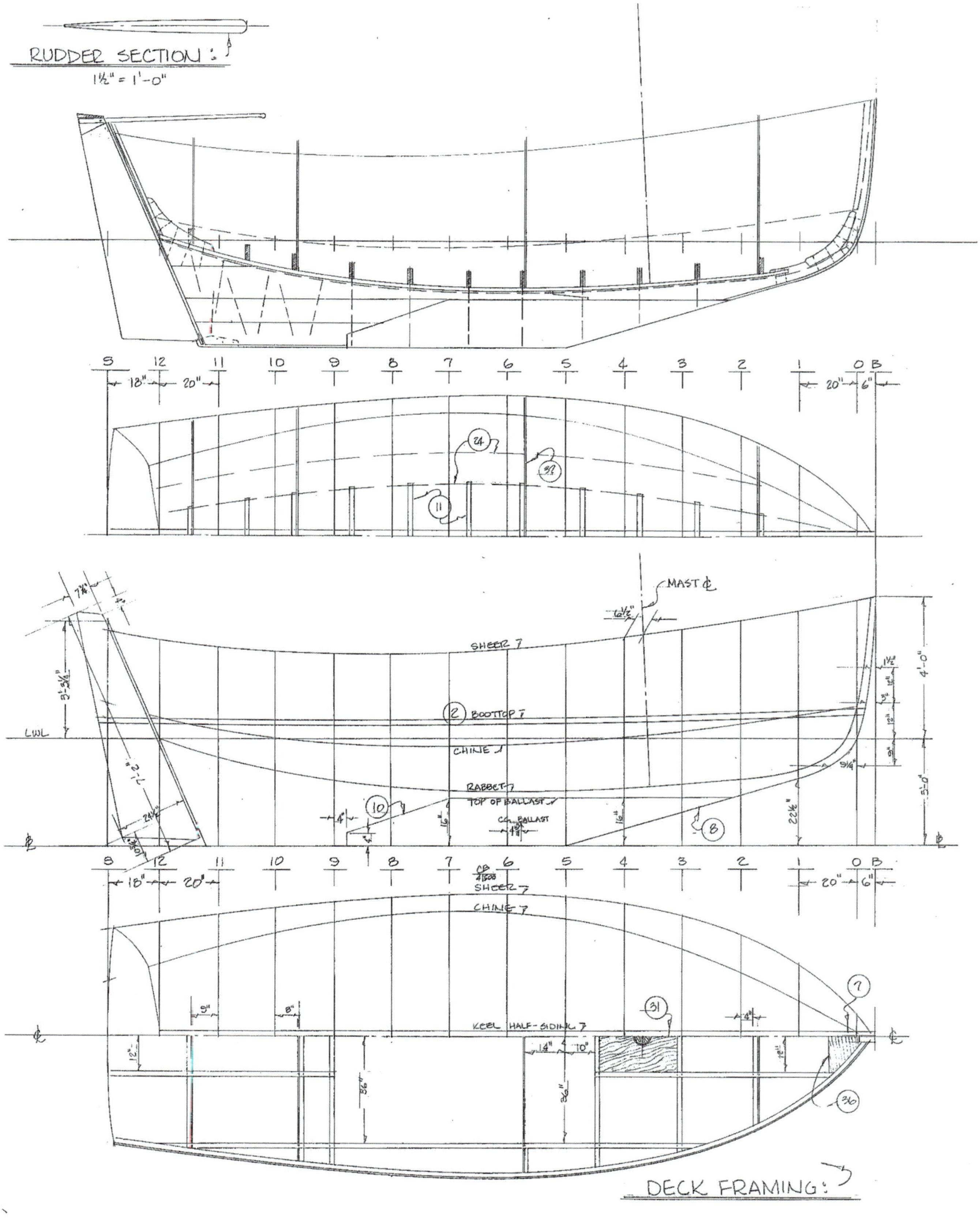

STATION	S	12	11	10	9	8	7	6	5
	HEIGHTS								
℄ TO RABBET	(3-6-7)	3-0-0	2-5-6	2-1-2	1-10-0	1-7-6	1-6-4	1-6-1	1-6-3
℄ TO CHINE	(4-0-0)	3-7-2	3-3-1	3-0-1	2-10-3	2-9-4	2-9-3	2-9-6	2-10-6
LWL TO BOOTTOP	0-7-0	0-6-6	0-6-4	0-6-3	0-6-2	0-6-2	0-6-3	0-6-5	0-6-7
LWL TO WALE									
LWL TO SHEER	(3-0-2)	2-9-2	2-6-7	2-5-3	2-4-5	2-4-3	2-4-6	2-5-7	2-7-6
	HALF-BREADTHS								
℄ TO CHINE	(1-6-5)	2-0-3	2-6-2	2-11-0	3-2-5	3-5-1	3-6-2	3-5-6	3-3-6
℄ TO SHEER	(3-0-0)	3-2-4	3-5-2	3-7-3	3-9-2	3-10-7	3-11-7	4-0-0	3-11-0

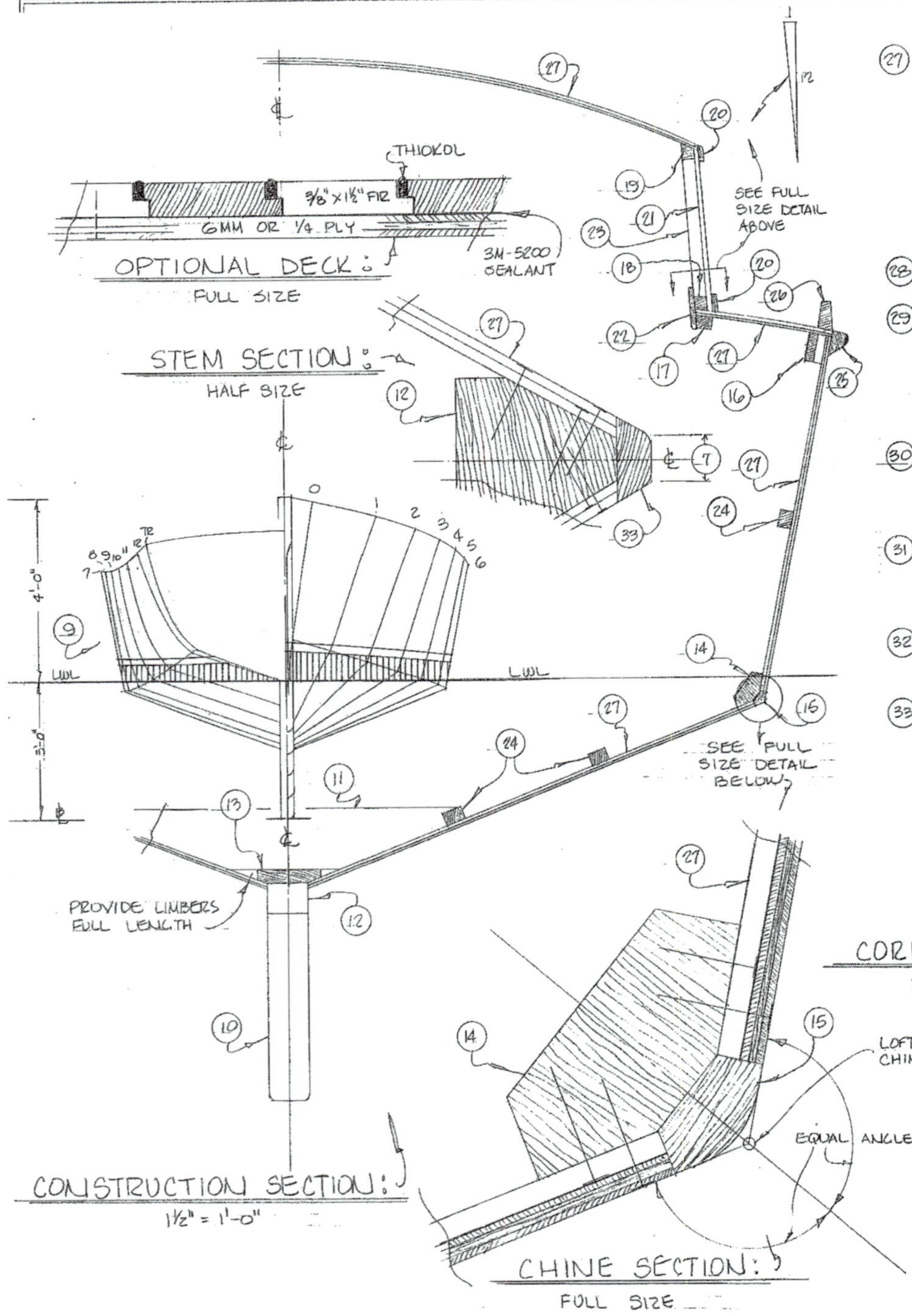

(27) PLANKING, DECK & HOUSETOP: 2 LAYERS OF 6MM BRUYNZEEL OR 1/4" FIR MARINE PLYWOOD CLOUT NAILED & GLUED TOGETHER — STAGGER BUTTS & JOINTS BETWEEN LAYERS — NAIL (ANCHORFAST) TO APRON, LONG'L. FRAMES, CHINE, CLAMP, CARLIN & HOUSETOP FRAME

(28) TRANSOM: SAME AS PLANKING (27)

(29) TRANSOM FRAME: FROM 1 1/8" FIR OR W. OAK SAWN TO 1 1/8" x 1 1/2" — ALL AROUND EDGE OF TRANSOM — ALSO IN WAY OF OUTBOARD MOTOR BRACKET, IF USED. (1 1/8" x 3 1/2" ON ℄)

(30) DECK BEAMS: 1 1/2" x 2 1/4" D. FIR OR W. OAK SAWN OR LAMINATED TO 4" IN 8'-0" CAMBER — NOTCH FOR CARLINS & LONG'L DK. FRAMES

(31) MAST DOUBLER: ADDITIONAL LAYER OF 6MM PLY UNDER DECK FITTED SNUGLY TO FRAMES.

(32) BULKHEADS: 10 MM (3/8") PLY ON 1 1/8" x 1 1/8" D. FIR OR W. OAK FRAMES

(33) STEM CAP & KEEL SHOE: FROM 3/4" IRONBARK, SHAPED AS SHOWN

CORNERPOST SECTION:
FULL SIZE

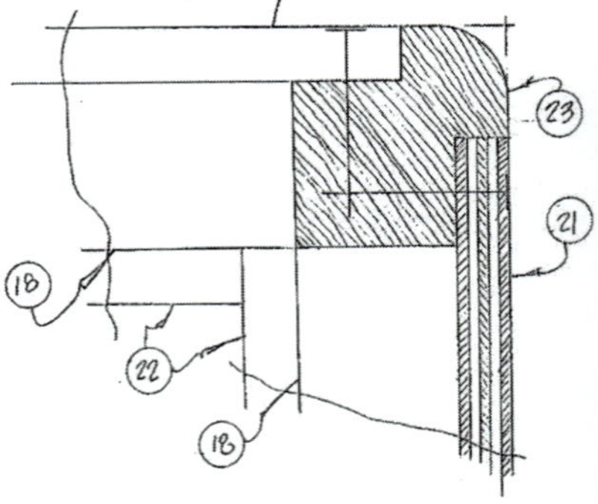

4	3	2	1	0	B
1-7-2	1-8-5	1-10-2	2-1-4	3-10-6	
3-0-2	3-2-2	3-4-6	3-7-4	3-10-6	
0-7-2	0-7-6	0-8-4	0-9-2	0-10-1	
2-10-0	3-0-6	3-3-6	3-7-1	3-10-6	4-0-0
2-11-4	2-5-3	1-9-3	0-11-7	0-1-6	
3-8-5	3-4-6	2-10-2	2-0-0	0-7-3	

NOTES:

(1) LINES & OFFSETS TO OUTSIDE OF PLANKING IN FEET – INCHES – EIGHTHS. DEDUCT FOR PLANKING & FRAMING AS SPECIFIED TO GET MOLD FRAME SIZES.

(2) BOOTTOP: OFFSETS TO TOP EDGE OF BOOTTOP — STRIPE IS 2" HIGH IN PROFILE FULL LENGTH.

(3) LINES MUST BE LOFTED & FAIRED FULL SIZE.

(4) ANY ALTERATION FROM THESE PLANS RELIEVES THE DESIGNERS FROM ANY FURTHER RESPONSIBILITY.

(5) THESE PLANS ARE THE PROPERTY OF THE DESIGNER & MAY BE USED ONLY AS AUTHORIZED BY THE DESIGNER IN WRITING.

(6) IT IS UNDERSTOOD THAT NO MORE THAN ONE BOAT WILL BE BUILT FROM THESE PLANS WITHOUT WRITTEN PERMISSION FROM THE DESIGNERS.

(7) HALF-SIDING OF STEM, KEEL & STERN-POST IS 1 3/4" — HALF-SIDING OF CUTWATER IS 1/2"

(8) LEADING EDGE OF KEEL TO BE RADIUSED FROM LWL TO KNUCKLE AT STA. 5, & THEN FLAT ACROSS BOTTOM TO STERN

(9) OUTWARD CURVATURE OF TOPSIDES (AFT) TO BE MEASURED THUS:

STATION	CURVATURE
8	1/4"
9	1/2"
10	1"
11	1 1/2"
12	2"
TRANSOM	2 1/4"

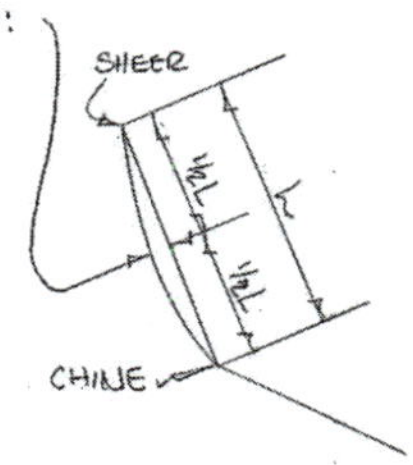

(10) BALLAST: TO BE 1060 LBS OF SCRAP LEAD & 200 LBS CONCRETE CAST TO SHAPE SHOWN — PROVIDE 100 LBS OF LEAD FOR TRIM BALLAST — WHEN CASTING KEEL, PROVIDE HOLES FOR KEEL BOLTS & SLIGHT RELESSES IN BOTTOM FOR BOLT HEADS (7 BOLTS - 1/2" ⌀ — THROUGH FLOOR TIMBERS, LOCATED AS SHOWN)

(11) FLOORS: SIDED 1 1/2" × 5 1/2" DEEP D. FIR OR Y. CEDAR — LOCATED AS SHOWN

(12) KEEL, STEM & STERNPOST & KNEES: SIDED 3 1/2" D. FIR OR GUM OR W. OAK

(13) APRON: 1 1/8" × 5 1/2" D. FIR OR W. OAK

(14) CHINE LOG: FROM 1 1/2" × 3 1/2" D. FIR OR W. OAK SHAPED AS SHOWN

(15) CHINE BATTEN: FROM 3/4" × 1 1/2" D. FIR OR GUM

(16) CLAMP: 2 PIECES 3/4" × 2 1/2" D. FIR OR W. OAK LAMINATED TOGETHER

(17) CARLIN & LONG'L. DECK FRAME: 1 1/8" × 1 1/2" D. FIR OR W. OAK

(18) LOWER HOUSE FRAME: FROM 1 1/8" × 1 1/8" D. FIR OR W. OAK — THRU-BOLT TO CARLIN

(19) UPPER HOUSE FRAME: 1 1/8" × 1 1/2" D. FIR OR W. OAK BEVELLED AS SHOWN

(20) HOUSE TRIM: 1/2" × 1 1/2" TEAK, MAHOG, OR D. FIR

(21) HOUSE SIDES: 10 MM BRUYNZEEL OR 3/8" FIR PLYWOOD

(22) FACIA: 3/8" × 3 1/2" TEAK, MAHOG, OR W. OAK — TO PROJECT ABOVE FRAMING AS SHOWN TO FORM HAND GRAB

(23) CORNER POSTS: FROM 1 1/2" × 1 1/2" D. FIR OR W. OAK (NEEDS LARGER PIECE IF NOT 90° CORNER.)

(24) LONG'L. FRAMES: 1 1/8" × 1 1/2" D. FIR OR W. OAK — SPACE EQUIDISTANT BETWEEN CLAMP & CHINE & BETWEEN APRON & CHINE — TO RUN CONTINOUS FULL LENGTH

(25) SHEER GUARD: FROM 1 1/8" × 2 1/8" W. OAK OR IRONBARK — TAPER OUTBOARD TO 1" & CAP WITH 1" BRZ. OR GALV. HALF-OVAL — MAKE SAW CUT FOR DRIP GROOVE AS SHOWN — BOLT GUARD THROUGH CLAMP & SCREW ON HALF-OVAL

(26) TOE RAIL: FROM 1 1/4" × 2 1/2" TEAK OR MAHOG. OR D. FIR — THRU-BOLT TO CLAMP

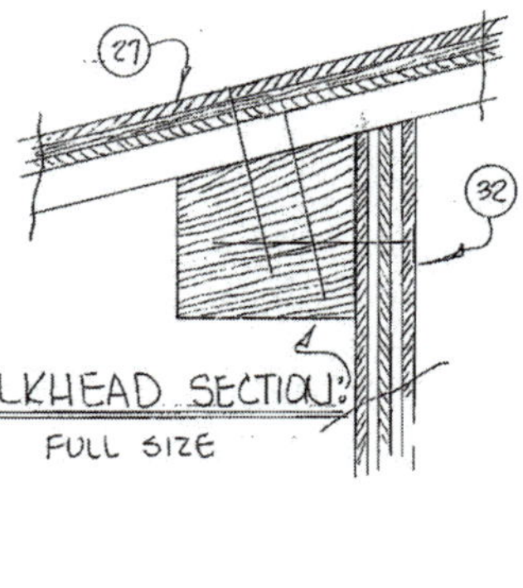

BULKHEAD SECTION
FULL SIZE

22' GUNKHOLER
FOR: PETER A. ZIMMER
DATE: 11-10-76
SCALE: 1/2" = 1'-0" & AS NOTED

V-BOTTOM VERSION #2

LOA	22'-0"
LWL	20'-0"
BEAM	8'-0"
DRAFT	3'-0"
FREEBOARD:	
FWD.	4'-0"
LEAST	2'-4"
AFT	3'-0"

JAY R. BENFORD
605 S. TALBOT ST, SUITE 1
ST. MICHAELS, MD 21663
(410) 745-3235
13-15 ~JRB

REVISED: 1-14-77 ~JRB
2-16-77 ~JRB

(18) LIMITS OF WELL OPENING

(19) ARRANGE COCKPIT WELL & FLATS AROUND ENGINE WELL TO ALL DRAIN INTO OUTBOARD WELL

(20) STAGGER KEEL BOLTS P. & S. OFF ℄ AS SHOWN.

(21) HATCH: OVER OUTBOARD WELL — HINGED AT AFT END — PROVIDE VENTS ON TOP OR PROP OPEN WHILE ENGINE IS IN OPERATION — PROVIDE LATCH OR CORD TO HOLD OPEN WHILE STARTING & WORKING ON ENGINE.

(22) MAST STEP: FROM 1½" x 5½" D. FIR OR W. OAK — TAPERED AT ENDS TO ¾" x 5½"

(23) BULWARK: STRIP-PLANKED FROM ¾" HONDURAS MAHOGANY OR D. FIR WITH ¾" x 1½" H. MAHOG. OR TEAK CAP

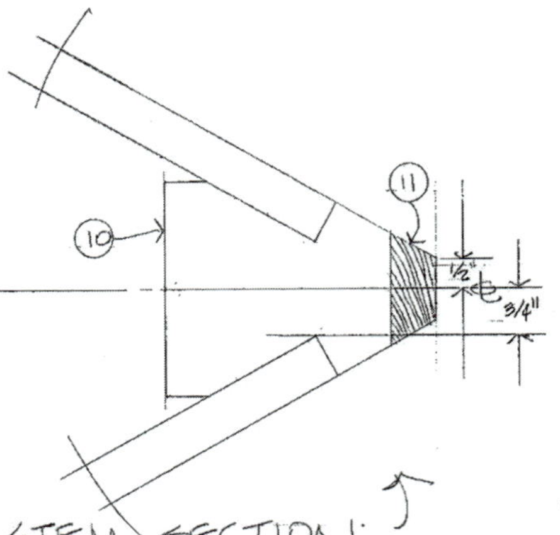

STEM SECTION: 10" = 1'-0"

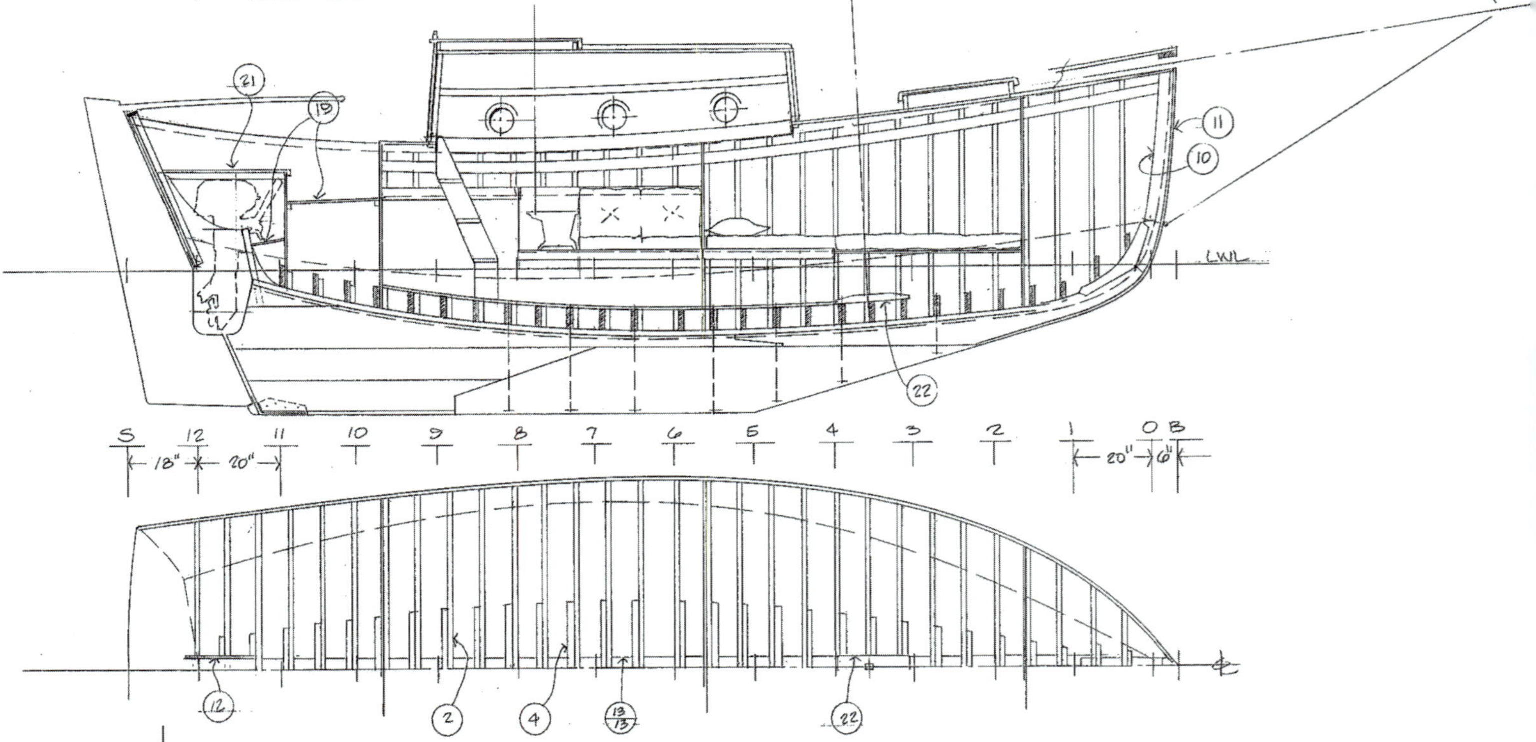

LONG'L. SECTION THRU OUTBOARD WELL: 5
3" = 1'-0"

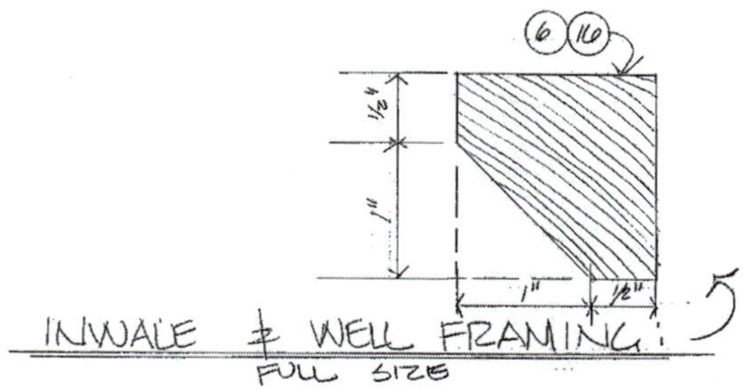

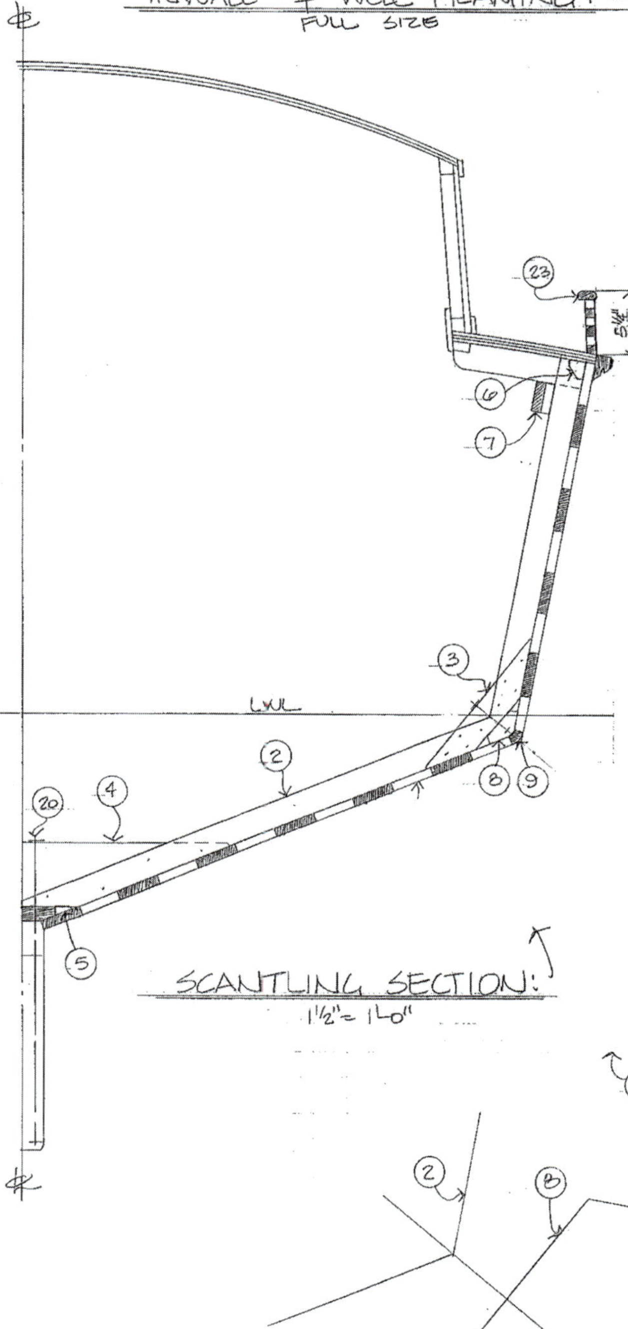

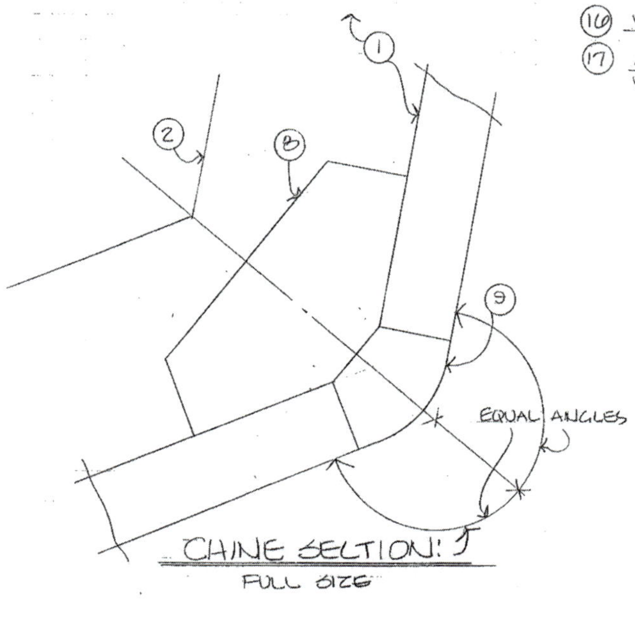

NOTES:

* FOR SCANTLINGS & DETAILS NOT SHOWN ON THIS SHEET, SEE DWG. № 13-15, "V-BOTTOM VERSION #2".

(1) PLANKING: 3/4" RED OR WHITE CEDAR, WITH CAULKED SEAMS.

(2) FRAMES: 1⅛" x 2⅛" D. FIR OR YELLOW CEDAR ON 8" CENTERS.

(3) CHINE GUSSETS: 3/8" PLY FWD. & AFT OF FRAME WITH 1⅛" D. FIR FILLER BLOCK BETWEEN TO TAKE CHINE LOG BOLT.

(4) FLOORS: FROM 1⅛" x 5¼" Y. CEDAR OR D. FIR — INCREASE TO 1⅝" x 5¼" IN WAY OF KEEL BOLTS.

(5) LIMBER THROUGHOUT BY RUNNING FLOOR STRAIGHT OFF TOP OF APRON & CUTTING OFF BOTTOM OF FRAME AS SHOWN.

(6) INWALE: FROM 1½" x 1½" D. FIR — LOWER, INBOARD CORNER (1" x 1" TRIANGLE) BEVELED OFF AS SHOWN.

(7) CLAMP: 1½" x 2½" D. FIR — MAY BE LAMINATED FROM 2 PIECES 3/4" x 2½" (OR W. OAK)

(8) INNER CHINE: FROM 1½" x 3½" D. FIR OR W. OAK

(9) CHINE BATTEN: FROM 1½" x 1" W. OAK OR GUM

(10) STEM: FROM 3½" W. OAK OR GUM OR D. FIR — NOTE MAXIMUM DEPTH OF RABBET SHOWN ON DETAIL.

(11) STEM CAP: FROM 3/4" W. OAK OR GUM OR IRONBARK — FROM STEMHEAD TO BALLAST — CONTINUE AFT OF BALLAST AS WORM SHOE

(12) WELL SIDES: 3/4" PLY FASTENED TO APRON & ℄ TRANSOM FRAME & TO WELL

(13) COCKPIT SIDES: 1/2" PLY

(14) TRANSOM & FRAME: 1½" x 5½" D. FIR FULL HEIGHT OF TRANSOM

(15) OUTBOARD MOTOR TRANSOM: 3/4" PLY DOUBLED IN WAY OF MOTOR — POSITION & ANGLE TO SUIT ENGINE CHOSEN (EVINRUDE 6 SHOWN)

(16) WELL FRAMING: SAME AS (6)

(17) APRON: FROM 1½" x 5½" D. FIR OR W. OAK

22' GUNKHOLER
FOR: REFF REINHART
DATE: JULY 14, 1983
SCALE: ½" = 1'-0" & AS NOTED
CARVEL CONST.

LOA	22'-0"
LWL	20'-0"
BEAM	8'-0"
DRAFT	3'-0"
FREEBOARD:	
FWD.	4'-0"
LEAST	2'-4"
AFT	3'-0"

JAY R. BENFORD
605 S. TALBOT ST, SUITE 1
ST. MICHAELS, MD 21663
(410) 745-3235
14-15
JRB

Miss Becky *in port, getting ready for another cruise.*

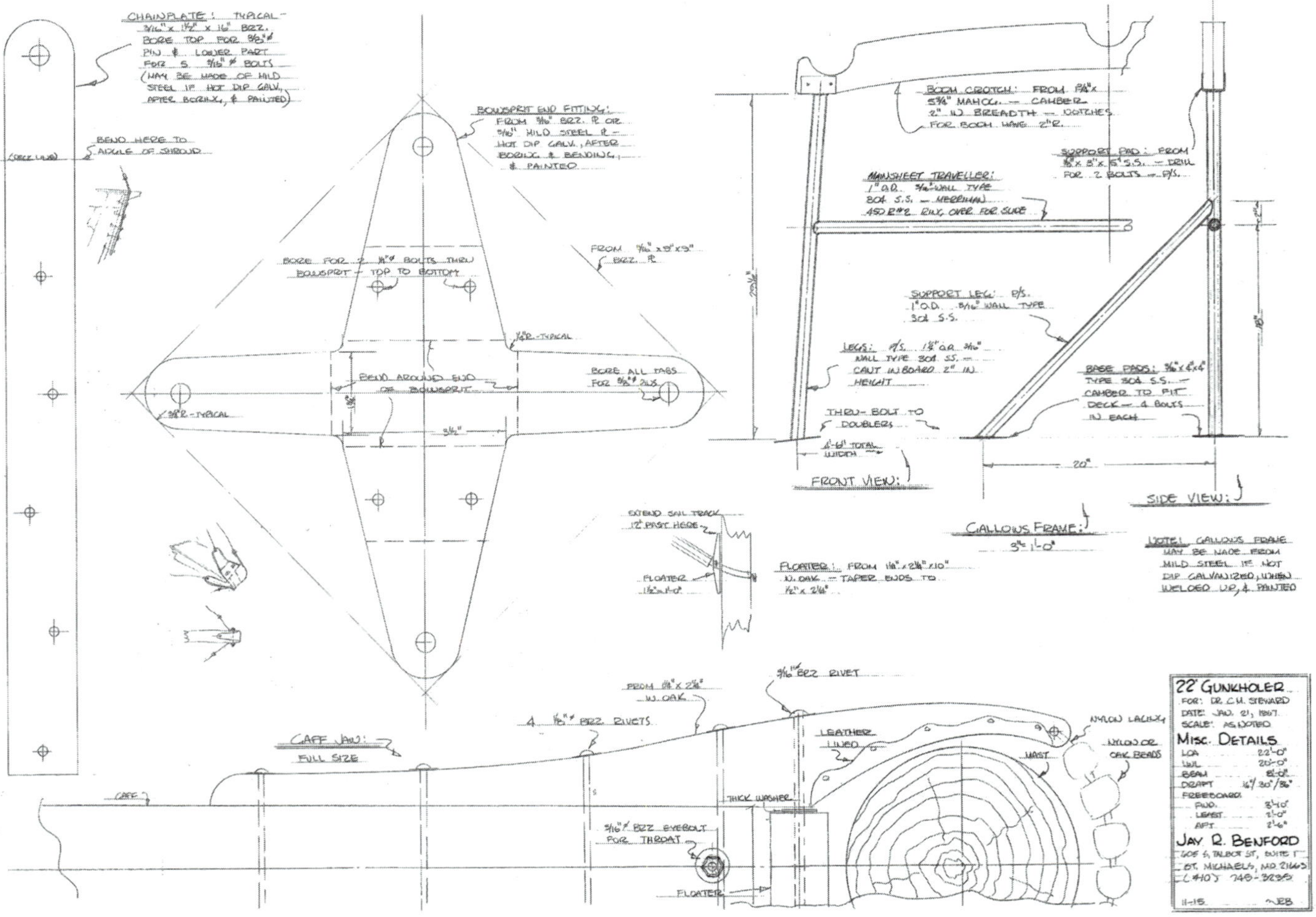

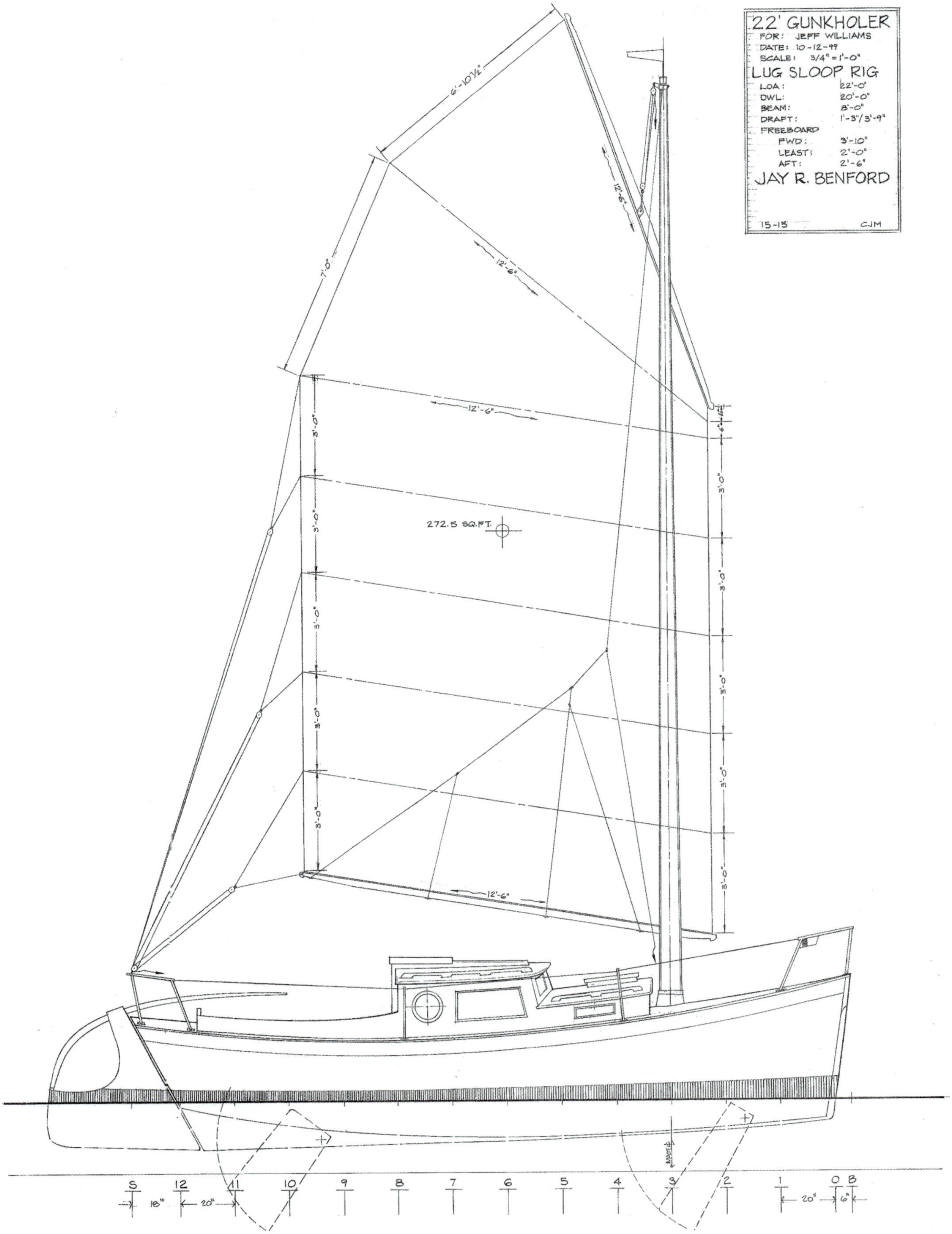
22' GUNKHOLER
FOR: JEFF WILLIAMS
DATE: 10-12-99
SCALE: 3/4" = 1'-0"
LUG SLOOP RIG
LOA: 22'-0"
DWL: 20'-0"
BEAM: 8'-0"
DRAFT: 1'-3"/3'-9"
FREEBOARD
FWD: 3'-10"
LEAST: 2'-0"
AFT: 2'-6"
JAY R. BENFORD
15-15
CJM
6'-10½"
7'-0"
12'-6"
272.5 SQ.FT.
3'-0"
S 12 11 10 9 8 7 6 5 4 3 2 1 0 B
18"
20"
6"

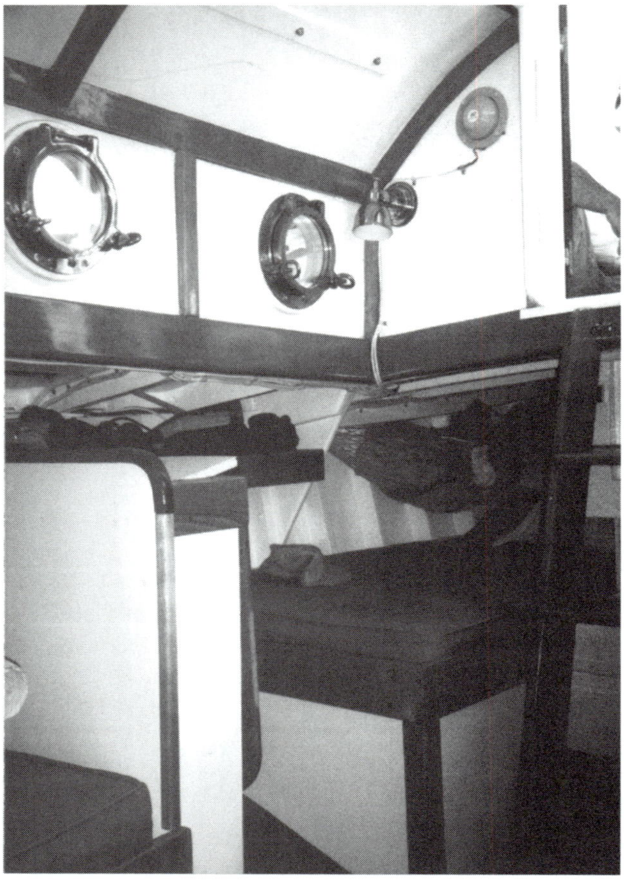

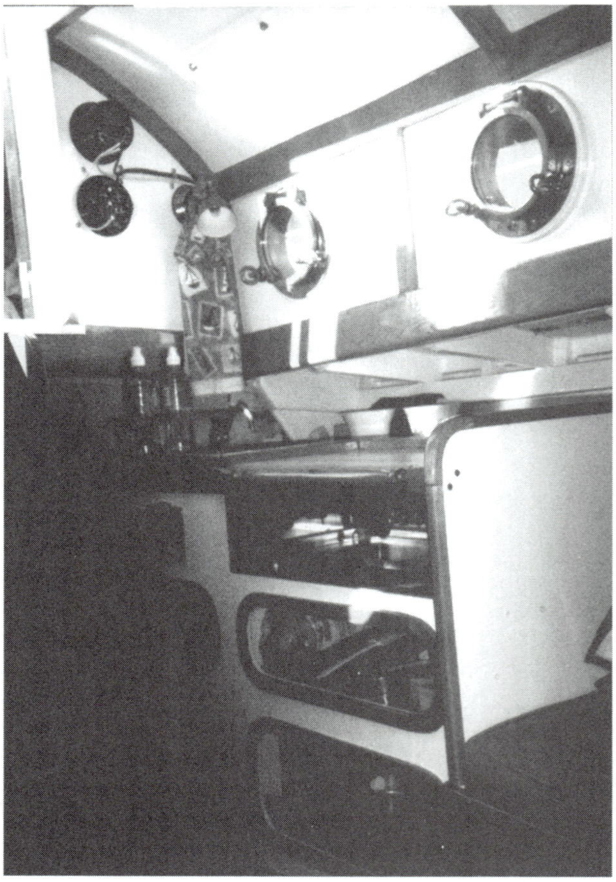

Miss Becky's *interior. Top left is looking aft to starboard giving a view of her querter berth. Top right is looking aft to port, showing her galley area and stowage spaces. Below is a view looking forward, showing the head in the bow and settee-berths either side of the mast.*

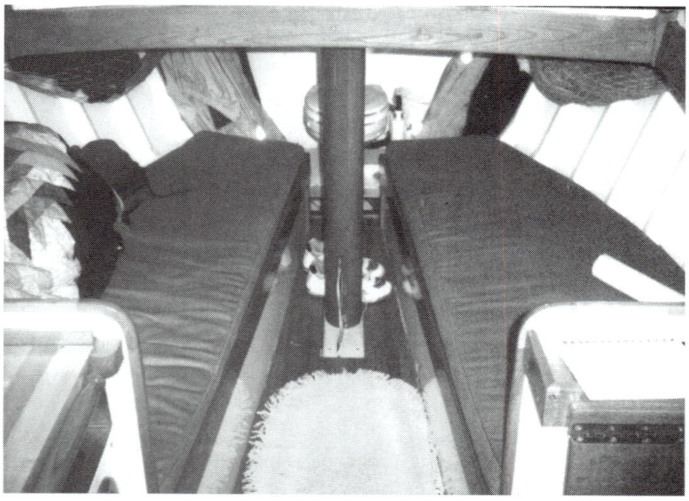

Miss Becky's *owner/builders enjoying her sailing ability.*

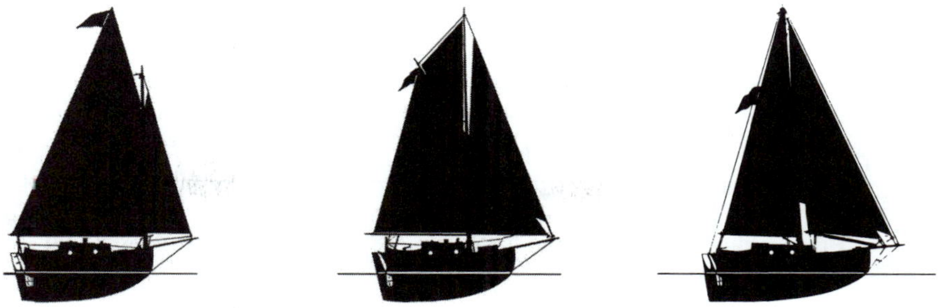

Chapter 6

22' Cruising Catboat or Cutter

Design number 128
1975 & 1988

A larger version of our popular 17' Cruising Catboat, this big sister has almost twice the room – useful living volume – in her proportionately longer hull. She's the answer for those who are in love with the 17' Catboat, but who'd like something "just a little bigger". She has 6'-2" standing headroom and the practicality of the traditional catbcat hull about her.

Her accommodations are great for a couple to spend long periods of time aboard. There is a private stateroom in the bow, port and starboard seats in the saloon, a roomy galley, and an enclosed head, which could have a shower in it. The alternate accommodations for the cutter shows a chart table, which is a treat to have. Of course, if you plan on having guests fairly often, keeping the longer port settee might be a good idea.

The catboat and topsail cutter rigs give the boat some options for interesting and fun sail handling.

The topsail cutter is the designer's favorite, with the variety of sails making her versatile and easily handled in a variety of conditions with a high fun quotient. Reefing can be done, in most cases, by simply dropping or furling one of her sails. And her mainsail is a size that is more easily handled. . . .

Her heavy displacement gives a great comfort to being aboard, and her hull form with the ballast on the keel gives her great stability for carrying sail in a strong breeze. She has very high sail area to wetted surface and sail area to displacement ratios, to assure good sailing performance in light airs too.

The last couple pages in this chapter show a lengthened version of this hull design, with a cutter rig. Her accommodations are a different approach to providing good sea berths and good space for socializing.

Particulars:		**Imperial**	**Metric**
Length overall		22'-0"	6.71 m
Length designed waterline		20'-0"	6.10 m
Beam		9'-8"	2.95 m
Draft		4'-6"	1.37 m
Freeboard:	Forward	4'-3¼"	1.30 m
	Least	2'-7 3/8"	0.80 m
	Aft	3'-0"	0.91 m
Displacement, cruising trim		8,800 lbs.	3,992 kg.
Displacement-length ratio		491	
Ballast		2,200 lbs.	998 kg.
Ballast ratio		25%	
Sail area: Cutter/Catboat		603/560 sq. ft.	56.02/52.03 m^2
Sail area-displacement ratio:		22.64/21.02	
Wetted Surface (W.S.)		211 sq. ft.	19.60 m^2
Sail Area/W.S. Ratio		2.86/2.65	
Prismatic coefficient		.573	
Pounds per inch immersion		633	113 kg./cm
Entrance half-angle		28°	
Water tankage		55 Gals.	208 liters
Fuel tankage		27 Gals.	102 liters
Headroom		6'-2"	1.88 m
GM		3'	0.91 m

***CAUTION:** The figure for displacement quoted here is for the boat in cruising trim. That is, with the fuel and water tanks filled, the crew on board, as well as the crews' gear and stores in the lockers. This should not be confused with the "shipping weight" often quoted as "displacement" by some manufacturers. This should be taken into account when comparing figure and ratios between this and other designs.

The stability curve for the 22' Cruising Cat, below, shows that she has very good form stability. The computer run was done by Bill Plice in 1975 when he was getting going with what has become an industry-standard stability program, GHS. This calculation did not include the deckhouse volume and this accounts for the drop off from 90-degrees onward. We have a similar curve for the 20' Catboat ***Joy*** *from the same era and it shows similar characteristics.*

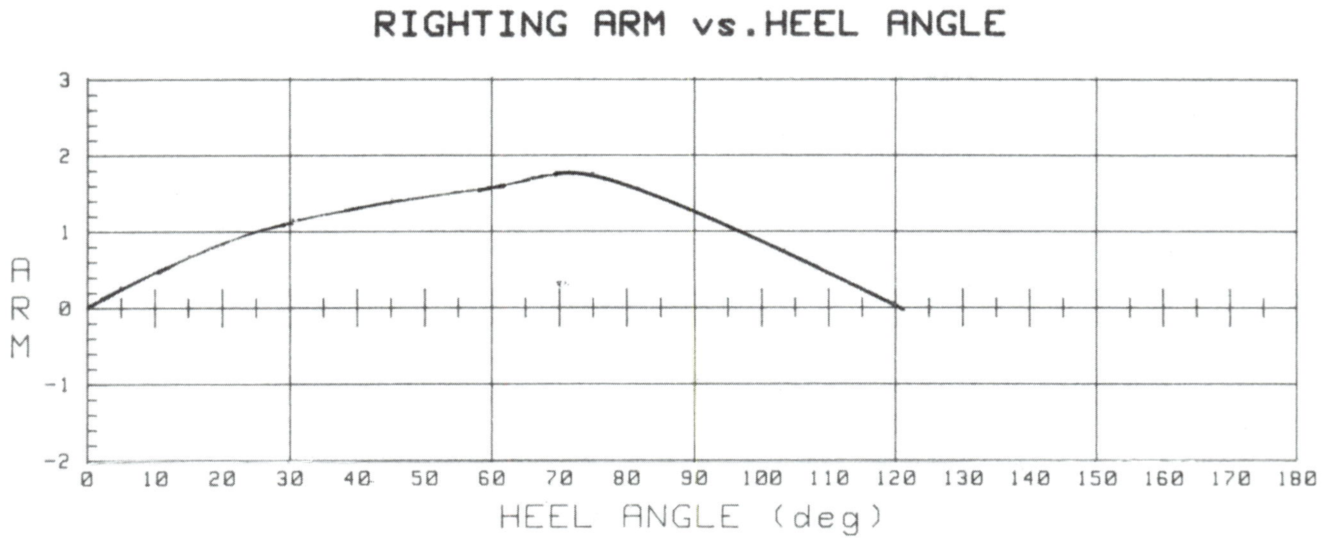

Just above the transom are the construction lines for developing the transom shape accurately. This is done small scale on the lines plan and repeated full size on the loft floor. Along a line perpendicular to the rake of the transom, the 15' radius point is located and the arc swung. From that arc are created the points that precisely define the transom.

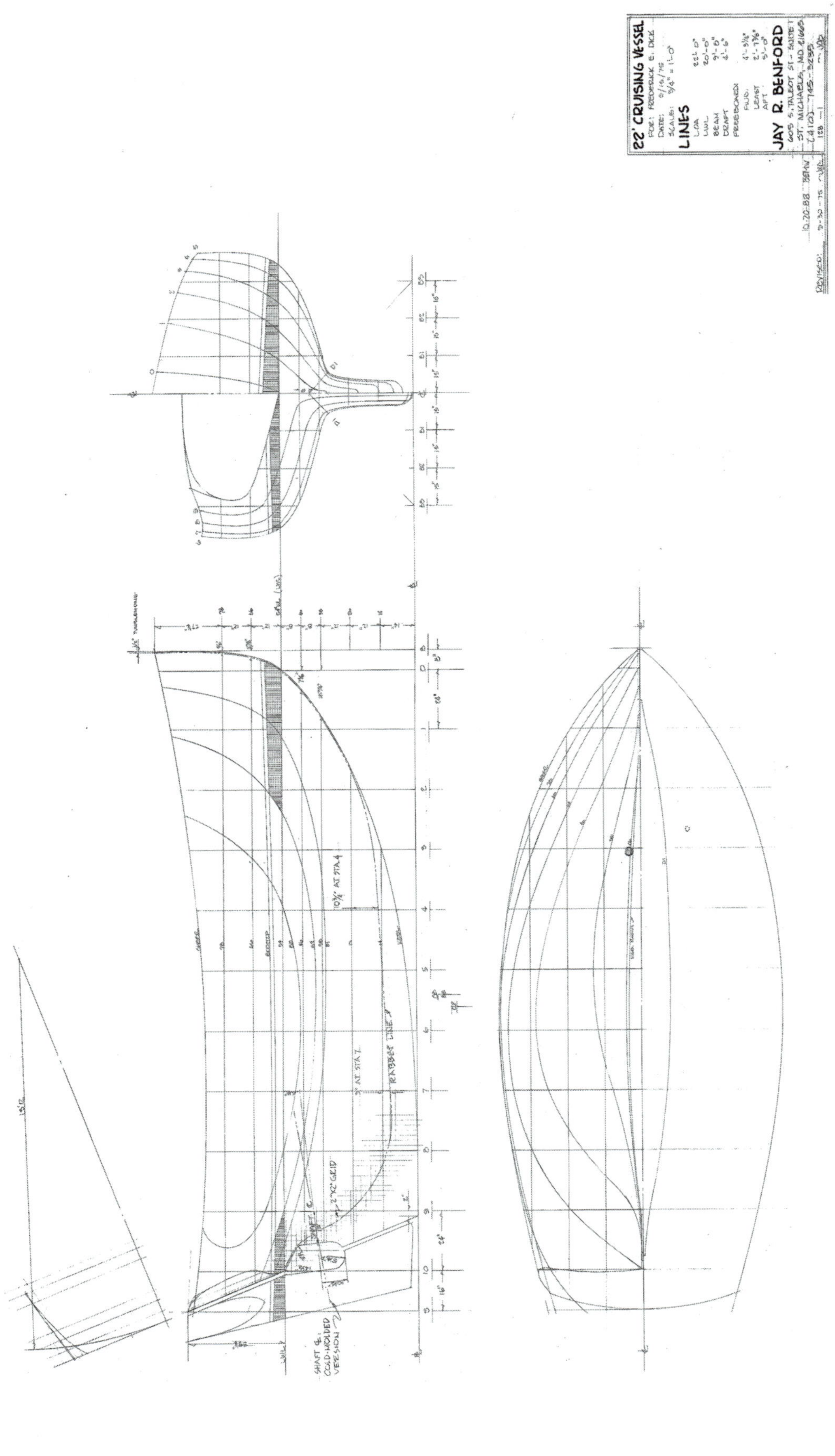

Three alternative layouts for the forward part of the interior are shown on this drawing. Think about how you actually use the boat and choose the one that best suits your cruising lifestyle.

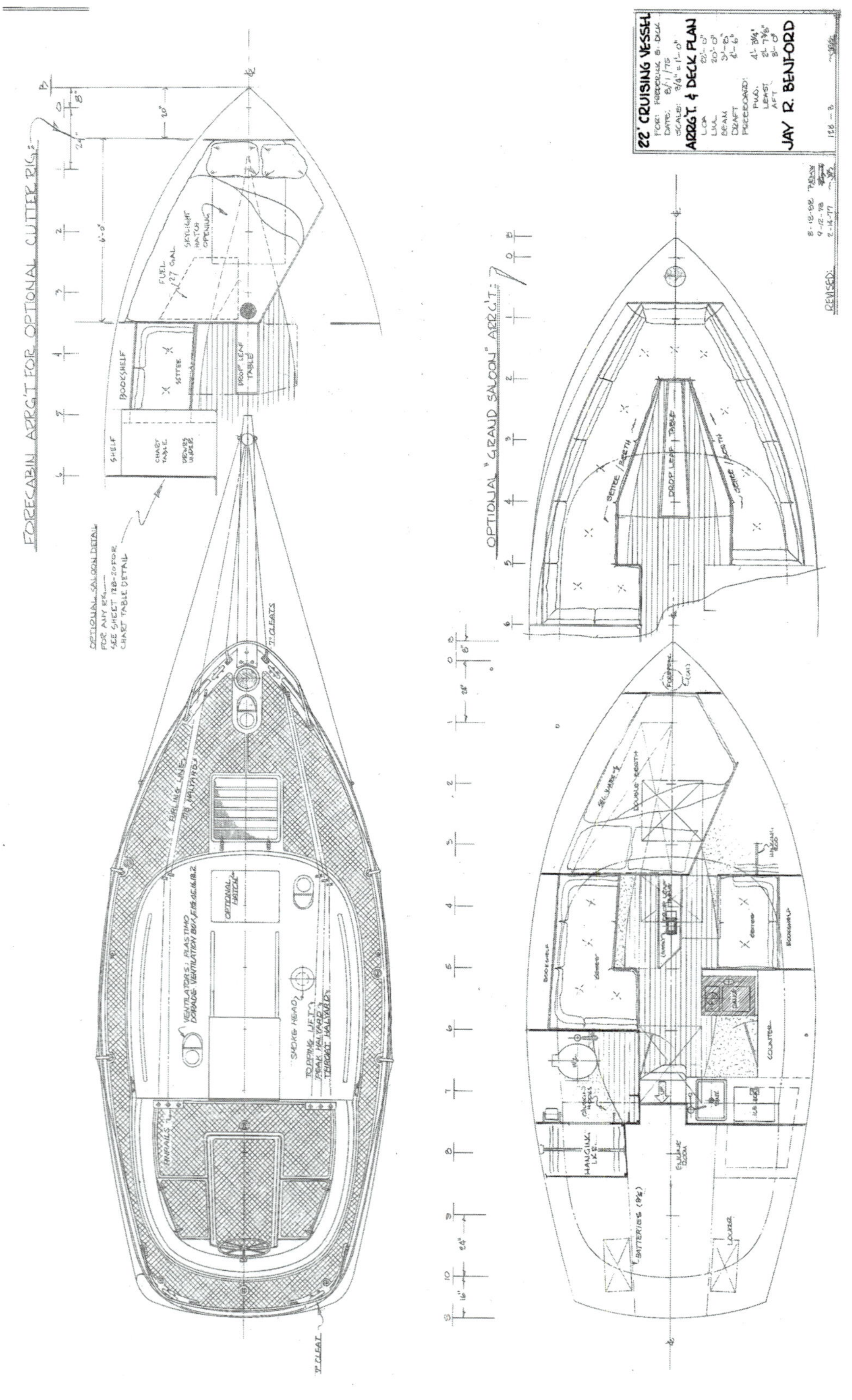

The engine shown here, the Sabb 10 hp with variable pitch propeller, is not widely available, and there are a number of good alternatives. Most of them will take up less space. Call or write the designer with ideas or questions.

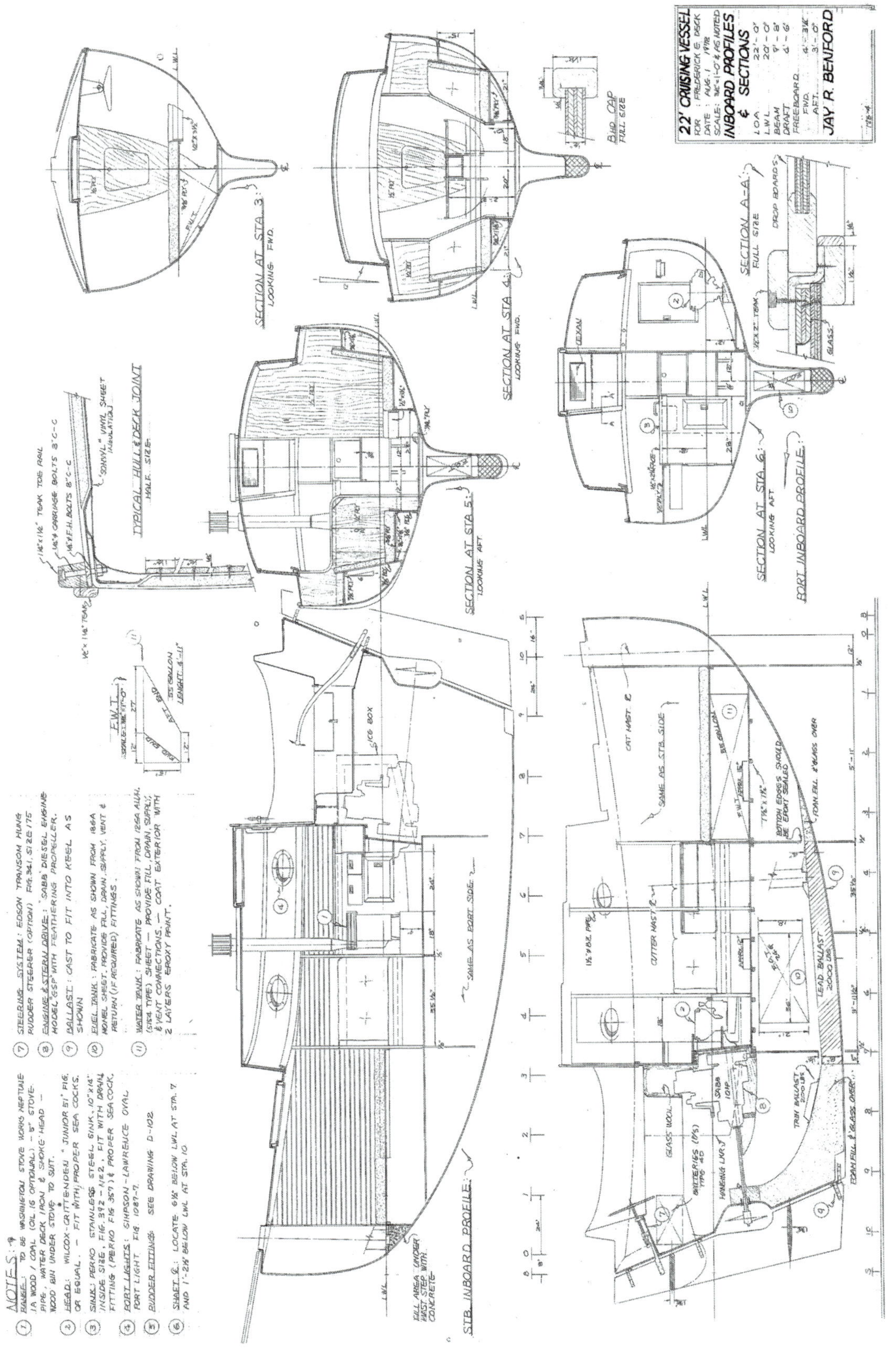

This is a preliminary study for a Marconi cutter version. It might benefit from having a bit more sail area for cruising areas with predominately lighter airs.

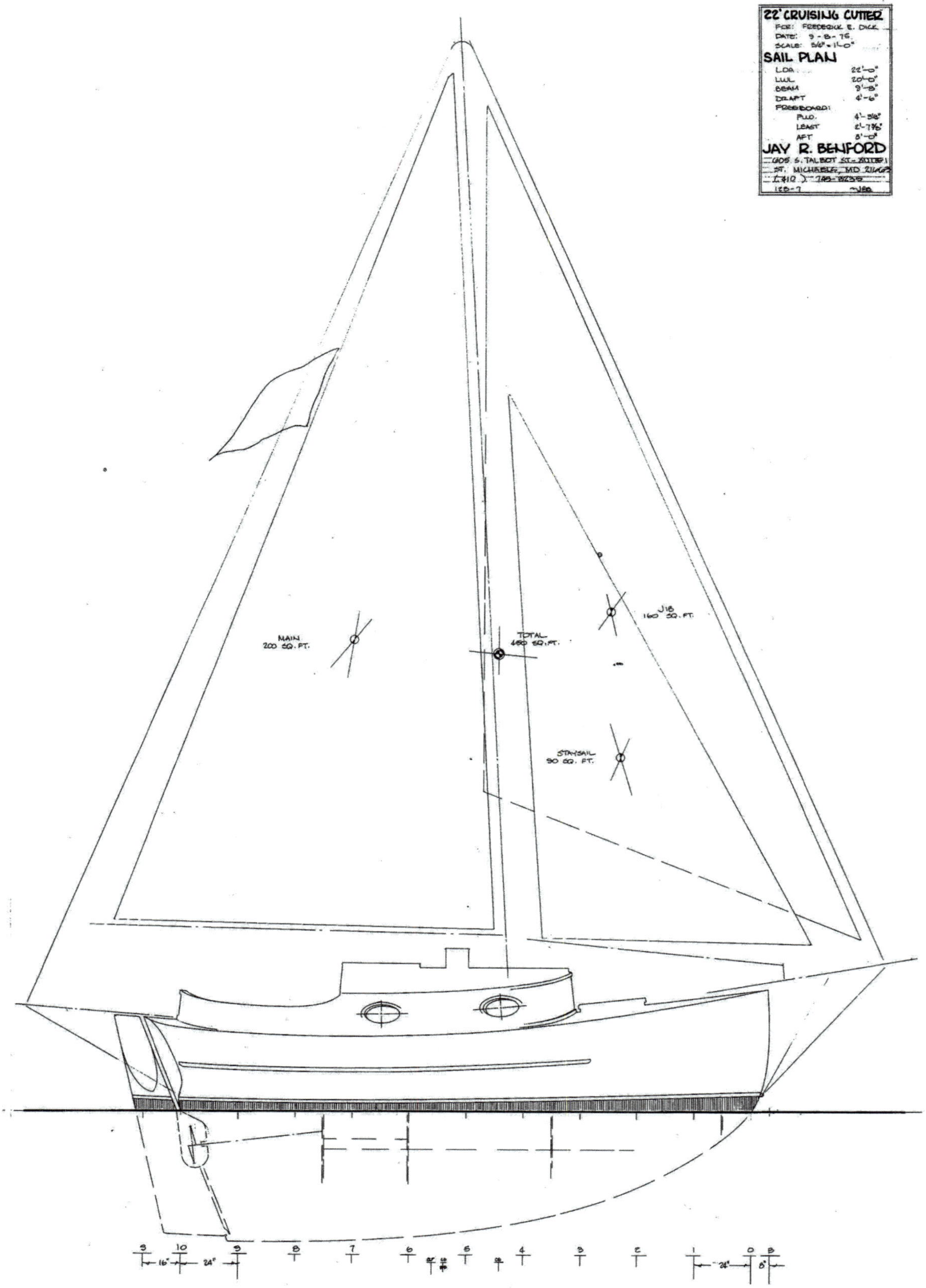

The cat has a large main and small jib set on the bowsprit, with roller furling gear. Keeping good tension in the jib halyara is essential to keep the weight of the mainsail from drooping the boom too close to the crew's heads in the cockpit. So don't omit that detail when building her!

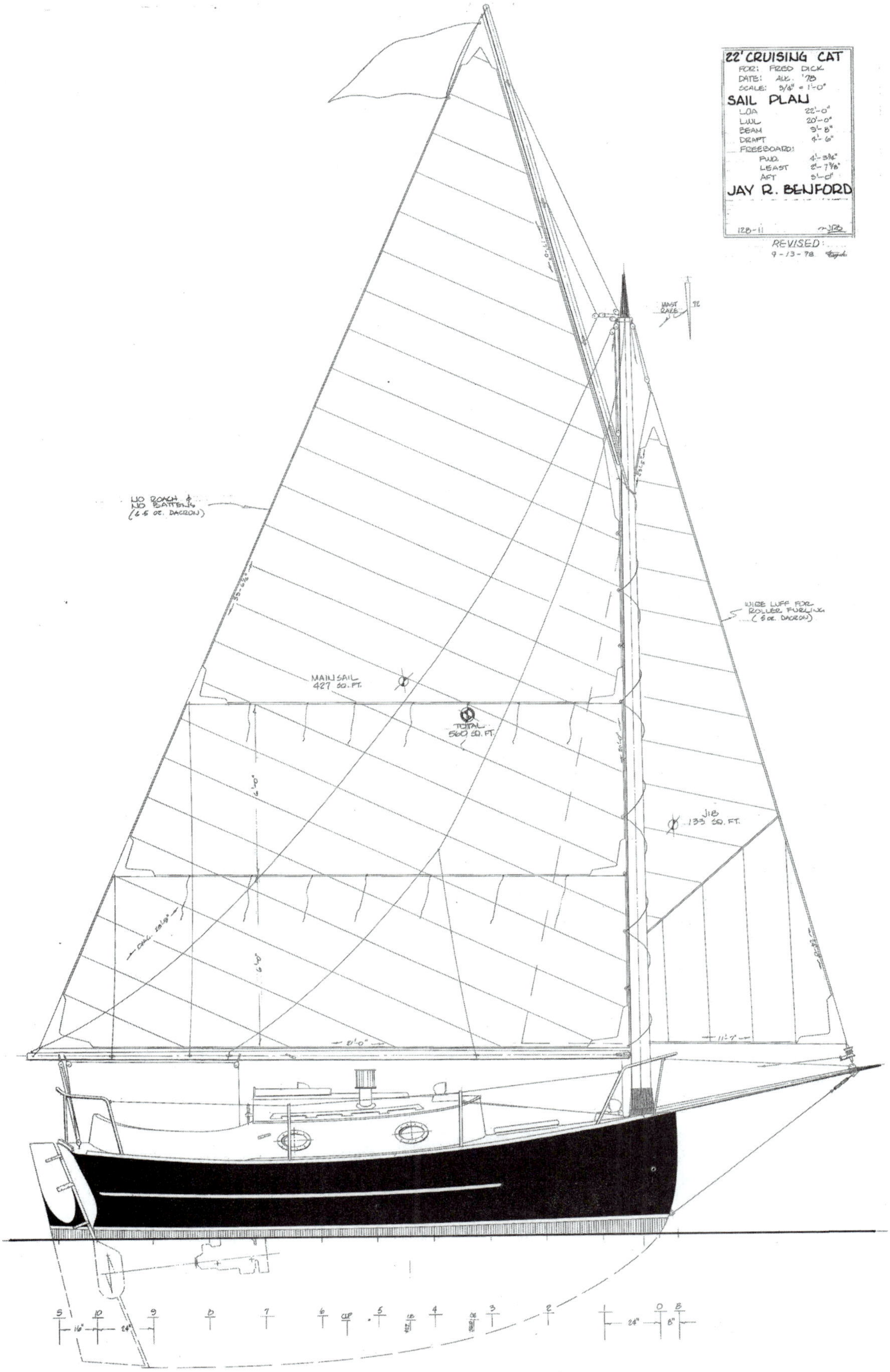

The Rigging plan shows how to build some of the parts for putting the cat rig together. Note that the gaff jaws have to have enough depth that they are still well around the mast when the gaff is raised. Also, the luff lacing around the mast needs to reverse direction around the mast each time it meets the luff. Otherwise, the lacing will act as a choker and you won't be able to raise or lower the sail. . . .

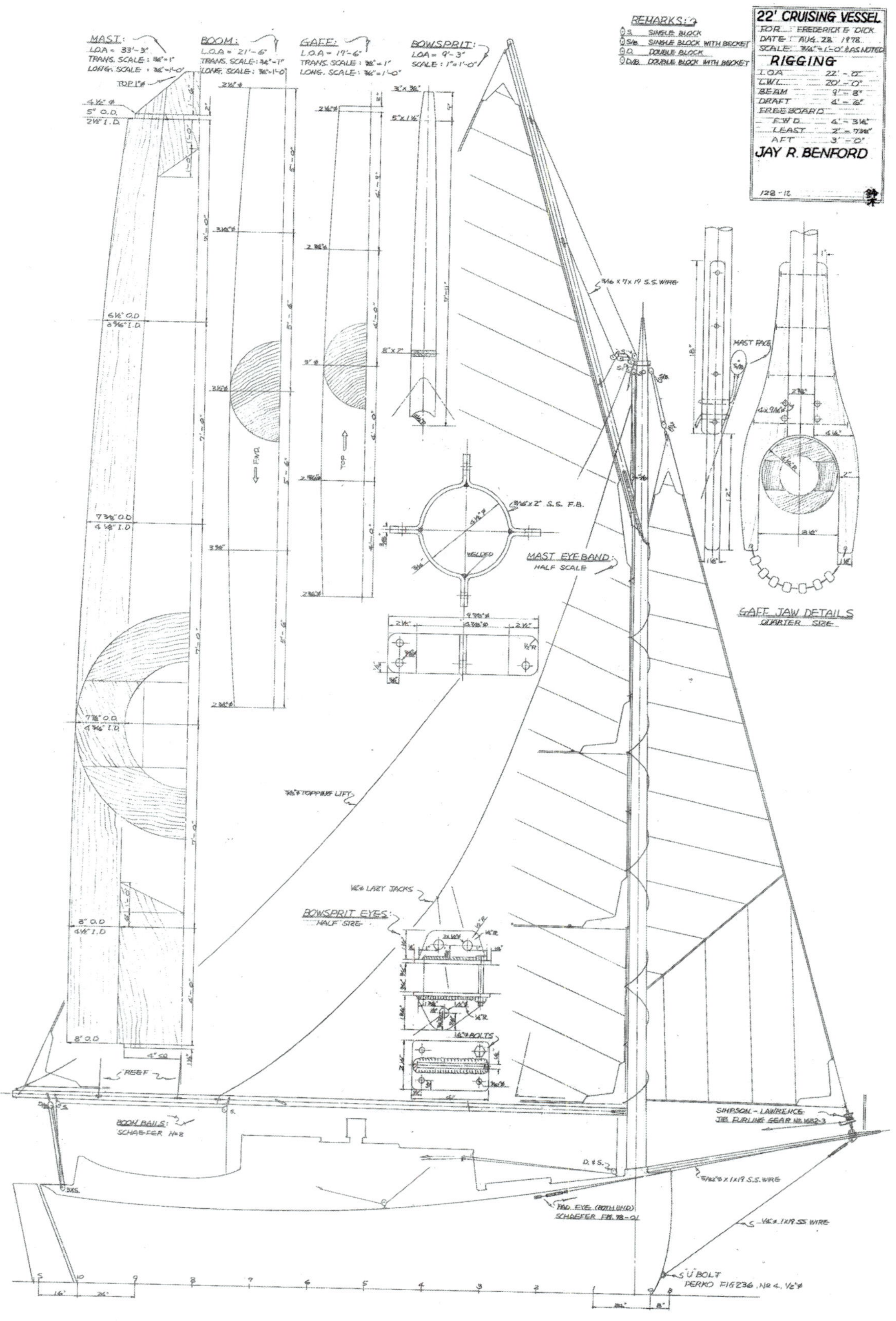

The Airex® construction plan is typical of our work with cored fiberglass. The body of the hull, including the living spaces, is all cored for insulation and stiffness. The keel is solid 'glass and this is a pretty conservative layup.

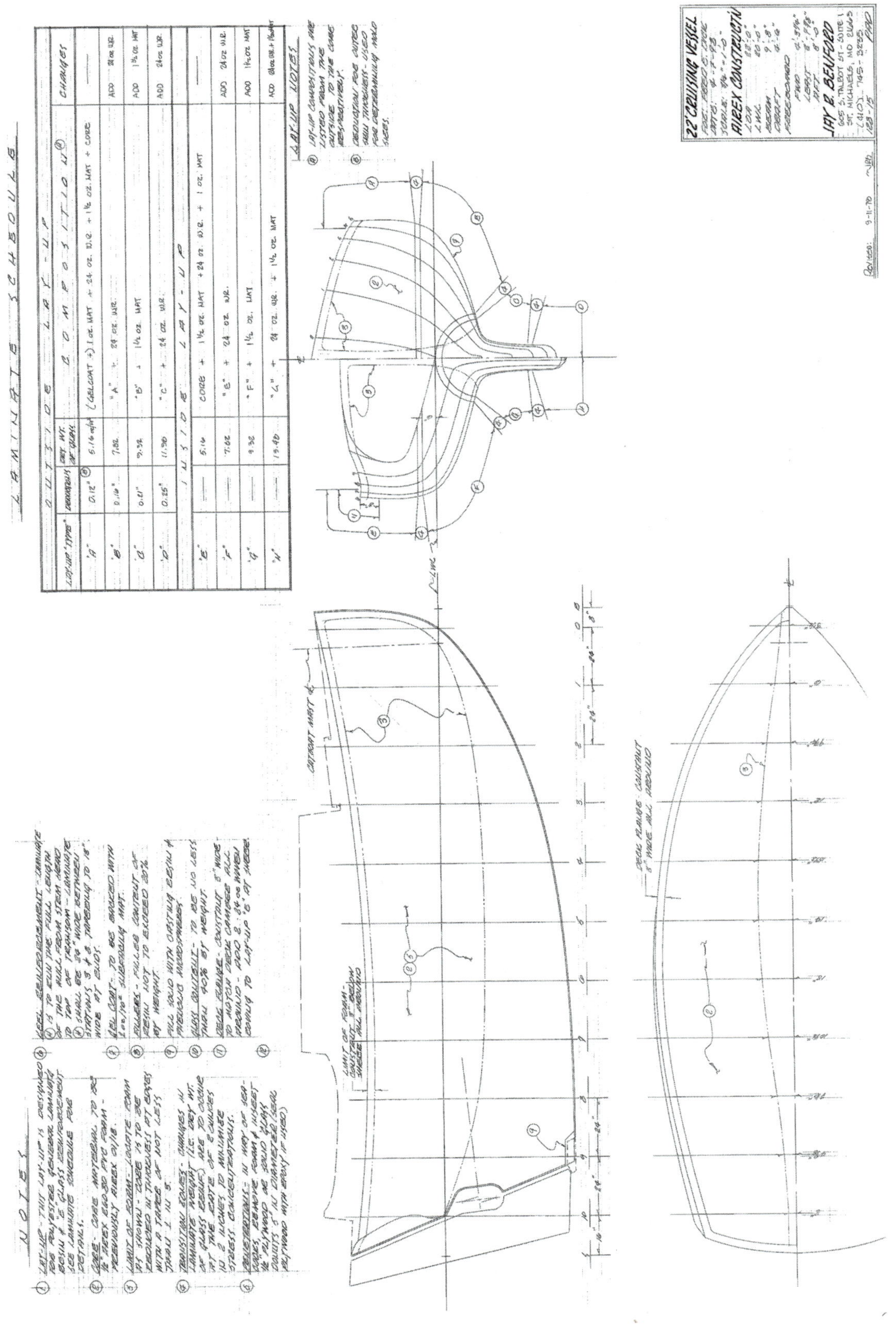

The Deck Lines & Construction show how the shapes for a molded deck are laid out, so that there is "draft" in the shape so that the part can be pulled out of the mold. The cockpit has good depth and the seatback slope will make for crew comfort. And the hatch scuppers are shown in the sections, arranged to drain water into the foot well and not let it drop into the lockers.

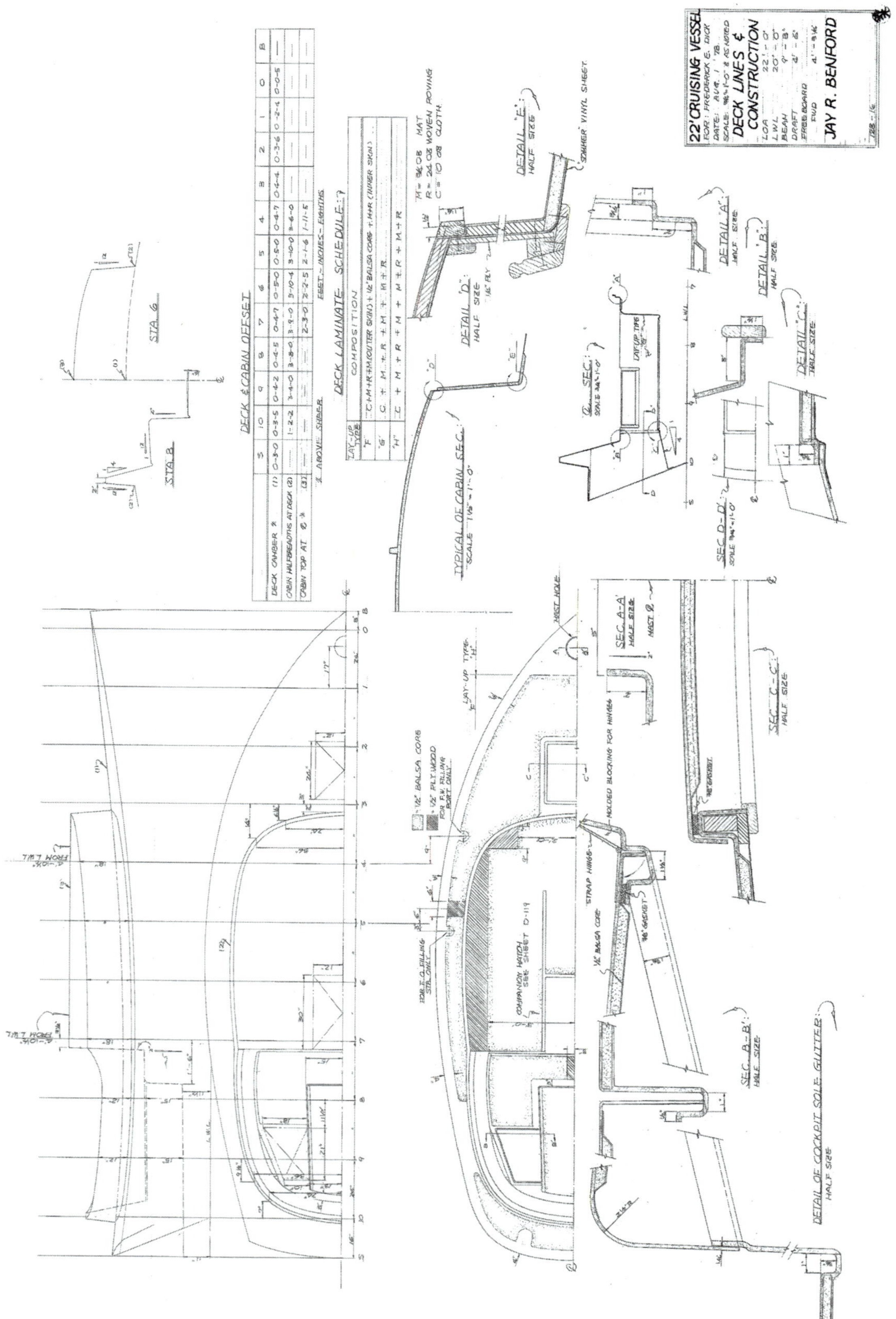

This is the solid fiberglass construction plan, showing the laminate schedule. This would be more economical for a builder to do, but the added insulation and stiffness of the Airex® cored hull would be more useful for the people living and cruising aboard. The engine beds shown should be modified for use with a different engine. Keep the same shaft line, if possible.

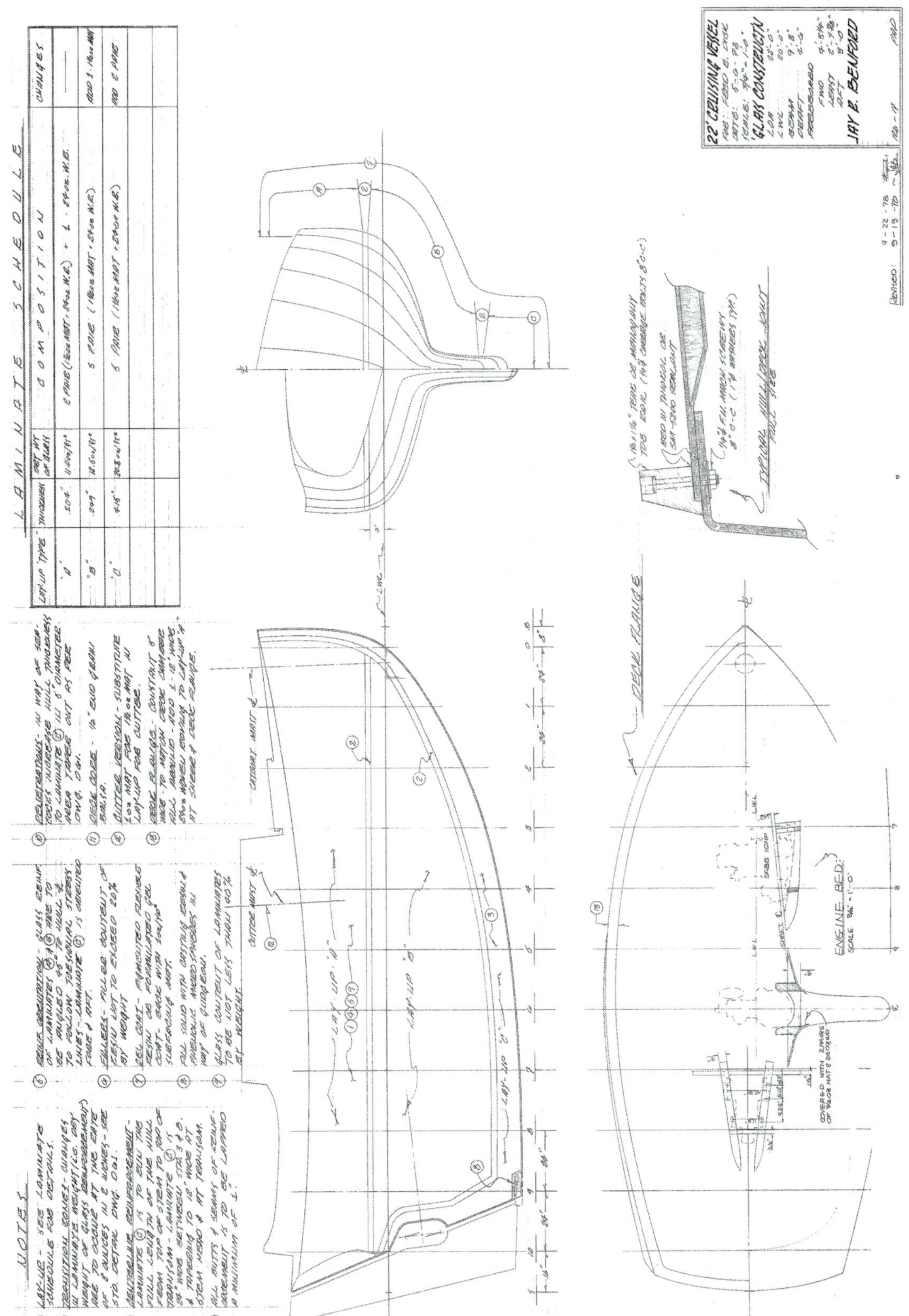

Note the differences between this deck plan and the one for the catboat version. That big skylight will let a wonderful lot of light and fresh air into the foc'sle. Having slept under one like this for a decade, I can testify to the usefulness of it. Like an instant weather check without getting out of bed. . . .

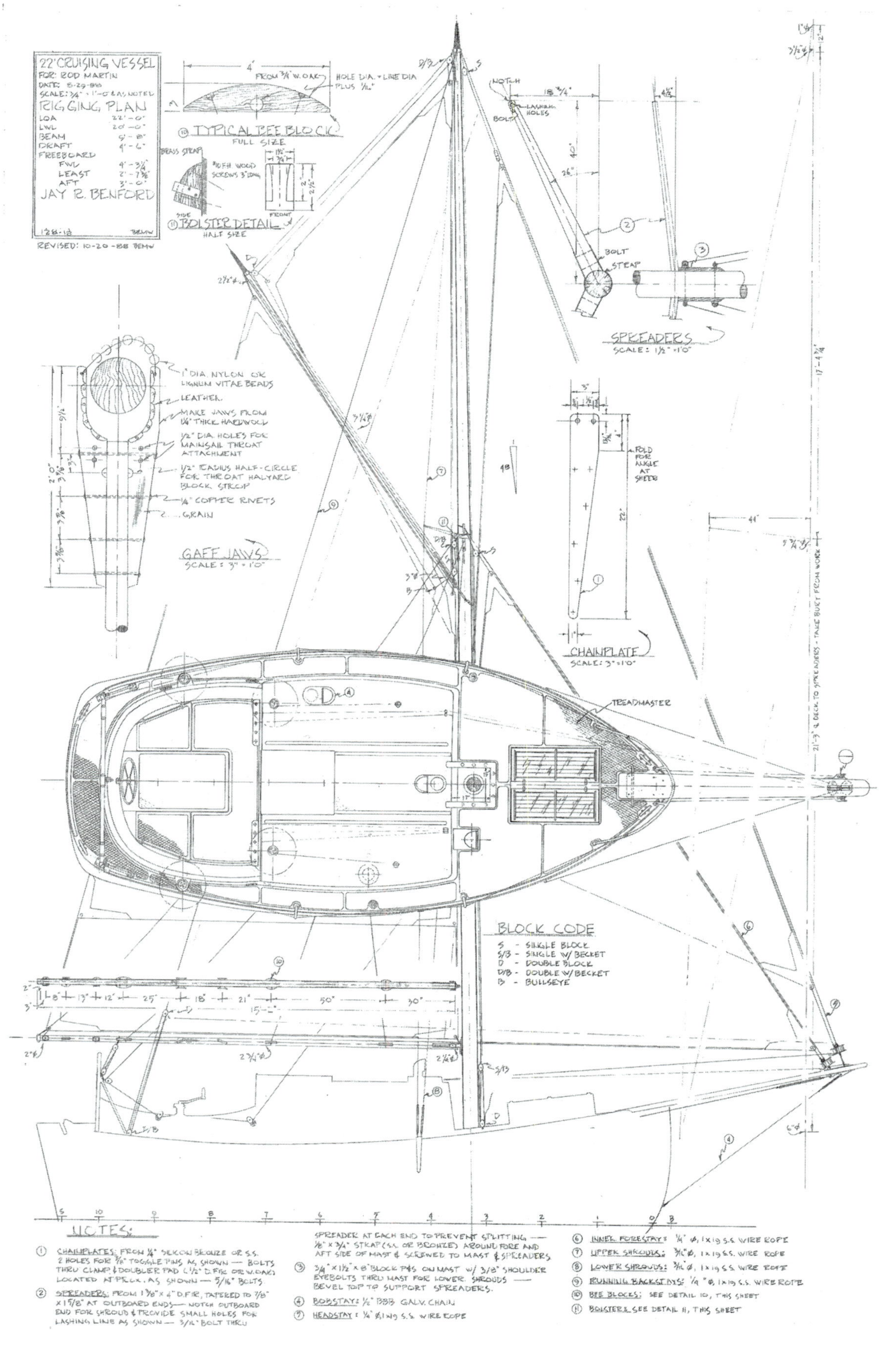

The topsail cutter rig has her sail area more evenly divided between her sails and much of the reefing down can be accomplished by just taking in one of the sails.

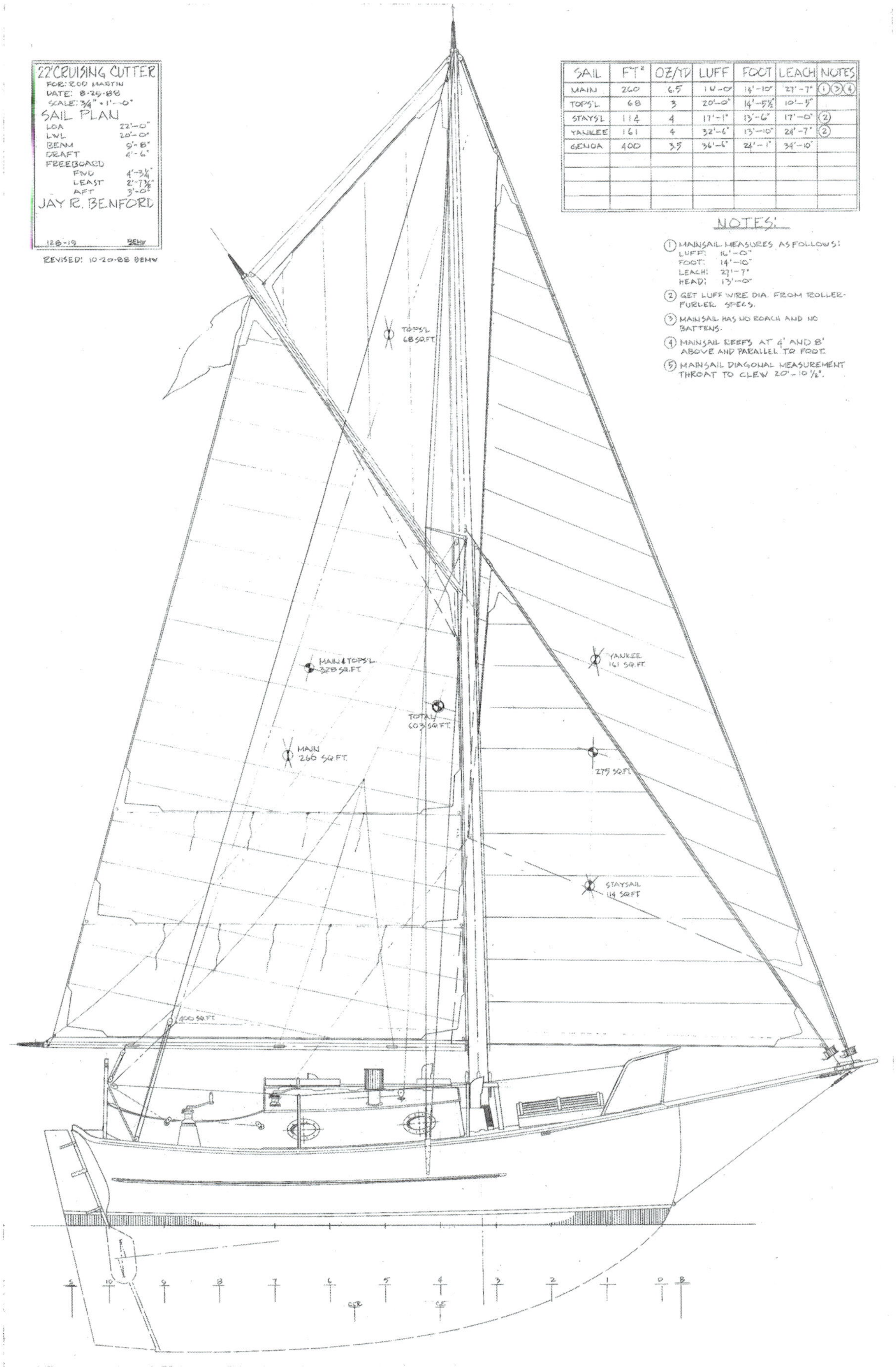

The cold-molded construction is probably the most suitable method for someone building their own boat. Starting with strip planks over the mold frames and bulkheads and then putting the diagonal veneers over them means there are no ribbands to be removed before the interior is built. Also visible on this drawing are the batteries for starting the engine and house usage aft under the companionway steps, the fuel tank under the forward berth, and the water tanks in the bilge.

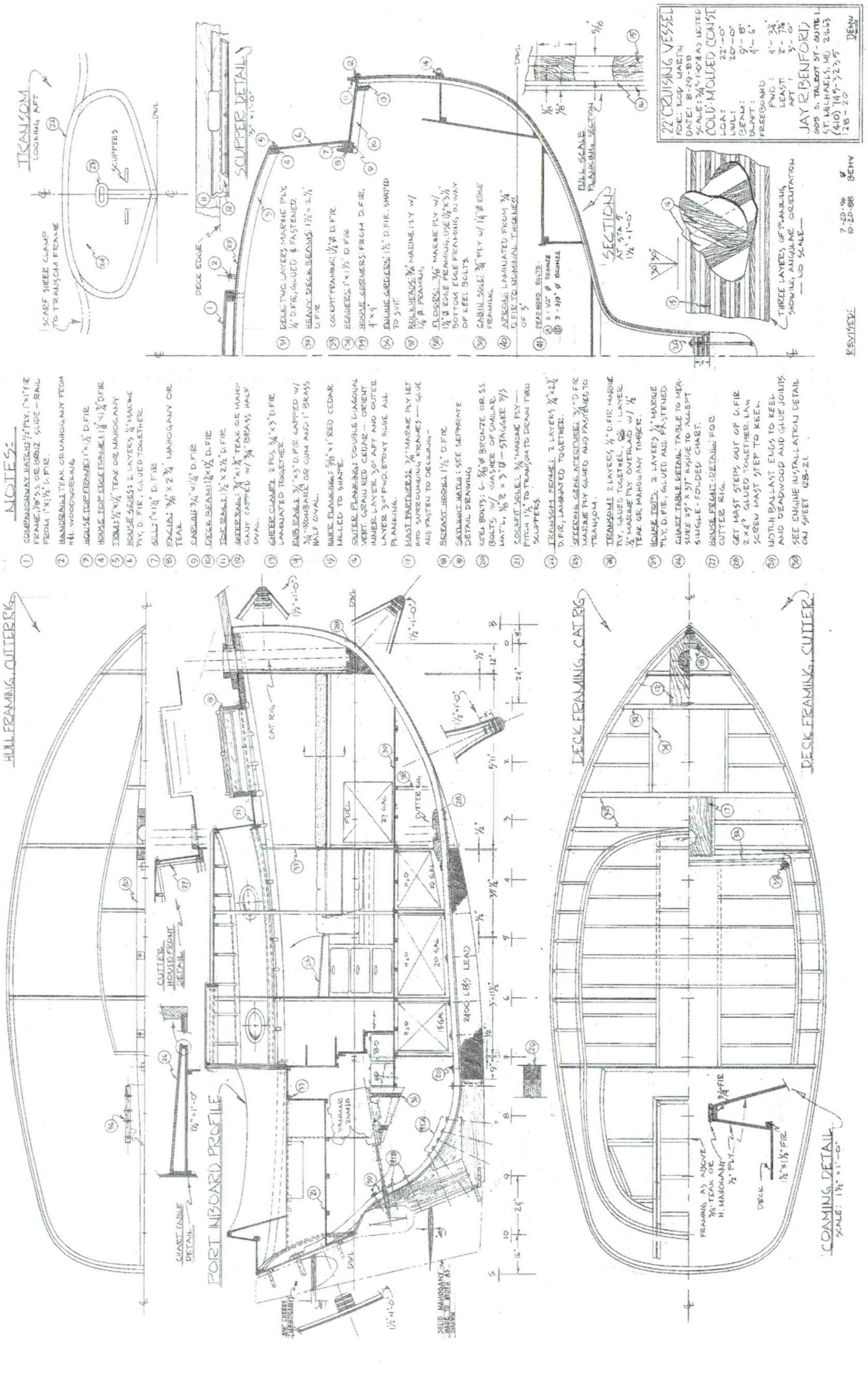

This sheet has a myriad of useful details on it for outfitting the boat, useful for both the cat and cutter versions. And a nice stern view drawing, showing that shapely transom. . . .

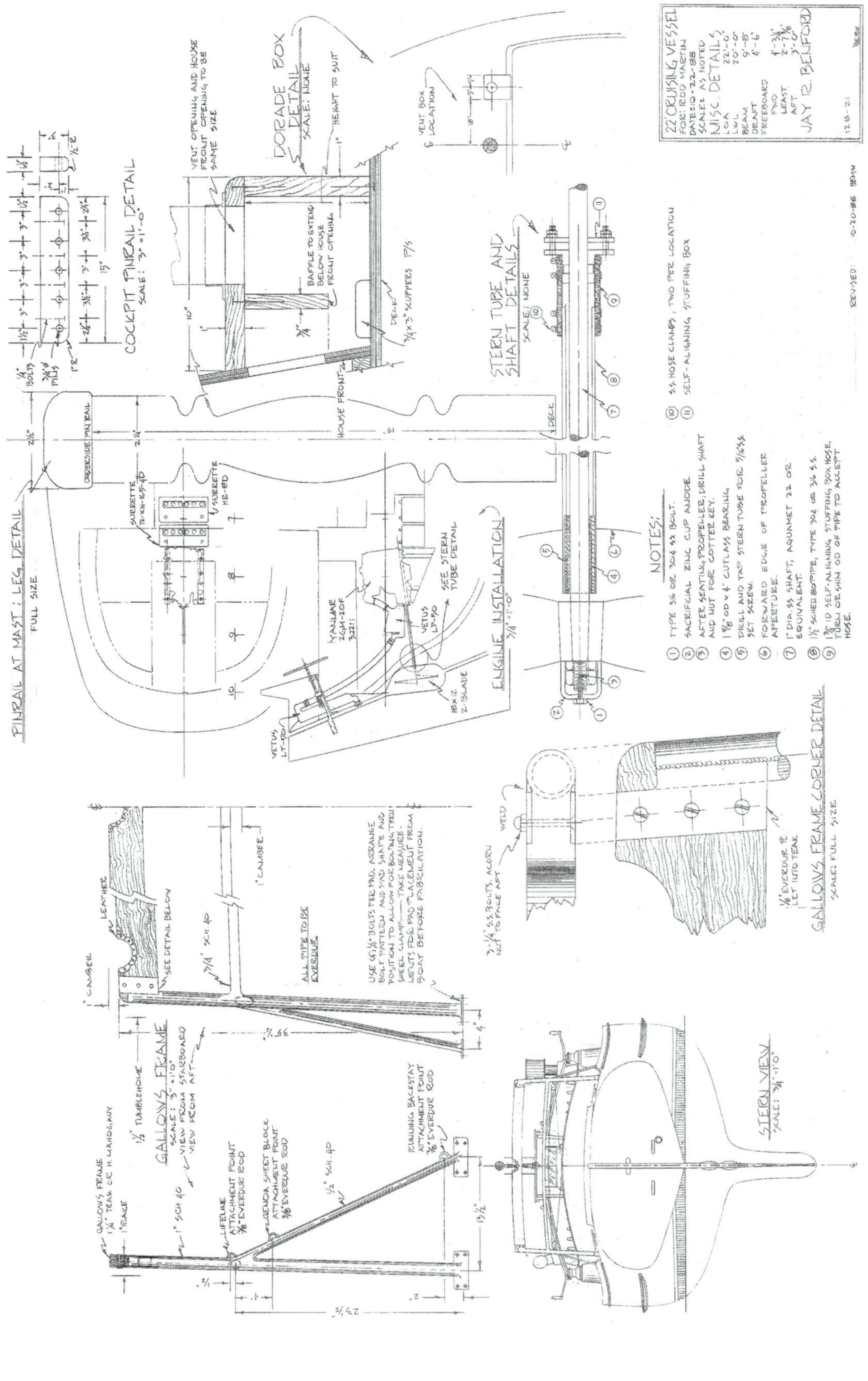

This is a longer version of the 22' Cruising Cat. Just as the 19' **Scrimshaw** *is a longer version of the 17'* **Liberty**, *this was achieved by judicious re-spacing of the design stations. This must be done* **very** *carefully by the designer, so that the fairness of the hull shape is maintained, not causing lumps in the resulting hull form. Trying to make a boat longer by just adding a few feet or a duplicate frame in the middle of the boat is pretty much a guaranteed way to have an unfair hull.*

Again, this is a shape like the ones Lyle Hess did in the designs made famous by Lin and Larry Pardey; firm bilges, long sweeping keel short ends and an overall shape that leads to a great sailing boat with wonderful seakeeping abilities.

Particulars	**Imperial**	**Metric**
Displacement	11,000 lbs.	4,990 kg.
Displ.-length ratio	355	
Sail-area-displ. ratio	18.12	
Ballast	3,300 lbs.	1,497 kg.
Ballast ratio	30%	
Prismatic coefficient	.562	
Pounds per inch immersion	736	131.4 kg./cm.
Entrance half-angle	21°	

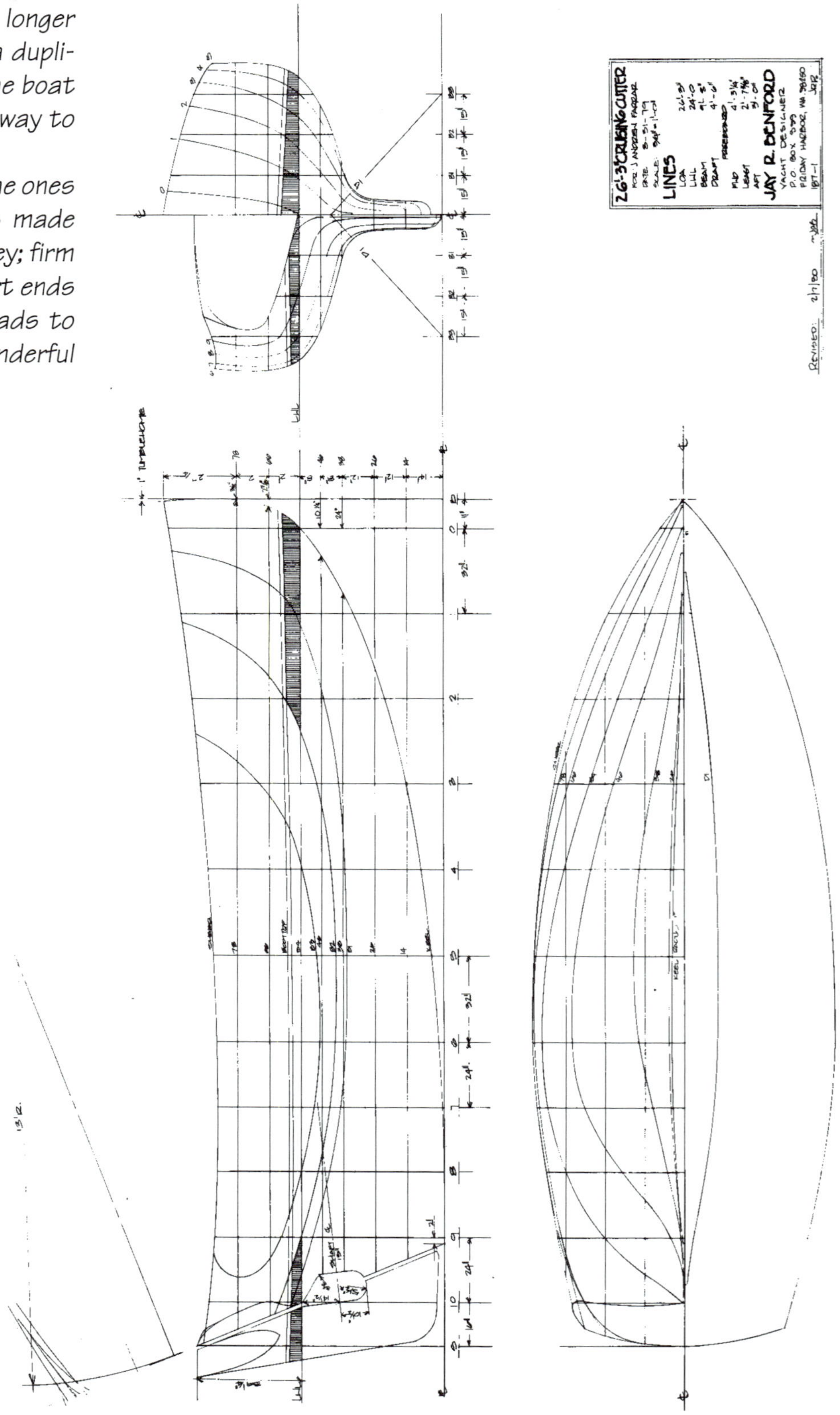

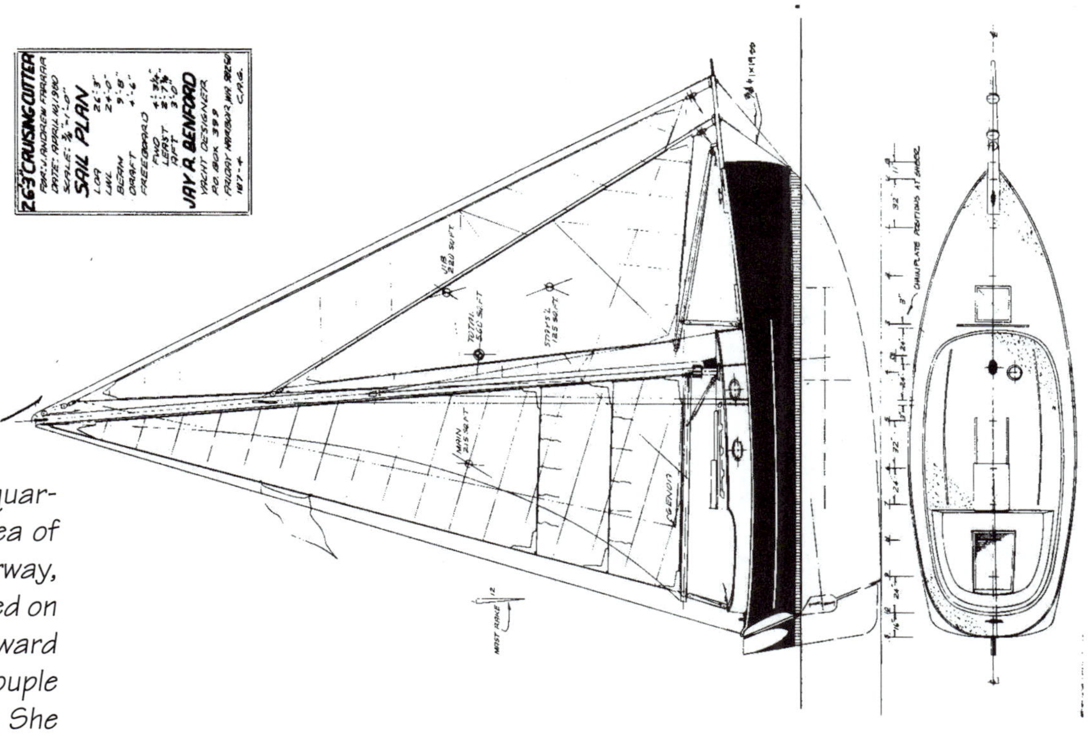

The sail plan is a compact cutter, with a modest length bowsprit, with 560 square feet in her working rig. The cockpit is a size that is comfortable for two couples. Her different accommodations have the main sleeping areas aft, as a double and wide single quarter berth. This is the area of least motion when underway, so they will be appreciated on a trip. The settee is forward and can serve as a couple extra berths, if desired. She has 6'-2" headroom under the cabin. The head has a shower in it – a "wet head". Immediately forward of the head is the large, one fold, chart table which can be used standing in the passageway or seated on the port settee. Just to starboard of the mast is the oil furnace, which will make cruising in cooler climates a joy. The galley has a place for the cook to stand out of the traffic and prepare meals. Tankage is 20 gallons of fuel and 40 gallons of water.

The engine is off center because of the asymmetric berth layout. This gives good working room around the engine. Angling the prop shaft like this can offset some of the prop torque, if the correct rotation engine is used. . . .

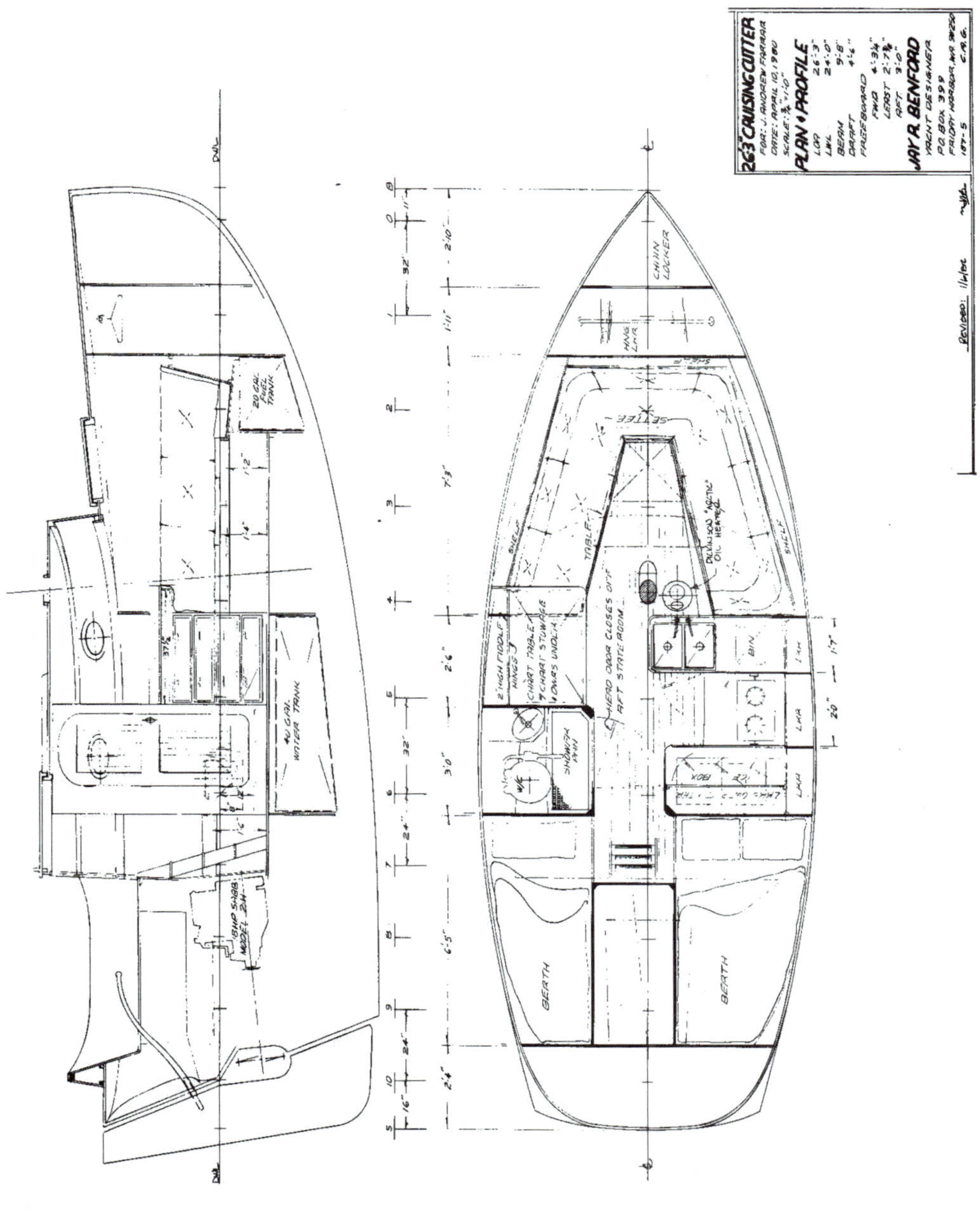

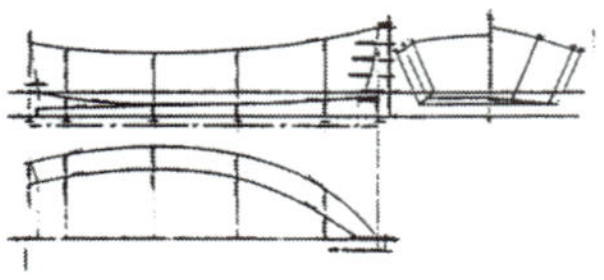

7'-3" Dinghy
7'-3" x 6'-9" x 3'-9¼" x 0'-4"

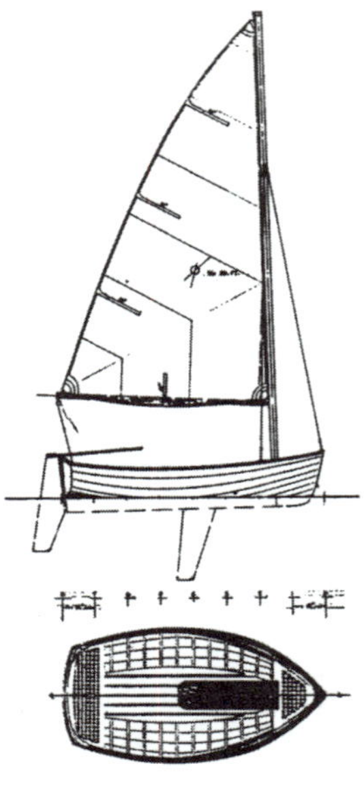

8' Portland Yawlboat
8' x 7'-6" x 4' x 0'-5"/2'-6"

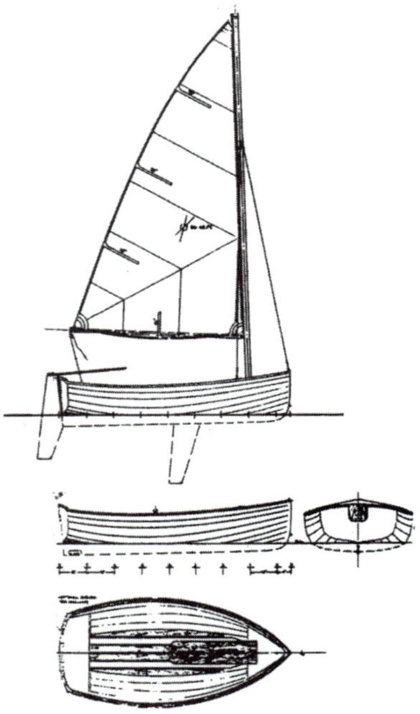

8½' Dinghy
8'-6" x 8'-2½" x 4' x 0'-9"/2'-6"

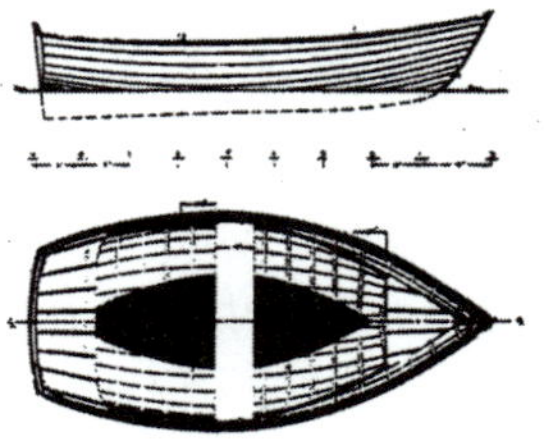

9½' Dinghy
9'-6" x 8'-3" x 4'-6" x 0'-7"

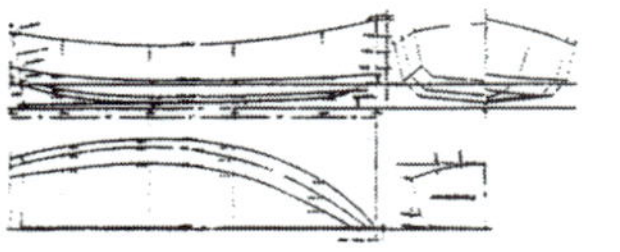

7½' Dinghy
7'-6" x 7' x 3'-9" x 0'-5"

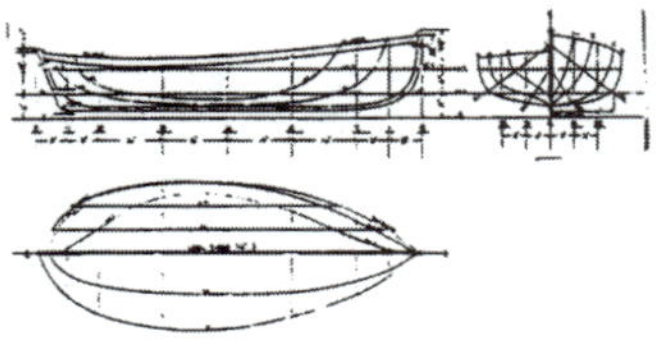

8' Double-Ended Dinghy
8' x 7'-3¾" x 3' x 0'-4"

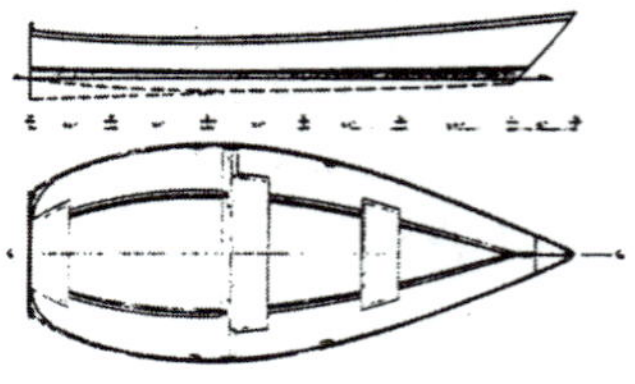

11'-4" Dinghy
11'-4" x 10'-1" x 4'-6" x 0'-3"

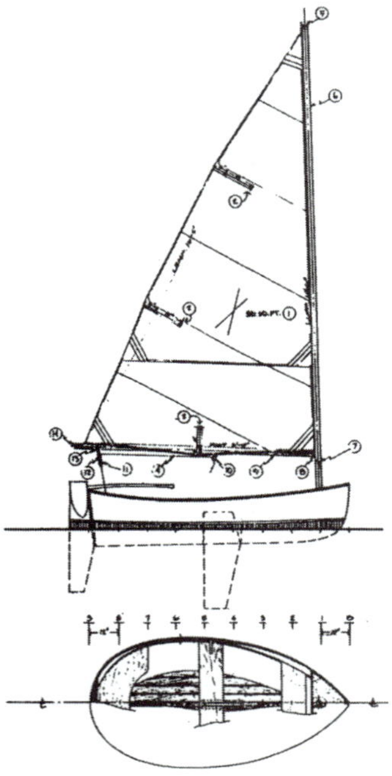

9' Pacific Peapod
9' x 8'-6" x 4'-6" x 0'-7"/2'-9"

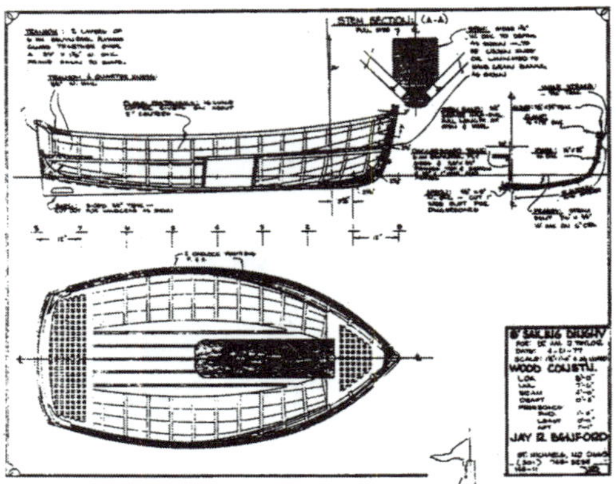

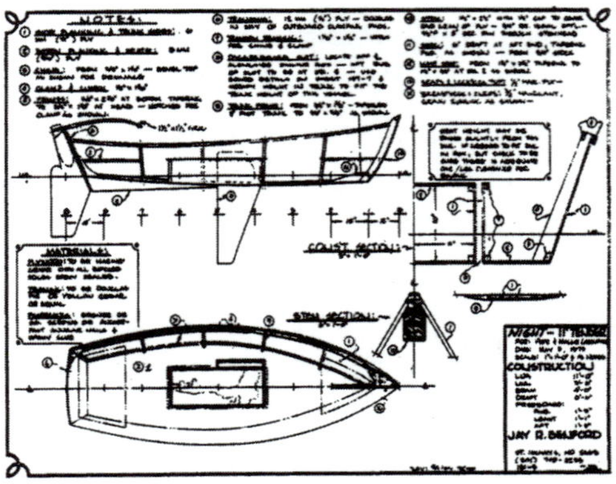

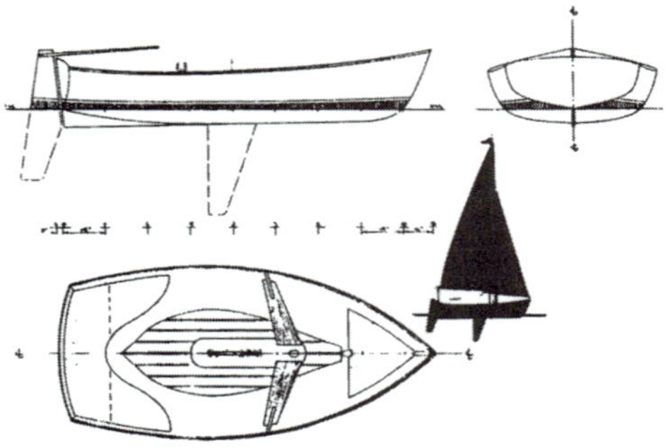

11' Dinghy
11' x 10' x 5' x0'-6"/3'

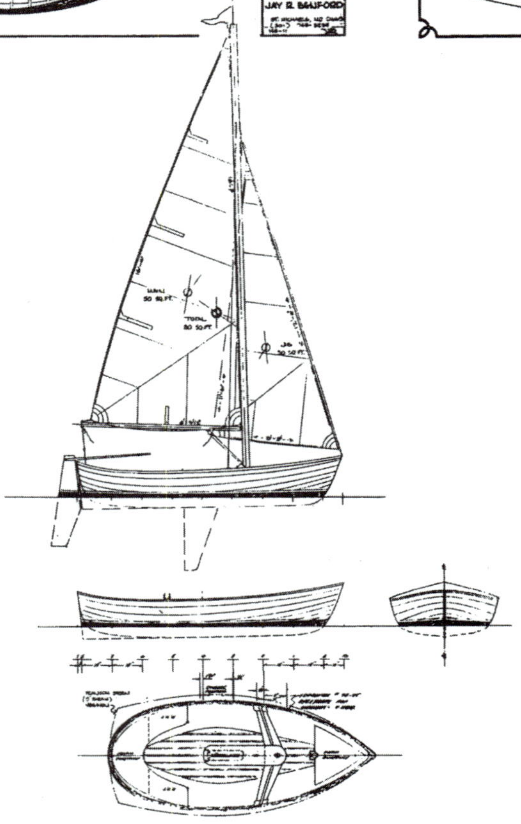

11' Oregon Peapod
11' x 10' x 4'-6" x 0'-6"/3'

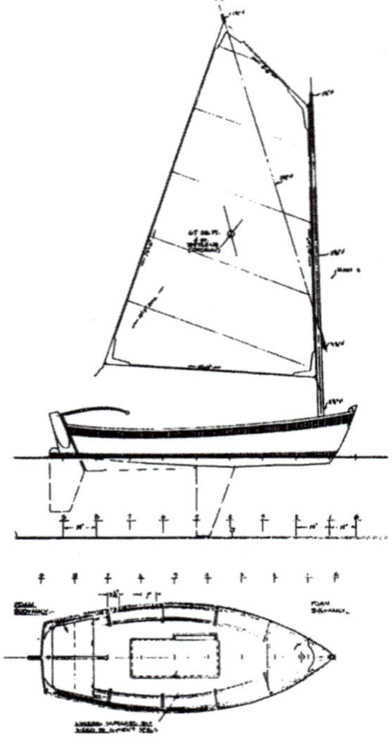

11' Dinghy ***Night***
11' x 9'-6" x 4' 0'-6"/2'-11"

Small Craft Plans

"Let's get it straight up front: This book is more than a catalog of designs. The 96 pages between its covers contain plans for 15 small craft. Designer Jay Benford knows that readers will build directly from the book — bypassing the formality of ordering plans from his office. In fact he encourages the process by including full working drawings and tables of offsets for all of the designs." **WoodenBoat**

The 15 sets of complete plans for skiffs & tenders from the Boards of the Benford Design Group (as shown on these two pages) will help you build your own dinghy, small craft, or tender. Plans for 7'-3" to 18-footers in cold-molded, carvel, lapstrake, plywood, and fiberglass construction are included, along with photos of some completed boats. 96 pages, softcover.

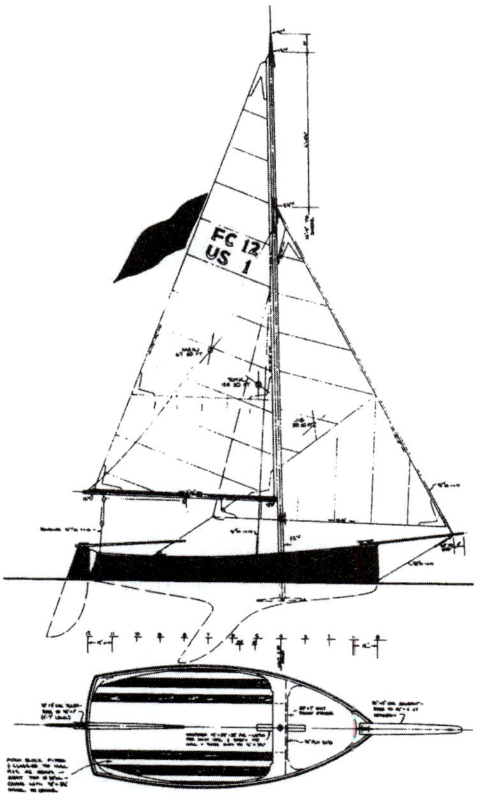

12' Keelboat
12' x 12' x 5' x 3'-4½"

To order this book of plans, using a credit or debit card, call 1-800-6Tiller or log on to www.tillerbooks.com.

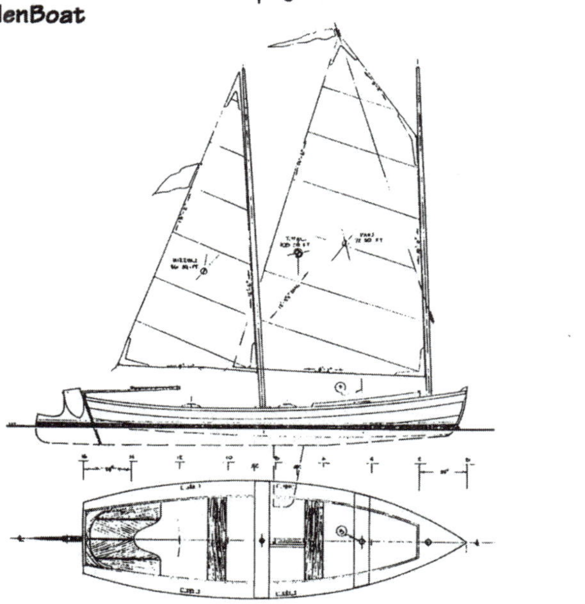

16' *Conch Ad Libitum*
16' x 15' x 5' x 0'-9"/3'-3"

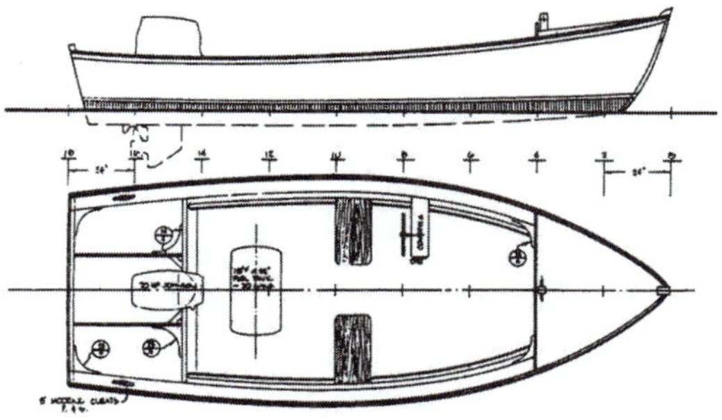

18' Texas Skiff
18' x 16' x 6'-8½" x 0'-5"

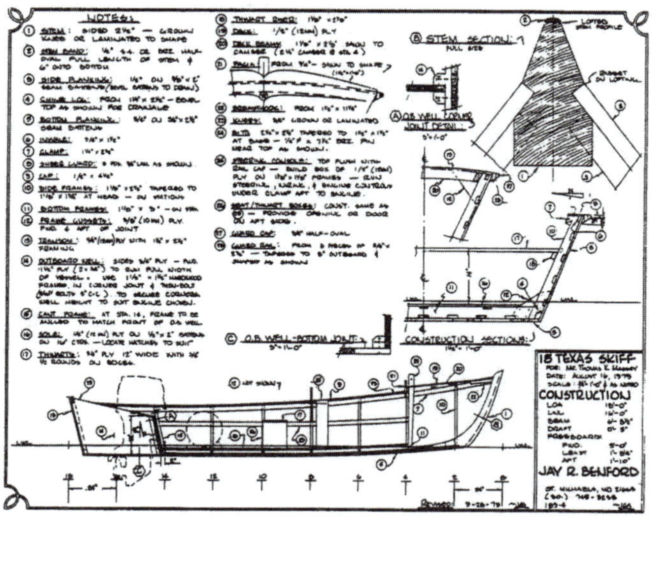

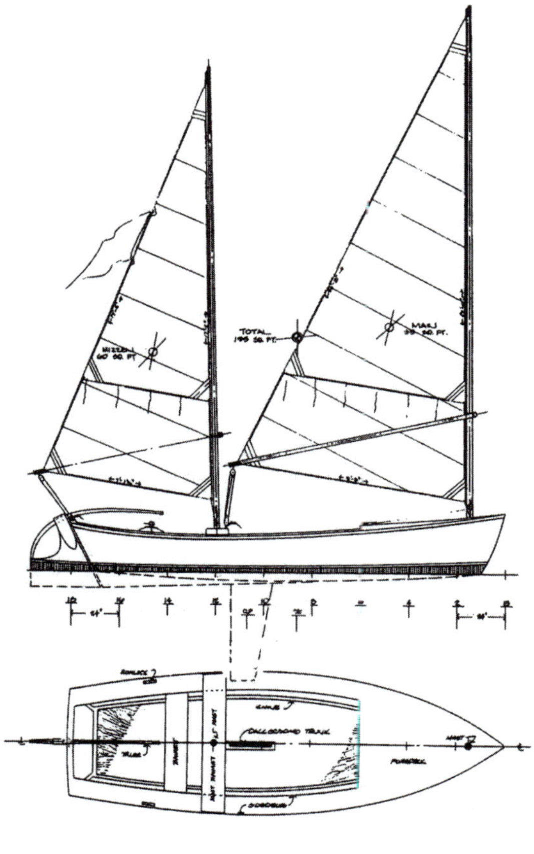

18' Cat Ketch
18' x 16' x 6' x 0'-8"/4'-6"

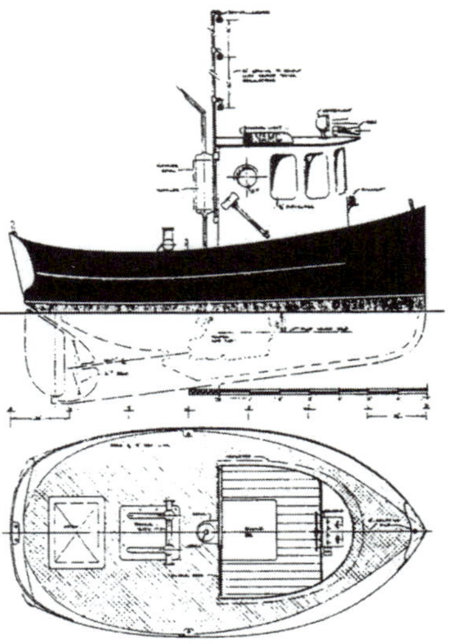

14' Tugboat
14' x 13' x 7' x 3'

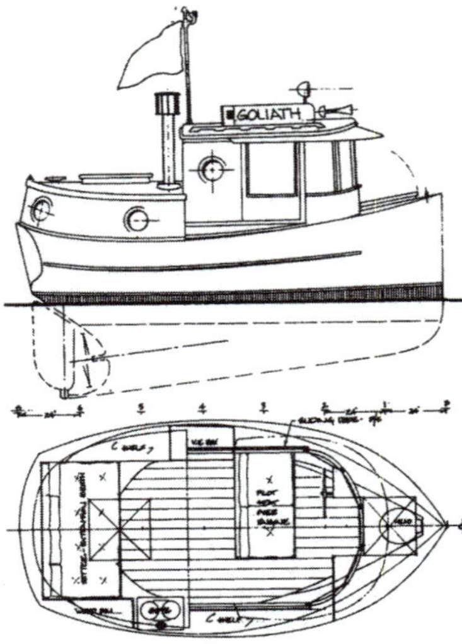

14' Tug-Yacht *Goliath*
14' x 13' x 7' x 3'

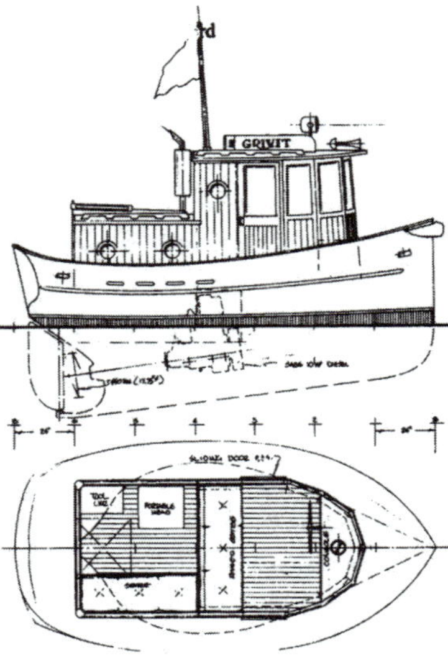

14' Tug-Cruiser *Grivit*
14' x 13' x 7' x 3'

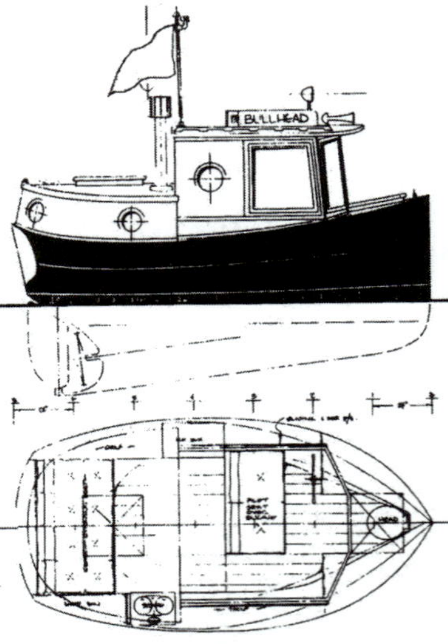

14' Trawler Yacht *Bullhead*
14' x 13' x 7' x 3'

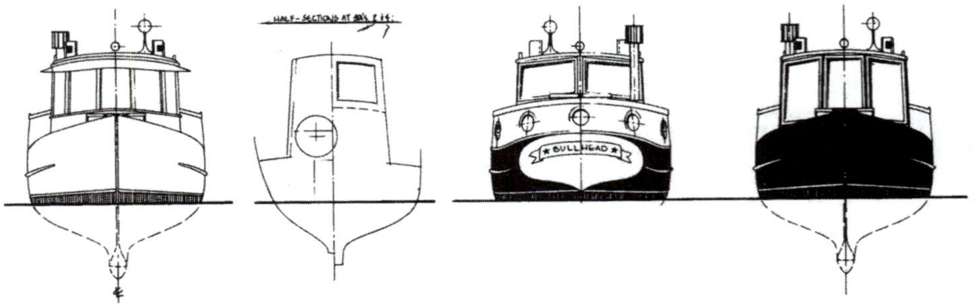

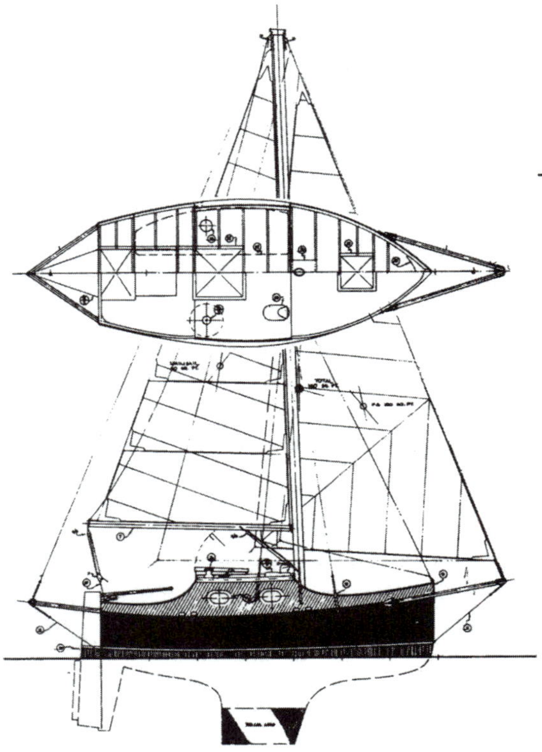

14' Cruiser *Happy*
13'-10" x 18'-8" x 6'-3" x 3'-7"

Pocket Cruisers & Tabloid Yachts
— Volume 1 —

Complete building plans for the small cruising boats shown on these two pages are in our book, *Pocket Cruisers & Tabloid Yachts, Volume 1.* These boats are some of the most popular plans from the boards of the Benford Design Group. They're shown in full detail in this 96 page, 8½" x 11" softcover book.

"All of the plans exhibit the professional presentation and well-conceived accommodations typical of the author's work"

—**WoodenBoat**

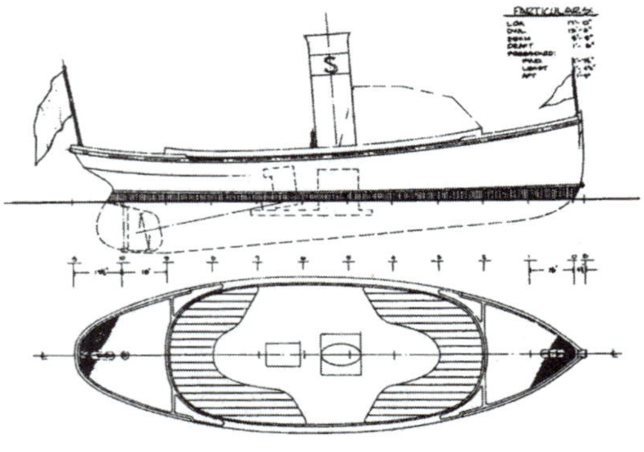

17' Fantail Steam Launch *Myf*
17' x 15'-6" x 5'-3" x 1'-8"

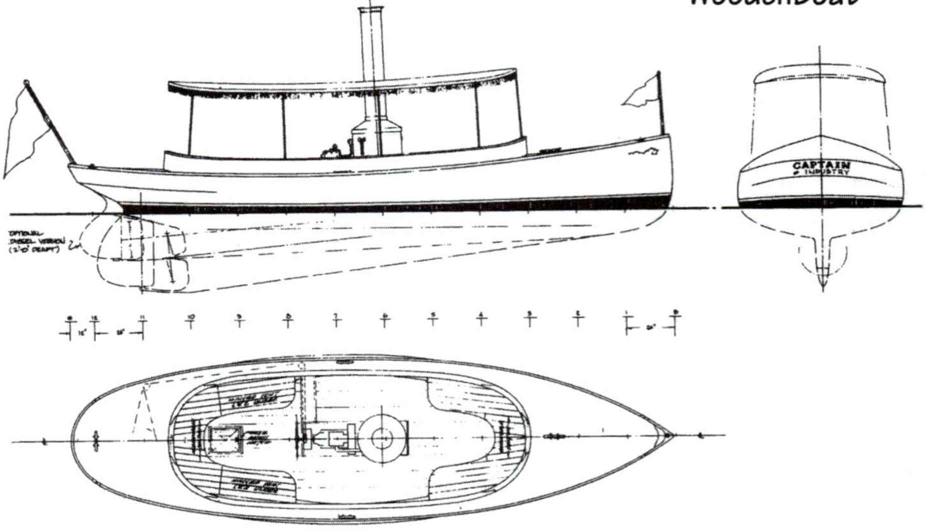

25' Fantail Steam Launch *Beverley*
25' x 22'-6" x 7' x 3'-3"

20' Supply Boat
19'-11½" x 19' x 7'-11½" x 2'-3"

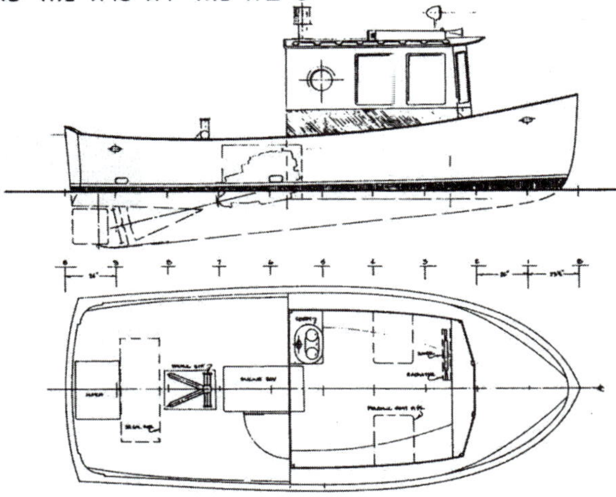

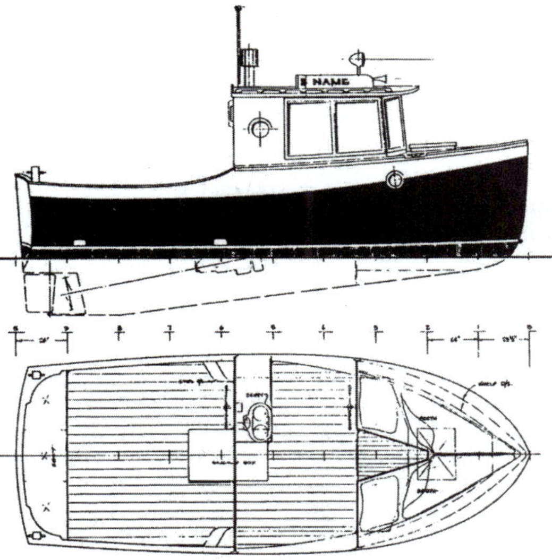

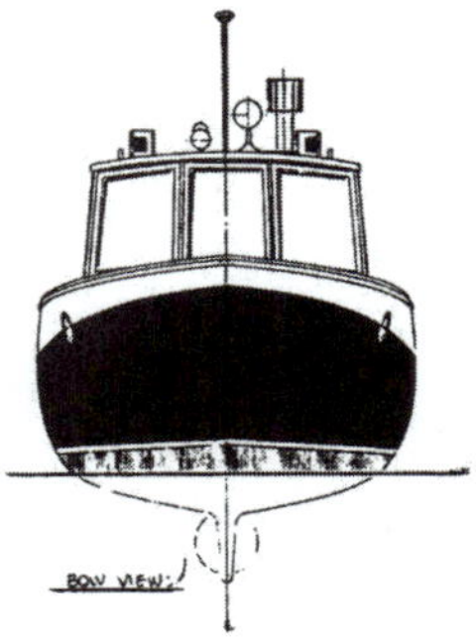

20' Supply Boat Båten
19'-11½" x 19' x 7'-11½" x 2'-3"

20' Lake Union Cruiser
19'-11½" x 19' x 7'-11½" x 2'-3"

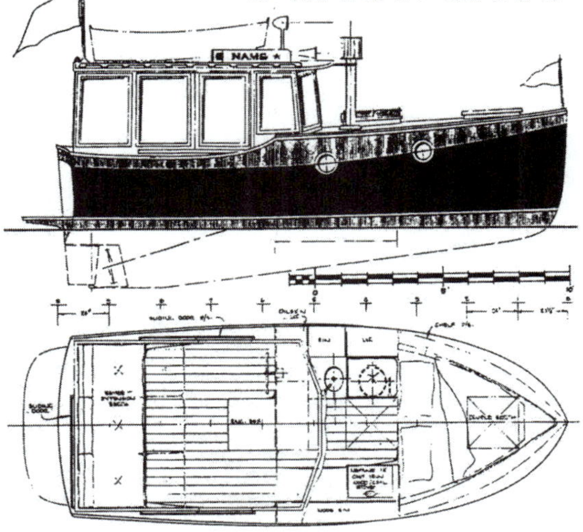

20' Catboat Joy
20' x 18' x 8' x 3'-6"

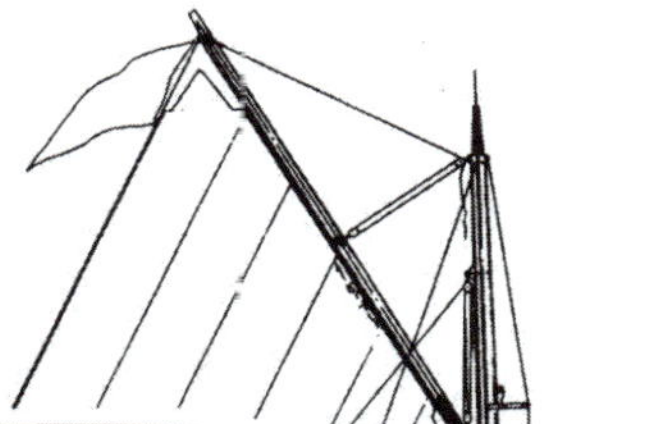

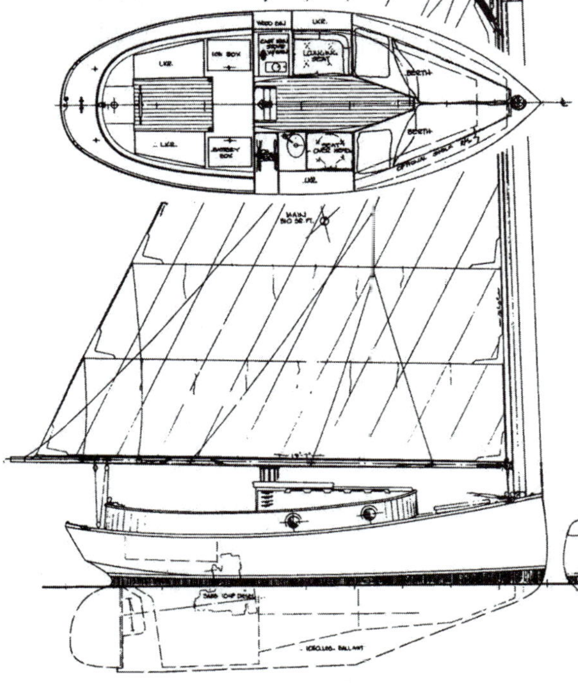

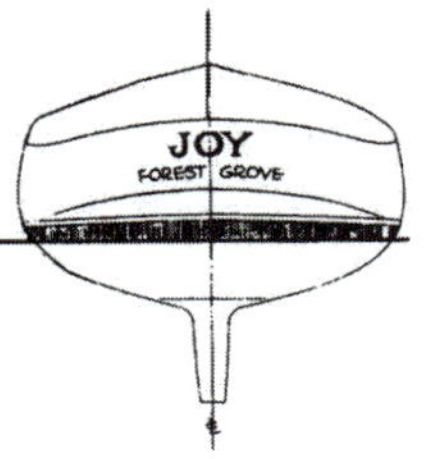

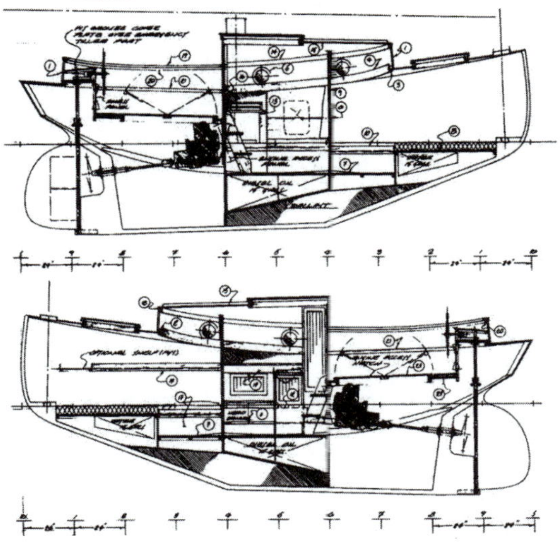

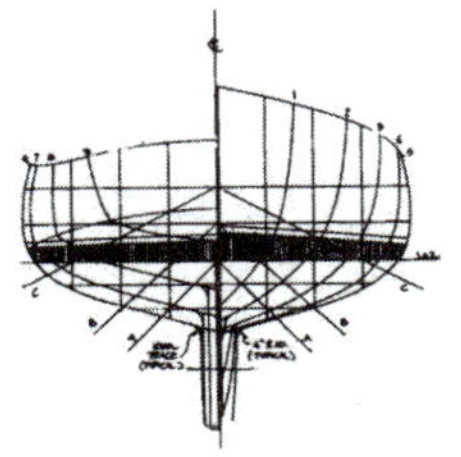

To order this book
of plans, using a credit
or debit card, call
1-800-6Tiller
or log on to
www.tillerbooks.com.

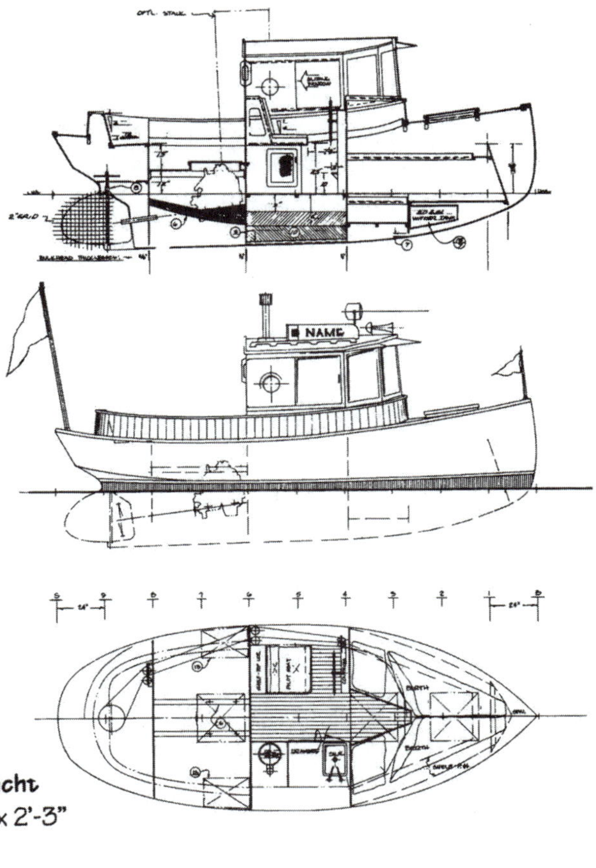

20' Tug Yacht
20' x 18' x 8' x 2'-3"